ELECTRICITY IN THE AMERICAN ECONOMY

Recent Titles in
Contributions in Economics and Economic History

ELECTRICITY IN THE AMERICAN ECONOMY

Agent of Technological Progress

SAM H. SCHURR
CALVIN C. BURWELL
WARREN D. DEVINE, JR.
SIDNEY SONENBLUM

*Published under the Auspices of the
Electric Power Research Institute*

CONTRIBUTIONS IN ECONOMICS AND ECONOMIC HISTORY,
NUMBER 117

GREENWOOD PRESS
NEW YORK • WESTPORT, CONNECTICUT • LONDON

Library of Congress Cataloging-in-Publication Data

Schurr, Sam H.
    Electricity in the American economy : agent of technological
progress / Sam H. Schurr . . . [et al.].
        p. cm. — (Contributions in economics and economic history,
ISSN 0084-9235 ; no. 117)
    "Published under the auspices of the Electric Power Research
Institute."
    Includes bibliographical references.
    ISBN 0-313-27512-2 (lib. bdg. : alk. paper)
    1. Electric power consumption—Economic aspects—United States.
2. Electric utilities—United States.   3. United States—Industries—
Energy consumption.   I. Electric Power Research Institute.
II. Title.   III. Series.
    HD9685.U5S348   1990
    333.79'32'0973—dc20   90-38411

British Library Cataloguing in Publication Data is available.

Library of Congress Catalog Card Number: 90-38411
ISBN: 0-313-27512-2
ISSN: 0084-9235

First published in 1990

Greenwood Press, 88 Post Road West, Westport, CT 06881
An imprint of Greenwood Publishing Group, Inc.

Printed in the United States of America

∞

The paper used in this book complies with the
Permanent Paper Standard issued by the National
Information Standards Organization (Z39.48-1984).

10 9 8 7 6 5 4 3 2 1

Contents

Figures and Tables

Figures

Tables

Preface

A strong and persistent rise in the importance of electricity is one of the major long-term trends in the history of energy in the United States during the twentieth century. In a remarkable profusion of applications, electricity has penetrated deeply and brought important changes into virtually every corner of American life, whether in industry, in the home, or in the rapidly growing commercial and service sectors.

This study treats the ever-widening scope of these applications of electricity and their linkages with technological progress in specific industries and in the economy at large. We draw on two categories of data: (1) case studies of the ways in which technological progress in particular industries and economic sectors has been dependent upon the adoption of electrified production methods; and (2) aggregative long-term national economic statistics that enable us to measure the changing relationships over time between increases in the use of electricity and other factor inputs (labor, capital, and nonelectric energy) and the growth in industrial productivity.

Eleven of the book's thirteen chapters cover the case studies, while the remaining two chapters (and a statistical appendix) contain the broad quantitative findings and supporting data. We use these two levels of analysis ("bottom-up" and "top-down," so to speak) in order to obtain a more reliable and more understandable picture of the underlying trends, and the forces at work in producing them, than either approach could yield by itself.

The book's attention, then, is focused on the proliferating applications of electricity and their interactions with technological progress throughout the economy. The well-documented technological advances that have been achieved *within* the electric power industry are not covered here. It needs to be understood, though, that long-term improvements in the efficiency of producing electric power have provided the fundamental preconditions for the developments that constitute the subject matter of this study.

One important measure of progress in producing electricity is in the striking improvements that have been achieved in the efficiency of converting fuels into electric power in the course of the twentieth century. Measured by the relationship between the thermal content of the electrical energy produced and that of the fuels consumed in its production, there has been an increase in the average thermal efficiency of converting fuels into electricity from roughly 5 percent early in the century to about

33 percent in recent years. There have also been tremendous increases in the scale of power plant equipment, typified by an increase in "large" generating units from a capacity of less than 5 megawatts (MW) at the turn of the century to recent levels of 1,000 MW, or so. And with increases in size came great economies in the cost of building power plants. As a result of improvements such as these, the cost of electricity (measured in constant dollar prices to residential customers) fell by roughly a factor of fifty from the turn of the century to the early 1970s.

By all accounts improvements in the efficiency of the electric power industry slowed to a crawl (if not a complete halt) during the 1970s. Why this happened remains a subject of controversy. Some place the main blame on such external cost-increasing factors as burgeoning environmental regulations; the much higher interest rates that have prevailed in recent years; and the post-1973 upheavals and uncertainties in the cost and supply of fuel. Others assign major importance to shortcomings within the management of the electric utility industry. Yet, despite the unfavorable developments affecting the costs of producing electricity in recent years, the studies in this book show that there continues to be a strong and growing dependence of technological progress throughout the economy on the use of electrified techniques of production. All the more reason why achieving solutions to the problems that have interfered with technological progress within the electric power industry should be a matter of great national concern.

The present study represents a cooperative effort of the Energy Study Center of the Electric Power Research Institute (EPRI) and the Institute for Energy Analysis (IEA) of Oak Ridge Associated Universities. The major EPRI participants were Sam H. Schurr, who directed the study, and Sidney Sonenblum, who served as a consultant. The major IEA participants were Calvin C. Burwell (now a private consultant) and Warren D. Devine, Jr. (now with Science Applications International Corporation). Other EPRI staff members who made important contributions were Milton Searl and Blair Swezey. At the Institute for Energy Analysis, other participants included E. Brent Sigmon, Ned L. Treat, and David B. Reister.

The authors gratefully acknowledge the technical review of various chapters of the book and of underlying draft reports provided by Martin N. Baily, Brookings Institution; Charles A. Berg, Northeastern University; Ernst R. Berndt, Massachusetts Institute of Technology; C. Roger Carnes, Martin Marietta Energy Systems; T. J. Grant, American Petroleum Institute; Walter R. Hibbard, Virginia Polytechnic Institute; Carl Hibscher, Toledo Engineering; Kenneth Iverson, NUCOR; Dale W. Jorgenson, Harvard University; Ethan B. Kapstein, Harvard University; H. G. MacPherson, IEA; J. Patrick McHugh, University of Tennessee; Larry Penberthy, Penberthy Corporation; W. G. Pollard, IEA; Nathan Rosenberg, Stanford University; Philip S. Schmidt, University of Texas; S. J. Shepard, Ball Corporation; E. G. Silver,

Oak Ridge National Laboratory; Bernard I. Spinrad, Iowa State University; Ray Squitieri, EPRI; J. K. Stone, private consultant; Alvin Weinberg, IEA; and David O. Wood, Massachusetts Institute of Technology.

The manuscript has profited greatly from the fine editorial hand of Mary Wayne. We want to thank her for giving the individual chapters a consistency of style and readability that was previously lacking. We want also to thank Sally Nishiyama for keeping our efforts reliably organized even though she had to cope with several different authors and an almost unending succession of draft manuscripts.

Finally, we wish to acknowledge the guidance provided by Chauncey Starr and Alvin Weinberg. When this study was conceived, and during much of the time that it was under way, Dr. Starr directed EPRI's Energy Study Center, and Dr. Weinberg was director of the Institute for Energy Analysis. We want also to thank Floyd L. Culler, president of EPRI at the time, who saw the need for the research and was instrumental in arranging for its support.

Sam H. Schurr

ELECTRICITY IN THE
AMERICAN ECONOMY

Overview

Electricity now powers American life to an unprecedented degree. In the year 1900, electricity's share of the nation's energy use was negligible. As the last decade of the century begins, that share has risen to almost 40 percent. Just within the past twenty-five years, electricity's importance in total energy has doubled.

The question is why. Because electricity is itself produced from fuels through a costly conversion process, its price per thermal unit has always been well above that of coal, oil, or natural gas. So something other than price must account for the long-term growth in electricity's market share.

Much of what we think accounts for this persistent growth is conveyed in the book's subtitle: "Agent of Technological Progress." The link between any energy form and the services it provides is the technology for its use. Simply put, electricity can be used in ways that no other energy form can. Technological progress leading to radically improved ways of organizing productive activities, to new products and new techniques of production, and to a variety of other advances has over and over again been heavily dependent upon the use of energy in the form of electricity.

The purpose of this book is to explore why and how electricity has functioned as an agent of technological progress in twentieth-century America and with what results. Our analysis proceeds in two complementary ways: through descriptive case studies, both historic and contemporary, for various industries; and through quantitative studies at a very high level of aggregation, based on long-term economic statistics for the total national economy and for the manufacturing sector as a whole.

The results flowing from these two levels of analysis serve to reinforce each other. Without the supporting documentation provided by the case studies, much uncertainty would surround the meaning of the broad quantitative findings because the underlying forces at work would be unclear; and without the confirming evidence of the aggregate results, the specific findings of the case studies might be regarded as little more than anecdotal. Together they tell a more convincing story than either stream of information could yield by itself.

Although the case studies come first in the book, it is helpful to begin this overview with a brief look at broad national trends. Much recent energy research has disregarded the long historical record, focusing instead on the years since 1973, when world energy conditions were turned upside down by the international oil crisis. As a result, it has been difficult to distinguish between temporary aberrations and long-term patterns. The present study, because of its evolutionary perspective, permits identification of certain persistent trends in the nation's energy use.

This study places its main emphasis on three long-term trends. Over the course of the twentieth century there has been (1) a strong increase in electricity's share of total energy use; (2) a substantial growth in the productive efficiency of the economy; and, at least since 1920, (3) a persistent decline in energy intensity — or to put it another way, a persistent improvement in the overall efficiency of energy use. These trends are shown in Figure O.1 for the manufacturing sector, for the private domestic business economy, and for the total economic activity of the nation.

Our historical research shows that these three trends are interrelated and that they reflect the following key influences:

Electrification of plant and equipment, which has contributed to technological progress and which helps account for the rising use of electricity relative to total energy;

Technological progress, which has been a major factor supporting productivity and output growth; and

Productivity growth, which in turn has yielded important savings in total inputs relative to output, including savings in inputs of energy.

The interrelationships are depicted in the following diagram:

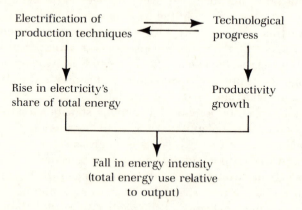

Because these interrelationships are critically important to an understanding of electricity's role in the economy, they will be discussed more fully later in this overview, where the results of our studies of long-term quantitative trends are summarized. First, however, we turn to a sequential review of the descriptive case studies.

Figure O.1 Electricity Share, Productivity Growth, and Energy Intensity, 1899–1985

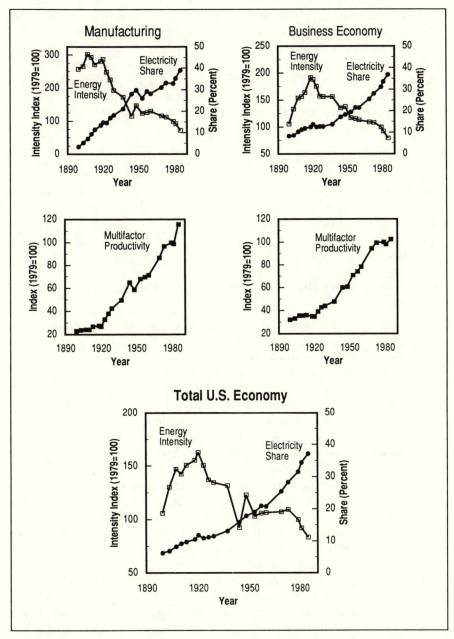

NOTE: To eliminate the effects of short-term economic fluctuations, the points plotted represent business cycle peak years.

THE CASE STUDIES

The opening case studies address electricity's role in manufacturing. We consider the effects of electrification on the organization of production first (Part I) because this generic force has affected the production framework throughout the manufacturing industries. In this capacity, electrification has exerted a powerful and pervasive across-the-board influence on the efficiency with which all manufacturing operations are performed. We then turn in Part II to more specific examples of how electricity, as an input into materials processing, has lowered costs and raised efficiency in many basic industries and has helped to create products and industries that did not previously exist.

A further distinction is made in both Parts I and II between events that began late in the nineteenth century and those changes that have begun to unfold in the more recent past. Examination of the temporal dimension reveals that even though the innovative role of electricity has extended through all of the twentieth century — and still shows no sign of abating — the key electric technologies, and the productive operations affected, have varied greatly over time.

Electricity and the Organization of Production

Part I begins by focusing on a major development in American technological history: the electrification of machine drive. At the beginning of the twentieth century, the location of machinery and the organization of production activities on the factory floor were severely constrained by the elaborate systems of shafts and belts needed to power the machines. Chapter 1 describes how the use of electricity removed these constraints in a transition that spanned several decades. At first, electric motors were simply substituted for steam engines in turning long line shafts; then, machines were grouped along shorter shafts, and each group was powered by a separate smaller motor; finally, shafting was eliminated altogether and each machine was run by its own electric motor.

Distributing power by wire to individual machines allowed a total reorganization of activities on the factory floor. Machinery could now be arranged according to the logical sequence of manufacturing operations. Materials handling could be reduced and the tempo of machine operations could be better coordinated, enabling work to proceed from one operation to the next in a continuous sequence or flow.

Although organizational changes within the factory associated with the electrification of mechanical drive started on a piecemeal basis much earlier, the real transformation of manufacturing operations took place largely in the years between World War I and World War II. We believe that it was the widespread changes in the organization of factory work that occurred during those years that drove the unprecedented productivity growth experienced at the same time and helped to initiate the long-term decrease in energy intensity.

Chapter 2 highlights a further transformation in manufacturing that began after World War II: the application of electricity and electronic devices to the automatic control of production machinery. By the 1950s, many factories utilized special-purpose equipment for rapid production of large numbers of identical products. However, the most significant change came with the advent of electronically controlled machine tools during the late 1950s and 1960s, which permitted manufacturers to produce smaller batches of products automatically with a variety of designs.

Today electronic digital computers greatly extend this capability because of the speed and ease with which computers can handle vast quantities of information and provide instructions to machines. Moreover, computer-controlled machines are sometimes linked together, forming whole systems that can manufacture a diversity of products with little human intervention. But controlling and linking machines is only part of what electricity and the digital computer can do in manufacturing. Electronic information management also allows manufacturers to schedule and coordinate to an unprecedented degree the use of machines, inventory, and even office resources, thus enhancing the speed and flexibility with which an enterprise can respond to the business environment.

Part I, then, deals with two eras in the history of manufacturing. One is historic and concerned with driving machinery, the other contemporary and concerned with handling information. These periods are similar, however, in this important respect: systemic changes in the way in which manufacturing activities are organized, with important consequences for the overall efficiency of production, are rooted in the use of electricity and electrified technologies. Just as the electrification of mechanical drive triggered the reorganization of production operations within the factory, so also is the electrification of control and information management facilitating the reorganization of the entire manufacturing production system, a system that extends well beyond the factory floor.

Electricity in Materials Processing

In Part II we turn to electricity's role in the processing of materials — production activities that are generally far more intensive than the rest of manufacturing in their use of energy. Currently the materials processing industries consume more than 70 percent of manufacturing's energy purchases, while accounting for only about 20 percent of the value added in all manufacturing. For the most part, the processes used in these industries depend upon heat. They have, therefore, tended to be based on the direct use of fuels rather than on electricity.

Even so, as described in Chapter 3, electric processes have for a long time provided the fundamental technological basis for producing particular metals and chemicals. Electrically based techniques for processing bulk materials first became commercially significant about a century ago, when electricity was just arriving on the industrial scene. As far back as the 1890s, electric technologies were introduced for manufacturing aluminum, chlorine, and silicon carbide, products that had not previously been produced commercially for lack of practical production techniques. Electroprocesses also took over in the production of a few materials that were already available — phosphorous, for example, and caustic soda, coproduced with the new product chlorine. In these cases, electroprocessing allowed the use of cheap raw materials rather than expensive manufactured substances as inputs and yielded products of high purity and uniformity.

Electricity's growing use in materials processing in recent years is due to a number of new influences. As described in Chapter 4, these factors have included relative energy price movements, growing constraints on the availability of certain raw materials, and concerns about environmental pollution. That chapter sets the stage for a more detailed discussion of the growth in the use of specific electroprocessing technologies in the chapters immediately following it.

Chapter 5 examines the use of electricity for high-temperature processing. The steelmaking and glassmaking industries are profiled, and modern processing options within both of these industries are described. The emphasis is on the unique advantages of electricity for high-temperature processing and on the consequent trend toward electrification in such industries.

An outstanding example is provided by the steel industry. Electrothermal processing is at the heart of a completely new configuration of production, utilizing an altogether different technology and even different input materials than in traditional steelmaking. The ferrous raw materials for the electric arc furnace consist entirely of scrap. Thus, coke ovens and blast furnaces, which are the technological core of the traditional steel industry, are no longer required to produce molten steel. Electric steel mills are, therefore, not constrained by the location of coal and iron ore, nor are they subject to the economies of scale associated with conventional steelmaking technologies. The overall advantages of electric steelmaking are reflected by the growth of electric steel from less than 10 percent of U.S. steel production in the 1960s to almost 40 percent in 1987.

Glassmaking exhibits still another facet of electricity's growing importance in high-temperature applications. Here, electricity is being used in conjunction with natural gas to boost the production efficiency of furnaces that can be fueled with either or both energy sources in varying proportions. In such a situation, comparative movements of electricity and gas prices have been important in determining manufacturer preferences for the different sources of heat.

Chapter 6 shows how sensitivity to comparative price movements has also influenced energy choices for low-temperature thermal processes, such as those employed in the paper, chemicals, and petroleum refining industries. Direct fuel use is generally more economical than the use of electricity at the temperatures required in these industries. However, there are situations in which the use of electrically driven mechanical processes are replacing or enhancing fuel-fired thermal processes. For example, mechanical fiberization in pulpmaking, mechanical separation of mixtures in chemicals manufacture, and use of oxygen (produced electromechanically) to enrich air for fuel combustion all yield substantial advantages in raw materials efficiency. The growing need for pollution abatement has also tended to favor the use of electrical processes.

Part II concludes with a brief sampling of some of the newest developments in electroprocessing. Some of these emerging technologies simply fill special niches in the production process. Others, because of their vast superiority over traditional production techniques, could dramatically transform the industries affected.

Electricity in Nonmanufacturing

In Part III the case studies profile various nonmanufacturing activities. Wide differences have existed among these sectors in the importance of electrification as a means of achieving technological progress. One important reason is that the flexibility associated with delivering energy by wire can become a major disadvantage where mobility is required, as is the case with transportation and field operations in agriculture. In such activities, requirements for freedom of movement and range of travel have most often been met by the internal combustion engine, powered by an on-board supply of liquid fuels.

But this is not universally the case in mobile applications. In coal mining, the subject of Chapter 8, electricity became the essential basis of mechanization in underground operations beginning late in the nineteenth century, and continues to be to this day. Only limited movement is possible underground, and the use of electricity eliminates the great hazards associated with other means of powering machinery in that environment. Surface mining, on the other hand, is unhampered by space constraints and is not subject to the severe health and safety risks inherent in underground operations. Consequently, electricity has been far less important than diesel fuel to mechanization of the earth-moving equipment used in surface operations. Nevertheless, electricity has advantages as a source of power for large draglines and shovels and so is also widely used in surface mining.

Chapter 9 is devoted to transportation, a major area of energy use in which the use of electricity is minimal. Transportation illustrates how sharply the pattern of energy dependence can change with the passage of time. Early in the century, when the automobile was just coming into use, electric passenger cars and gasoline-powered vehicles started out on a roughly equal footing, but the gasoline vehicle soon surged ahead. The electric automobile could, however, still stage a comeback if and when an adequate storage battery is developed.

Electricity also provided the original means for mechanizing urban transport very early in the century when electric streetcars, interurbans, and trolley buses were in widespread use. But only scattered fragments of such networks remain today, having fallen victim to the much greater mobility provided by the use of gasoline-powered automobiles and motor buses after World War I. Railroads illustrate a situation in which steam transport was replaced by the use of diesel-electric locomotives, a technology that gave mobile units many of the advantages of electrification at a lower cost.

Agriculture, discussed in Chapter 10, presents a mixed picture. The need for mobility in field operations has created a virtual monopoly in such applications for equipment based on the internal combustion engine powered by liquid fuels. However, in irrigation and in livestock and poultry operations, roughly half the purchased energy is used in the form of electricity. The largest single use of electricity on farms is to pump irrigation water, an important source of increased agricultural productivity.

Chapter 11, which concludes Part III on nonmanufacturing, is a case study of electricity use in the home. It reveals how completely technological progress in household operations has been tied to electrified equipment, beginning with lighting and continuing midst a great — and still proliferating — variety of consumer products that have totally transformed daily life. The scope and diversity of these electrical applications in the home is now so great that residential electricity use has grown from 15 percent of total utility sales in 1930 to over 30 percent in the 1980s, a share equal to that of industry.

THE QUANTITATIVE STUDIES

The foregoing case studies provide a detailed view of changing conditions that are analyzed in Part IV within a broad quantitative framework. This analysis covers the same ground from a different perspective, allowing the identification of general trends that become evident only when individual industries are grouped together and aggregate measures are derived.

Electricity Use and Productivity Growth

Chapter 12 presents a long-term quantitative view of electricity use and productivity growth in manufacturing as a whole. Following an approach widely used in economic studies, we define growth in productivity as an increase in productive output minus increases in the major inputs (i.e., the weighted sum of the individual inputs of capital and labor). Using standard data sources, we estimate that multifactor productivity thus defined has accounted for over half of the growth rate in manufacturing output since the turn of the century, with the remainder due to increased inputs of labor and capital (Figure O.2).

Technological progress represents an important component of the growth in multifactor productivity. The central question for us concerns the role of electrification in achieving such technological advances. Having looked at particular manufacturing technologies in Parts I and II, we undertake, in Part IV, to assess electricity's role in technological progress for manufacturing as a whole. We do so by measuring the growth of electricity

Figure O.2 Average Annual Growth Rates in Manufacturing: Output, Productivity, and Input, 1899–1985

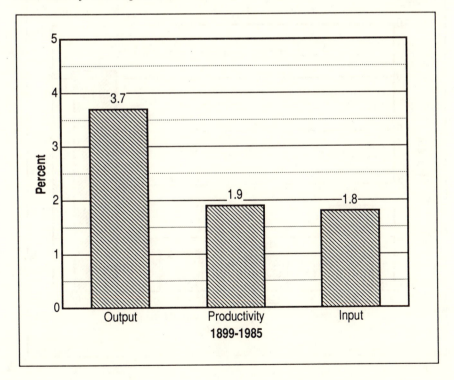

as a production input and comparing it with the growth of nonelectric energy and the major factor inputs of capital and labor.

The data show that from 1899 to 1985, the entire period we cover, the use of electricity in manufacturing grew at an average annual rate of more than 8 percent. This strong growth compares with much lower rates of growth in the other inputs: 1.5 percent for nonelectric energy, 1.3 percent for labor, and just over 3 percent for capital (Figure O.3).

Particularly revealing is the fact that the long-term growth in capital (i.e., plant and equipment) has been associated with a much steeper increase in electric than in nonelectric energy. Since changes in plant and equipment are the main vehicle for achieving technological improvements, electricity's very high rate of growth relative to capital signifies that technological progress in manufacturing over the course of the twentieth century has shown a strong affinity for energy in the form of electricity. This conclusion is strongly borne out by our descriptive case studies in Parts I and II.

Figure O.3 Average Annual Growth Rates in Manufacturing: Capital, Labor, Nonelectric Energy, and Electricity, 1899–1985

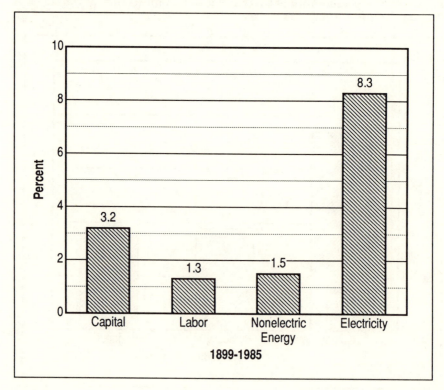

Variation in the ratios between electricity's growth and the growth in capital are also used in Chapter 12 to trace changes in the importance of electricity use to technological advance during several different periods. This analysis shows, for example, that the ties between electricity use and capital growth were strongest between 1920 and 1929, at the same time that multifactor productivity in manufacturing grew at an unparalleled rate of more than 5 percent per year. During this period, electricity use grew thirteen times faster than capital, while nonelectric energy use actually declined (Figure O.4). The surge in growth in electricity use relative to capital during these years resulted from the electrification of machine drive, which constituted the dominant technological thrust in manufacturing at that time. Other time periods, both before and after the 1920s, also show how evolutionary change in manufacturing technology has been reflected in varying productivity growth rates and in the changing relative importance of electricity as a production factor.

Figure O.4 Average Annual Growth Rates in Manufacturing: Capital, Labor, Productivity, Nonelectric Energy, and Electricity, 1920–1929

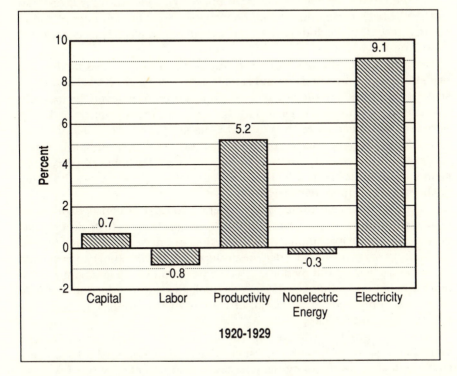

Electricity Use and Energy Conservation

What many would regard as the most surprising long-term trend is discussed in Chapter 13. This is the decline in the intensity of overall energy use over much of the twentieth century, coincident with a strong rise in the electricity fraction of total energy use. The energy intensity measure on which we focus is the ratio between energy consumed and output produced which, at the most highly aggregated national level, is represented by the ratio of energy to gross national product (GNP).

Government statistics show that between 1973 and 1985, when energy conservation became a major national policy goal, the British thermal units (Btus) of commercial energy used per dollar of GNP fell at an average annual rate of 2.2 percent. Although it is widely believed that such a decline in energy intensity is unprecedented, the data in Chapter 13 (summarized in Figure O.1) show that this is definitely not so. Energy efficiency grew through much of the twentieth century and for some periods at even faster rates than in 1973 to 1985.

This decline in the intensity of energy use is particularly puzzling because many more Btus are contained in the primary fuels used to generate electricity than in the electricity that is produced. This fact would suggest that as the electricity fraction of total energy use rises, overall energy use in relation to output (including the fuels consumed in electricity generation) should be more, rather than less, intensive.

However, just as the rise in multifactor productivity has resulted in the need for fewer inputs of labor and capital to produce a given level of output, so, too, have fewer inputs of energy been required. Whatever the losses in raw energy resulting from electricity's rising share in the energy total, they have over time been more than offset by the growth in productivity due to technological progress based on electrified production techniques. In this context, energy conservation may be viewed as a by-product of new, more economically efficient production methods rather than the result of any deliberate effort to save energy.

In view of the great importance assigned to energy conservation as a policy objective in recent years, the lesson taught by U.S. historical experience deserves strong emphasis. A reduction in energy inputs per unit of final output (as reflected in a decline in the energy-GNP ratio) can over time be successfully achieved by doing what comes naturally in a dynamic, competitive economy — that is, adopting technologies and production methods that seek to maximize the overall efficiency of production, not just the efficiency of energy use.

Thus, the aggregate statistics in Part IV of the book confirm what the individual case studies have suggested. As electric technologies have penetrated industry, productivity has increased, and less energy has been utilized relative to output. It is clear that these three persistent trends can and

do coexist not only in individual industries but throughout the manufacturing sector as a whole. Indeed, other statistics, also shown in Figure O.1, make it clear that these trends pervade the private business economy of the United States and the total national economy as well.

THE OUTLOOK

What is the outlook for the continuing electrification of American industry? Our major theme concerns its role in several long-term developments supporting growth and productive efficiency in the American economy: more efficient organization of factory operations, the emergence of new products and new industries, and various improvements in the production operations of the materials processing industries and of particular nonmanufacturing activities. All of these will almost certainly continue to be powerful forces in the future.

Certain other themes also emerge, but in patchwork fashion, at various points throughout the book. These merit notice because they provide additional clues as to the future direction and impact of electrification. One such theme is the importance of relative energy price movements. Reductions in the cost of electricity in comparison with competing energy sources can hasten the process of electrification, whereas increases in electricity's comparative cost can retard it. Yet, as the studies in this book show repeatedly, fuels and electricity often entail fundamentally different industrial technologies, with different levels of productivity. So the effects of relative price movements of electricity and other energy sources can go far beyond the primary choice of energy sources. To the extent that such price movements encourage or discourage electrification, they may also affect the nation's future productive efficiency.

Raw materials also figure heavily in the outlook for electrification. Electric-based technologies can sometimes, as in wood pulping or petroleum refining, utilize more plentiful, lower-grade raw materials than those required in conventional fuel-based processes. Sometimes, as with coke and iron ore in the steel industry, electroprocessing can completely eliminate the need for production inputs that were once essential. As a result of process modifications such as these, electrification can help to offset the effects of depletion of particular resources.

Pollution concerns constitute yet another theme. Increasingly stringent air quality and waste disposal regulations for industry favor the use of electricity over fuels in process industries. In addition, the use of electric automobiles in place of combustion engine vehicles could have very significant air quality benefits in crowded, smoggy cities. Electricity, being absolutely clean at the point of use, could come into increasing demand as the base for improved technologies to help preserve environmental quality.

Finally, there is the decentralization that characterizes some electricity-based operations. Steel production, once centralized in a few huge complexes near the required resources, now also thrives in the form of more widely scattered minimills. More generally, computerized information technologies can link small producing units across the country. And people can now work in increasingly remote office settings, or even in their homes. The productivity of today's far-flung electrified enterprises would amaze the early captains of industry, whose fuel-based technologies required the concentration of capital, labor, and energy in one central location.

Taking all of these themes into account, what can be said about the relationship among the three quantitative trends basic to our work in this book — a rising electricity share of total energy use, strong productivity growth, and a persistent improvement in national energy efficiency? Recent history can be misleading on this score. After many years of marching along together, these three trends parted company for a period of years as a result of the energy turbulence of the 1970s. As shown in Chapter 13, the rise of the electricity share continued, although at a reduced rate, while energy efficiency shot ahead and productivity growth tended to lag.

Yet these departures from the long-term pattern appear to have been only a temporary aberration. The force of soaring energy prices created a heavy emphasis on energy conservation as such, which disrupted the established relationships. The need to save energy often led to choices that reduced the intensity of energy use but failed to support overall productivity growth. Once the energy shortages eased and prices began to fall, the three trends appear once again to have fallen more closely into step.

At core, the book addresses three interrelated questions: Why has the electricity share risen so dramatically throughout this century despite electricity's relative price disadvantage? How has this rise been related to productivity increases? And how can it be consistent with steady improvements in the efficiency of energy use?

The answer to all three questions lies in technological progress strongly dependent on the use of electricity. The unique characteristics of electricity have for long periods allowed innovative electric-based technologies to produce more output than ever before with fewer combined inputs of labor and capital per unit of output and also with fewer inputs of energy.

As these advances continue, the historic rate of increase in electricity use could moderate somewhat because the energy requirements of some of the newest computer-related technologies are so minimal. Even so, dependence on electricity as an energy form will likely continue to grow. It will grow as computerized information management systems reshape the nation's productive capacity and as the heavily energy-intensive materials processing industries continue to shift from fuels to the use of electricity.

Part I
Electricity in Manufacturing:
Its Role in the Organization
of Production Systems

Introduction

In this part of the book we explore electricity's role as an agent of organizational change, in contrast to its role as an agent of physical and chemical change in materials (the subject matter of Part II). Chapter 1 describes the electrification of mechanical drive, a major transition in energy use that largely took place between 1890 and 1930. Chapter 2 discusses the use of electricity in the control of production processes through the management of information — a story that begins where Chapter 1 leaves off and continues through the present, into the midst of what some have called the "information revolution."

These two eras in the history of manufacturing — one historic, the other contemporary — appear quite dissimilar because the first focuses on methods of driving machinery, while the second focuses on methods of handling information. Moreover, in the first era, electricity itself was a new, almost magical tool, and its adoption was hastened by enthusiastic reports of the wide variety of ways in which it could be used to assist production. In the contemporary era electricity is taken for granted, and even though the "information revolution" is driven by electronic technologies, the role of electricity is, at best, accorded only implicit recognition.

Despite these obvious differences, the historic and contemporary eras are similar in one basic respect. In each era, systemic changes in the organization of manufacturing activities, with important consequences for the overall efficiency of production, spring from the use of electrified technologies. In Chapter 1 a fundamental change in the way in which energy is provided to machinery and in Chapter 2 a fundamental change in the way in which information is provided become the basis for a transformation in the organization of manufacturing. In both situations, those changes are completely dependent upon the use of electricity as an energy form.

What characteristics of electricity allow it to effect marked change in activities as diverse as machine drive, production control, and information management? The answer may lie in what is perhaps the most unique attribute of electricity: the exceptional precision — in space, in time, and in scale — with which energy in this particular form can be transferred. The very nature of electricity (electrons moving in a conductor) is responsible for this precision.

With wires, electricity can be provided anywhere in a factory — to motors driving huge machines as well as to tiny circuits within process controllers. By using simple devices such as switches and rheostats, electric current is provided on any time schedule. And in production, electric power is provided in any desired amount, from watts to megawatts. Such precision far exceeds that which is possible when energy is transferred mechanically (e.g., by the historic steam-based rotating shaft system described in Chapter 1) or when energy is transferred as heat (a subject taken up in Chapters 5 and 6).

Chapter 1 traces the historic shift from steam to electric power and its impact on factory organization. Today, the driving of production machinery is almost universally carried out via electricity. But for centuries machine drive shafts were turned by water or steam via a system of shafts and belts. When it became available, electricity proved to be a vastly superior means of driving production machinery. As a result, electric generators, wires, and motors have replaced virtually all of the fixed-in-place infrastructure required by the earlier system of power distribution.

In the late nineteenth and early twentieth centuries, the organization of factory production was dictated by the need to arrange machinery in long rows parallel to the constantly rotating iron and steel shafts that brought them their power. As recounted in Chapter 1, electrification changed all that, but not in one giant step. The first change was to turn the line shafts by using electric motors rather than steam engines. Although it appears unnatural today, even foolish, simply to replace a steam engine with an electric motor and leave the mechanical power distribution system unchanged, this was in fact the usual early stage juxtaposition of the beginnings of a new technological system upon the framework of an old one.

The potential for real change became far clearer as industry gained experience with electric motors and their intrinsic advantages were more clearly understood. In its initial form, electric drive, like its predecessor steam power, was centralized; later, electric drive was distributed separately to each machine. Decentralization was the decisive step. Only then, with electrified unit drive (i.e., a motor on each machine) did a wholly new system of mechanical power distribution come into existence, with its revolutionary consequences for the overall organization of factory production.

The new system would not have been possible but for two specific conditions: the ability to subdivide electric power into appropriately sized increments that could be delivered with precision and the availability of appropriately sized motors in which the electricity could be used. These two — the precision power source and the motor for its use — constituted a dynamic new production team in American industry.

Flexibility in the form of electrified unit drive was the new ingredient introduced by this innovative production team. Freed from the constraints imposed by power distribution via fixed shafts and belting, activities on the factory floor could be reorganized to bring about a much better integration of production through the coordination of related mechanized operations and the adjustment of their tempos to achieve a natural or continuous sequence. With better integration, a faster rate of throughput resulted, and this meant higher productivity and lower costs per unit of output.

Greater throughput was the focus of assembly-line production systems that began before World War I in the automotive industry and that flowered during the interwar years. Such manufacture increasingly utilized special-purpose equipment whose high initial cost could be spread out over a great number of identical pieces produced rapidly. However, this system also brought with it a degree of standardization that has become a source of concern in more recent years, as firms find it increasingly necessary to respond to changing product demands and the pressures of international competition.

Chapter 2 addresses that contemporary need for ever-greater flexibility in manufacturing production. The key is the use of machinery that can be controlled automatically and reprogrammed easily to produce a variety of products and a diversity of product designs. Here, once again, flexibility is being achieved through technology that is tied to the use of electricity. The new production team is electricity plus the digital computer, now playing the same critical role that was filled by electricity and the electric motor earlier in the twentieth century.

The mainframe computer of the 1960s, with its ability to acquire, store, process, and transmit large amounts of information, at first seemed to be an ideal controller for these complex processes. Because such computers were expensive, manufacturers would use one of them to control a number of machines. Firms at the forefront of computer control brought more and more activities under direction of a central computer. This tendency toward centralized control slowed, however, because of the complexity of the inter-actions and the vulnerability of the systems.

The failure of centrality and the development of microcomputers consti-tuted a point of demarcation in the evolution of electronic control and information management. Since the mid-1970s, control has evolved toward decentralized techniques, the same direction in which electrified mechani-cal drive evolved half a century earlier. In the case of computerized equipment, this decentralized organization is sometimes called "distributed intelligence."

Endowing an automatic machine with the information-handling ability of a microcomputer gives it the capacity for flexible operations. But it does more than this. Information can be shared automatically with computers controlling other machines, with supervisory computers, and even with computers supporting certain office activities.

The microcomputer has become a key unit in a new electricity-based production network that can accomplish far more than just controlling and linking machines. Not only fabrication, but also product design, engineering, and distribution can readily participate in this network. Electronic integration of a firm's diverse activities can free companies from the constraints of time-honored organizational structures. Factories and the office activities that support them, indeed entire production enterprises, can be reorganized so as to enhance the speed and flexibility required in constantly changing markets. Today, as in the past, electricity-based technology offers highly innovative ways to organize the work of manufacturing industries.

1 | *Electrified Mechanical Drive: The Historical Power Distribution Revolution*

Warren D. Devine, Jr.

THE HISTORICAL SETTING

Major changes took place between 1880 and 1930 in the forms of energy that were produced and used in the United States. These changes included switches from coal to oil and natural gas, and the shift from direct use of raw energy commodities (coal and water power) to the use of processed energy commodities (internal combustion fuel and electricity).

Perhaps the most rapid and complete transition in energy use was the shift from steam power to electric power for driving machinery. Steam power prevailed at the turn of the century, with steam engines providing around 80 percent of total capacity for driving machinery. By 1920 electricity had replaced steam as the major source of motive power, and by 1929 — just forty-five years after their first use in a factory — electric motors provided 78 percent of all mechanical drive.

In the following sections the evolution of this major application of electricity in production is viewed from the standpoint of engineers and entrepreneurs present at the time. These witnesses agreed that the substitution of electric power for steam power reduced the energy needed to drive machinery and that it sometimes yielded savings in the total cost of driving machinery. Much more important, however, they observed that this substitution went hand in hand with changes in factory organization. These changes considerably enhanced output relative to all inputs to production, including inputs of energy.

LINE SHAFT DRIVE

Until late in the nineteenth century, production machines were connected by a direct mechanical link to the power sources that drove them. In most factories, a single centrally located prime mover, such as a water wheel or steam engine, turned iron or steel "line shafts" via pulleys and leather belts.

These line shafts — usually 3 inches in diameter — were suspended from the ceiling and extended the entire length of each floor of a factory, sometimes even continuing outside to deliver power to another building.

Power was distributed between floors of large plants by belts running through holes in the ceiling; as these holes were paths for the spread of fire, interfloor belts were often enclosed in costly "belt towers." The line shafts turned, via pulleys and belts, "countershafts" — shorter ceiling-mounted shafts parallel to the line shafts. Production machinery was belted to the countershafts and was arranged, of necessity, in rows parallel to the line shafts. This "direct drive" system of distributing mechanical power is illustrated in Figure 1.1.

The entire network of line shafts and countershafts rotated continuously — from the time the steam engine was started up in the morning until it was shut down at night — no matter how many machines were actually being used. If a line shaft or the steam engine broke down, production ceased in a whole room of machines or even in the entire factory until repairs were made.

To run any particular machine, the operator activated a clutch or shifted the belt from an idler pulley to a drive pulley using a lever attached to the countershaft. Multiple pulleys offered speed and power changes. Drip oilers, suspended above each shaft hanger, provided continuous lubrication. Machine operators were usually responsible for the daily filling and adjusting of these oilers and for periodically aligning the belts. As the belts stretched and became loose, they had to be shortened slightly and the ends laced tightly together. These maintenance tasks took significant amounts of time, since a large plant often contained thousands of feet of shafting and belts and thousands of drip oilers.[1]

Electricity was probably first used for driving machinery in manufacturing in 1883 — the year after electric power was first marketed as a commodity by Thomas A. Edison.[2] (A chronology of the electrification of mechanical power in American industry is given in Figure 1.2.) Early electric motors operated on direct current (dc) and had usually less than 1 horsepower (hp) capacity.

The first reliable and efficient dc motors in capacities exceeding 1 hp were introduced in 1885 and were designed for use on the Edison dc circuits.[3] The Edison Electric Light Company encouraged the use of motors because daytime motor loads would complement nighttime illumination loads; since the marginal cost of serving these loads was relatively low, large profits were foreseen. By late 1886, 250 motors of 0.5 to 15 hp capacity were operating in a number of cities across the United States.[4] By 1889, total electric motor capacity in manufacturing exceeded 15,000 hp, with over one-quarter of this capacity in printing establishments.[5]

Figure 1.1   Evolution of Power Distribution in Manufacturing

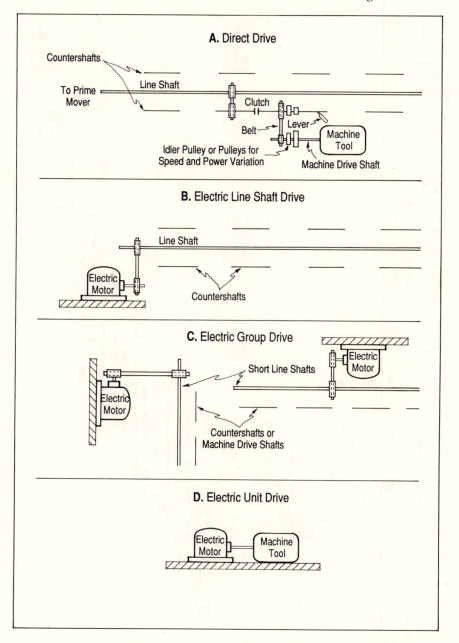

Figure 1.2 Chronology of Electrification of Mechanical Power in Industry (A) Methods of Driving Machinery; (B) Rise of Alternating Current; (C) Share of Power for Mechanical Drive Provided by Steam, Water, Electricity; (D) Key Technical and Entrepreneurial Developments. (The percentages shown in "C" do not necessarily add up to 100 because sources of mechanical power other than steam, water, and electricity are not shown.)

	1870	1875	1880	1885	1890	1895	1900	1905	1910	1915	1920	1925	1930

A.

Direct drive ———

Line shaft drive ——————————————————

Group drive ———————————————

Unit drive ——————————

B.

Dc transmission ——————

Dc motors in manufacturing ——————

"Battle of the currents" ——

Ac transmission ——————

Ac motors in manufacturing ——————

C.

Steam 52%		Steam 64%		Steam 78%		Steam 81%		Steam 65%		Elec. 53%		Elec. 78%
Water 48%		Water 36%		Water 21%		Water 13%		Elec. 25%		Steam 39%		Steam 16%
				Elec. <1%		Elec. 5%		Water 7%		Water 3%		Water 1%

D.

1870 Dc electric generator (hand-driven)

1873 Motor driven by a generator

1878 Electricity generated using steam engine

1879 Practical incandescent light

1882 Electricity marketed as a commodity

1883 Motors used in manufacturing

1884 Steam turbine developed

1886 Westinghouse introduces ac for lighting

1888 Tesla develops ac motor

1891 Ac power transmission for industrial use

1892 Westinghouse markets ac polyphase induction motor, General Electric Company formed by merger

1893 Samuel Insull becomes president of Chicago Edison Company

1895 Ac generation at Niagara Falls

1900 Central station steam turbine and ac generator

1907 State-regulated territorial monopolies

1917 Primary motors predominate; capacity and generation of utilities exceeds that of industrial estab.

By the early 1890s, dc motors had become common in manufacturing, though they were far from universal. Mechanical drive was first electrified in industries such as clothing and textile manufacturing, and printing, where cleanliness, steady power and speed, and ease of control were critical.[6] Here electric motors improved the quality of work, but they did not change the method. As illustrated in Figure 1.1, the only difference between direct drive and the earliest electric drive system was the type of machine used to turn the line shafts. In the first electric drive system — called "electric line shaft drive" — all countershafts, belts, pulleys, and clutches remained. Thus a single motor might have driven a few or several hundred machines; in textile mills it was not uncommon for large motors of several hundred hp capacity to drive well over 1,000 looms.

The costs of turning line shafts with steam engines and with electric motors had been thoroughly examined by 1891. Dr. Louis Bell reported that when small amounts of power were needed, it was usually cheaper to use electricity than steam.[7] This was so because small steam engines were much less energy efficient than large ones and because the price of small amounts of direct current electricity could be low if generated in large quantities in a central station. In plants that used large amounts of power, it was usually cheaper to drive machinery with steam engines than with electric motors. However, as Bell pointed out,

> The electric motor may be cheaper than steam even when the latter may be used on a large scale; the only condition being that we shall be able to take advantage of cheaper production (elsewhere) by the ability electricity gives us to transfer power from a distant point . . . we must look upon electricity as an enormously powerful and convenient means of transferring power from one point to another with the greatest simplicity and very small losses.[8]

This view of electricity as a means of power transmission was common in the early 1890s. It became apparent that large factories did not have to be located adjacent to sources of water power; moreover, they did not necessarily have to be designed around a large steam engine. Instead, power could be produced at good water power sites or coal depots some distance away and transmitted to the plant in the form of electricity.

A Columbia, South Carolina, textile mill built in 1893 had been located near water power, but mechanical transmission of power from the water wheels to the mill machinery proved impractical. Two electrical equipment manufacturers, Westinghouse and Siemens-Halske, proposed transmission systems that were in accord with the practice of the time: direct current would be generated at the river and transmitted across a canal to the mill, where large motors would turn the line shafts. In these proposals, electricity was simply a substitute for mechanical power transmission via a 1,000-foot cable.[9]

In 1895, Professor F. B. Crocker of Columbia University visited Baltic and Taftville, Connecticut, "to see the practical working of the well-known power transmission plant between these places."[10] Both water and steam power had previously been used to run the Ponemah textile mill at Taftville. A hydroelectric plant was then installed at a dam upriver at Baltic. The power was transmitted via overhead lines to Taftville, where "motors are located in the basement of the mill, near the engines which they replace. They are belted to pulleys which are connected to their respective shafts by friction clutches. . . . One of these motors drives 1,200 looms requiring an expenditure of about 155 horsepower. The other drives 500 looms."[11]

In both the above examples, electric power was preferred because it enabled distant, low-cost mechanical power to be transmitted to the mills relatively easily. The means of distributing power within the plants, however, remained unchanged. Steam engines and water wheels were simply replaced by electric motors, and these motors drove line shafts via belts and pulleys. Electricity was seen as a way to transmit mechanical power to factories but not yet as an agent for distributing power within factories.

Replacing a steam engine with an electric motor, leaving the power distribution system unchanged, was the usual juxtaposition of the beginnings of a new technological system upon the framework of an old one. But few had this perspective in the 1880s and 1890s. Shaft and belt power distribution systems were in place, and manufacturers were familiar with their problems. Turning line shafts with motors was an improvement that required modifying only one end of the system.

ELECTRIC GROUP DRIVE

As long as electric motors were simply used in place of steam engines to turn long line shafts, the shortcomings of mechanical power distribution systems remained. According to mechanical engineer H. C. Spaulding, the most serious problems were the large friction losses in the system and the necessity of turning all the shafting in the plant regardless of the number of machines in operation.[12] Spaulding realized that these problems continued to exist because manufacturers had not yet come to view electricity as a means of power distribution within their plants. Production machinery ought to be arranged in groups, he said, with each group of machines driven from a relatively short line shaft turned by its own electric motor. Such a group could be operated most efficiently if the machines ran at similar speeds and if the group exhibited little variation in load.

The first large-scale application of such a "group" approach to machinery was probably in the General Electric Company plant in Schenectady, New York, in the early 1890s. Forty-three dc motors totaling 1,775 hp turned a total of 5,260 feet of shafting. These motors were located in perhaps forty different shops or departments. Thus the line shafts must have been relatively short, perhaps averaging 100 to 150 feet of shaft per motor. These shafts turned countershafts that drove machinery in the usual manner.[13]

General Electric salesmen undoubtedly made good use of their company's experience with electric group drive. Early customers of the company were industries near water power, such as the Columbia, South Carolina, textile mill noted previously. The 1893 proposal of salesman S. B. Paine for powering the mill differed from those of his competitors in two important respects. Polyphase alternating current would be used rather than the much more common direct current, and seventeen 65-hp alternating current (ac) motors would drive groups of machines.

The use of electric group drive rather than electric line shaft drive would reduce airborne dirt and grime, which was particularly important in textile mills. The use of ac induction motors rather than dc motors would eliminate commutator sparking, which was a fire hazard in the lint-filled atmosphere of the mill. Finally, the motors were to be mounted on the ceiling so as to occupy no floor space.[14]

Paine's proposal was important because it implied electricity was more than a substitute for direct transmission of mechanical power to the plant and more than a means of power distribution within the plant: properly applied, it could improve the overall efficiency of production. The General Electric bid was accepted by the mill owners, even though it was the most costly of those submitted.

Other early adopters of group drive were plants located near Niagara Falls to take advantage of the low-cost hydroelectric power that became available there in 1895. One of these plants was the nation's largest nut and bolt factory, that of Plumb, Burdict, and Bernard, in North Tonawanda, New York.[15] Their facilities were typical of those designed around electric group drive instead of around line shaft drive: machines were grouped together and belted to countershafts turned by a motor-driven line shaft that served only those particular groups of machines (Figure 1.1).

Group drive offered more flexibility in locating machinery than line shaft drive, often increasing production efficiency. For example, machines that performed related operations — but that might have been located some distance apart with line shaft drive — could now be consolidated in an individual shop or department. The specialized shops of the North Tonawanda plant evidently represented a considerable advance over older plants.

Organizing manufacturing operations into specialized shops or departments was facilitated by group drive because the motor-driven line shafts and production machinery could be oriented in virtually any direction. In 1897, for example, the Keating Wheel Company built a new factory in Middletown, Connecticut, with shops in six wings perpendicular to the plant's main building.[16] This plant would probably not have been built this way without electric power distribution because it would have been quite troublesome to turn line shafts in the wings from the shaft in the main building by means of belts. Furthermore, motors turning line shafts in individual shops took up no more floor space than overhead mechanical power distribution. Motors were often mounted on platforms suspended from the ceiling or were attached to the wall; in one machine shop the motor platforms were above the tracks of an overhead traveling crane![17] Obviously, the structural design and internal organization of these and other plants was intimately associated with electric group drive.

As manufacturers gained experience with group drive, countershafts were eliminated and production machinery was belted directly to line shafts, reducing power losses between the motor and the machines and leading to greater consolidation and specialization. Line shafts became shorter; shafts thirty to fifty feet long were typical in many machine shops.[18] Some engineers even held "the extreme view . . . that a motor should be applied to every tool."[19]

The increasing use of electric drive in the late 1890s and first decade of the twentieth century was concurrent with the rise of on-site electricity generation via steam. From 1899 to 1909, electric drive increased from less than 5 percent of total capacity for driving machinery to 25 percent. During this decade 60 percent to 70 percent of the electric motor capacity in manufacturing plants consisted of motors powered with electricity generated by the manufacturing establishments themselves.

For plants employing steam engines, the use of electric drive involved three transformations of energy as opposed to only one with the use of direct drive. Even so, energy consumption was often less with electric group drive. Why?

The greatest reduction in energy consumption upon adoption of group drive came because any department, shop, or group of machines could be operated independently; the motor driving a group of machines could be stopped when the machines were not being used.[20] Conversely, a particular group of machines could be operated without rotating the shafting throughout the entire plant; energy savings could be significant if only part of a plant was working overtime or at night. The ability to shut down or start up selected equipment is taken for granted today, but such ability represented technical and organizational innovation in the 1890s.

Reductions in energy consumption also came with electric group drive because these systems contained less shafting and fewer belts and pulleys than line shaft drive systems. Thus, less power was lost to friction in turning shafts. Indeed, estimates of the power required to rotate the shafting in factories using direct drive range between about one-third and three-quarters of the power made available at the plant's steam engine.[21]

But energy use with electric group drive was not dramatically lower than with direct drive based on steam. A saving of 20 percent to 25 percent in the amount of coal used was typical.[22] However, even large energy savings would have had a rather minor direct impact on total production costs. This is so because the cost of fuel for steam generation was relatively small, usually between about 0.5 percent and 3 percent of the total cost of producing a unit of output.[23]

We have already seen that during the late 1890s and early 1900s electric group drive was used in a number of new factories that had been specifically designed around electric power distribution. For these and other firms, electrification of mechanical drive and factory reorganization went hand in hand. This reorganization often entailed creating individual shops or departments specializing in a particular operation or in production of a particular component.

Reorganization also frequently involved relocating the steam plant in the basement or in a separate building, thus freeing space in the main building for production and isolating power generation from power use. Regardless of where the steam plant was located, electric power distribution dramatically reduced opportunities for the spread of fire. This was so because shaft and belt holes in floors and walls were no longer needed; if necessary, production could be contained in individual, fireproof rooms. Thus, after installing group drive, many manufacturers obtained reductions in fire insurance premiums.[24] Use of electric motors to drive machinery in groups also meant that no single line shaft breakdown could affect the entire plant. Only machines in that shop or portion of the room in which the mishap occurred would be stopped, with the rest of the plant running as before.

Finally, as mentioned earlier, direct drive imposed certain constraints on the size and configuration of individual buildings. Concurrent with the rise of electric drive, very large factory buildings were constructed. Contemporary accounts imply that such large installations were uncommon at the time.[25]

Thus these various benefits of electric group drive appear to have been at least as important as energy efficiency in fostering electricity use in manufacturing around the turn of the century. Electricity was now generally seen as the preferred method of power distribution within new manufacturing plants. Group drive was a major form of electric drive through World War I and was vigorously defended as late as 1926.[26] Yet, as early as the turn of the century, a few innovative manufacturers decided it was best to eliminate shafting altogether and run each machine with its own electric motor.

ELECTRIC UNIT DRIVE

During the 1890s many engineers advised against running any but the very largest machines with individual electric motors. This was primarily because the power capacity required to drive a group of machines was much less than the sum of the capacities required to drive each machine separately. By the early 1900s, however, most observers agreed that individual (or "unit") drive would eventually replace other techniques for driving nearly all large tools. Some enthusiasts, such as engineer G. S. Dunn, had an even broader outlook; Dunn felt unit drive would soon be adopted for very small machines as well.[27] Why?

The Cost of Driving Machinery

With unit drive, a motor was usually mounted right on the machine being driven (Figure 1.1). Motor and machine drive shaft were often connected by a belt and pulleys or by gears. Sometimes the motor armature and drive shaft were directly linked via a key-and-slot coupling.

Unit drive used less energy than group drive for the same reasons that group drive used less energy than line shaft drive. Unit drive entirely eliminated power losses due to friction in rotating line shafts and countershafts, sometimes almost doubling overall, full-load efficiency of power production, transmission, and distribution.[28] More important, no energy was wasted turning shafts with some machines out of service.

A manufacturer's total cost of driving machinery — consisting not only of energy costs, but also of capital, labor, and materials costs — was often somewhat lower with unit drive than with group drive. For this to occur, savings in energy, labor, and materials had to offset any increases in capital costs.

Capital costs could be high with unit drive. Many more motors were required, and their total horsepower was often five to seven times the horsepower of a single motor for group driving of the same machines.[29] This was because with unit drive, each motor had to be of sufficient capacity to handle the maximum demand of its machine; with group drive, the motor could be sized to take advantage of load diversity. With adoption of unit drive the aggregate horsepower of electric motors in a plant increased dramatically, but the actual peak power need of the plant often decreased somewhat.

In principle, this permitted installing a proportionally smaller power plant, with capital cost savings. Furthermore, factory buildings could be of lighter and cheaper construction since their roofs no longer had to support heavy line shafts, countershafts, and pulleys. The elimination of this mechanical power distribution system effected the greatest saving, sometimes offsetting the first cost of additional motors and wiring.[30] Nevertheless, the cost of equipping a plant with electric unit drive was usually somewhat higher than installing electric group drive.[31]

Labor and materials costs of driving machinery, however, were generally lower with unit drive. There were no belts to tighten and adjust, and no drip oilers to fill. Thus, lower costs of energy, labor, and materials were probably often sufficient to offset the capital cost penalty of electric unit drive, giving this technique a slight cost advantage over group drive or line shaft drive.

But savings in the total cost of driving machinery were not terribly important. As noted previously, the cost of fuel for electric power generation or the cost of purchased electricity was a minor item, usually between about 0.5 percent and 3 percent of the total cost of producing a unit of output. Since the cost of energy has always been a major fraction of the total cost of driving machines,[32] it follows that the cost of mechanical drive was a small component of total production cost — certainly less than one-tenth and probably closer to one-twentieth of total cost per unit of output. Thus, even a large reduction in the cost of mechanical drive would have had a minor impact on production costs.

Electricity: A Lever in Production

By the early years of the twentieth century, manufacturers were beginning to recognize that savings in the cost of driving machinery were almost insignificant compared to other benefits from use of electric unit drive. According to American Institute of Electrical Engineers (AIEE) member Oberlin Smith,

> The problem talked much about until quite recently has been whether we should put in motors at all, because we did not know whether they were going to take more power or not . . . that is a point of very little importance, compared with the total expenses of the shop. It doesn't matter if it is 5 or 10 or 20 percent, considering the great advantages we are going to get in all these other ways.[33]

S. M. Vauclain, superintendent of the Baldwin Locomotive Works, reports his company's favorable experience:

> In conclusion, while the question of the saving in power which the adoption of electric motors permitted was of importance, it was by no means the deciding factor; I would have put in electric driving systems not only if they saved no power, but even if they required several times the power of a shaft and belting system to operate them.[34]

Electric equipment sales engineers began to shift their emphasis from energy savings to what an engineer with the Crocker-Wheeler Electric Company called "indirect savings":

> There were many factories which introduced electric power because we engaged to save from 20 to 60 percent of their coal bills; but such savings as these are not what has caused the tremendous activity in electric power equipment that is today spreading all over this country . . . those who first introduced electric power on this basis found that they were making other savings than those that had been promised, which might be called indirect savings.[35]

Thus, with the advent of unit drive, electricity was beginning to be seen as more than an economical means of power distribution within factories; to many, it was a "lever" to increase production.

Manufacturers often estimated the additional production they could ascribe to electric unit drive and reported their experiences at technical society meetings. At a meeting in 1901 Professor F. B. Crocker summarized a number of these reports:

> It is found that the output of manufacturing establishments is materially increased in most cases by the use of electric driving. It is often found that this gain actually amounts to 20 to 30 percent or even more, with the same floor space, machinery, and number of workmen. This is the most important advantage of all, because it secures an increase in income without any increase in investment, labor, or expense, except perhaps for material. In many cases the output is raised and at the same time the labor item is reduced.[36]

How did electric unit drive facilitate these increases in output and productive efficiency?

Increased Flow of Production

Unit drive gave manufacturers flexibility in the design of buildings and in the arrangement of machinery. No longer were machines grouped and placed relative to shafts. Machinery could now be arranged on the factory floor according to the natural sequence of manufacturing operations, minimizing materials handling. The tempo of machine operations could be better coordinated, enabling workpieces to proceed from one operation to the next in a continuous sequence, or flow. With this integration of production, greater throughput was possible using essentially the same production resources. This meant lower cost per unit of output.

The ability to arrange machinery irrespective of shafting made all space in the factory equally useful. Furthermore, a machine's position could be changed readily, without interfering with the operation of other machines. Engine-driven overhead cranes were used for factory assembly of certain large items before 1900, but overhead mechanical power transmission precluded cranes almost everywhere else.

By eliminating shafting, electric unit drive left clear and unobstructed passages and headroom and allowed the use of overhead traveling cranes in any part of a plant. One speaker at the 1895 AIEE meeting predicted that small electric cranes would revolutionize materials handling.[37] Nine years later electric cranes were being called an "inestimable boon" to production.[38] By 1912, the importance of clear headroom for cranes was "so generally recognized as to require no comment."[39]

But unit drive did more than permit easier moving of work to machines; it also made it possible to move machines to the work. Movable power tools could now be readily applied to any part of a large workpiece. As of 1912, such tools "played an active and extensive part in increasing the output in structural iron works, locomotive works, and modern shops of almost every description."[40] Finally, as group drive had reduced the effect of a motor malfunction or breakdown in the mechanical power distribution system to the affected group of machines, so unit drive further limited the disruption of production to the single malfunctioning machine.

Improved Working Environment

Without overhead mechanical power transmission, improvements were possible in illumination, ventilation, and cleanliness. Formerly, mazes of belts practically precluded shadowless lighting. With unit drive, lights could be provided in places formerly occupied by belts, pulleys, and shafts.

Some new buildings also incorporated skylights, thus improving ventilation as well as illumination. With line shaft or group drive, continuous lubrication of shafting added oil or grease to the working area and moving shafts and belts kept grease-laden dust circulating. Walls and ceilings became dirty rapidly and were rarely cleaned or painted because of the difficulty of getting around the shafting. Factories could be cleaner and brighter after adoption of unit drive, and many observers felt this had a very positive impact upon the quantity and quality of work.[41]

Improved Machine Control

Belt slippage, common with group drive, caused the speed of some machines to vary with the load, reducing the quantity and quality of output.[42] Furthermore, the two or three pulleys used on most drive shafts and countershafts limited the number of operating speeds. Often work was turned out at a slower than maximum rate.[43] In addition, valuable time could be lost during speed changes if the operator had to leave his work to shift the belt between pulleys.

Unit drive practically eliminated these problems. Individual motors, having a minimum of transmission apparatus, maintained relatively steady machine speed. Until after the turn of the century, dc motors provided almost all industrial electric drive, and with them operators could conveniently vary the speed of their machines. After 1904 the need for speed variation diminished; increasing use of constant speed ac motors and the rise of unit drive went hand in hand, enhancing the quantity and quality of output via steady speed and convenience of control.[44]

Ease of Plant Expansion

The first quarter of this century was a time of rapid growth in manufacturing. A number of rapidly growing firms felt that mechanical power distribution systems imposed constraints on the expansion of their plants.[45] With line shaft drive, the original power distribution system had to be designed with provisions for expansion, or it had to be replaced entirely; ad hoc additions to the system reduced efficiency and increased fluctuations in the speed of driven machinery in the new parts of the plant.[46] Even with electric group drive, plant expansion often required undue rearrangement of machinery.

Once a plant converted to electric unit drive, however, the power distribution system no longer hampered expansion of production facilities. Departments could be enlarged and buildings could be added readily. Costly hanging or rehanging of shafts was unnecessary, and production could continue even during construction of the new works. In a case study of the Scovill Manufacturing Company of Waterbury, Connecticut, E. B. Kapstein argues that the removal of constraints on the expansion of production was the primary reason for the company's switch to electric unit drive.[47]

In summary, although unit drive used less energy and sometimes cost less than other methods for the driving of machinery, as such, manufacturers came to find these savings to be far less important than their gains from increased productivity in their operations overall. In essence, electric unit drive offered the opportunity to obtain greater output of goods per unit of input employed in production. Electricity had come to be viewed as a factor in improving overall productive efficiency.

THE PENETRATION OF UNIT DRIVE

In the manufacturing sector as a whole, the trend toward freeing production from constraints imposed by mechanical power transmission and distribution culminated with the widespread use of unit drive. Although no comprehensive data were ever collected on power distribution systems in use, there is evidence that unit drive did not become the predominant form of electric drive until after World War I.

First, the merits of driving machines in groups or driving them individually were discussed in the technical literature throughout the first quarter of the twentieth century. Between 1895 and 1904, this subject was vigorously debated in meetings of technical societies.[48] Those who advocated unit drive were probably well ahead of established practice. Over twenty years later, group drive was still being strongly recommended for many applications. In 1926, F. H. Penney of the General Electric Company's Industrial Engineering Department reviewed the place of unit drive vis-à-vis group drive at the New Haven, Connecticut, Machine Tool Exhibition. He concluded,

> The experience of the author in the motor-application field inclines him to the belief that, unless all of the operating conditions are known, it is difficult to decide which would be the better of the two methods. . . . Generally, the

author thinks that at the present time individual drives seem to predominate — i.e., as far as newly installed equipment is concerned.[49]

Two textbooks printed in 1928 also make it clear that there were many situations in which group drive was justified but that the tendency during the 1920s was toward exclusive use of unit drive.[50]

Second, machines had to be made compatible with motors. The considerable time needed for this adaptation is best illustrated by a brief look at that class of machines known as "machine tools." Machine tools are powered devices, not hand-held, that cut, form, or finish metal workpieces. Included are lathes, milling machines, other boring, grinding, and threading implements, and pressing, bending, and die-casting devices. Virtually every manufactured good is produced with one or more machine tools or with machines made using machine tools.[51]

Machine tools and other production machinery had traditionally been built with drive shafts and pulleys for use with line shaft drive systems. In the early 1900s, machine tools were not generally built to be directly connected to an electric motor, and the control and performance of some machines were only marginally better with unit drive than with line shaft or group drive.[52] The situation improved a few years later as better quality steel became available and made possible changes in the design of machine tools. Sometimes these new machines provided for the mounting and direct connection of electric motors.[53] In 1904 several of the largest manufacturers of lathes adapted 30 percent of their product line for unit drive.[54] But according to engineer F. B. Duncan, more significant changes were needed:

> No permanent advance in electrical operation of machine tools will be made until the motor and the tool are designed for each other as much as the old cone pulley was designed for the machine on which it was used. . . . What is needed (and this cannot be emphasized too strongly) is a complete redesign of present machine tools with motor operation alone in view.[55]

Through 1904 the means of providing power to production machinery had changed significantly while the machines themselves had changed very little. Now the spread of the culminant form of mechanical drive and the development of new machines were closely related. Yet, progress toward the "complete redesign" of tools advocated by Duncan was not as rapid as he might have hoped. A 1928 textbook indicates only a "trend toward incorporating the motor as an integral part of the machine tool."[56] Machines designed specifically for unit drive were probably not in wide use until after World War I.

Consider, for example, the case of the milling machine — a machine that removes metal during movement of a workpiece past rotating multiple-point cutters. At the turn of the century, the Cincinnati Milling Machine Company experimented with ac motors that drove machine spindles through a friction clutch and an all-geared drive and offered their first ac motor-driven machine in 1903. But the placement of the motor and the design of the drive train were major technical problems. Subsequent models were offered with the motor in various locations — on top of the column or bolted to the frame, for example. Finally, the motor was placed on a mount cast integral to the base of the machine. The first commercially successful electrically driven milling machine — marketed by Brown and Sharpe in 1919 — was based on such a design.[57]

The third reason for believing that unit drive was not widespread until the 1920s is that electricity did not become widely available until the rise of the electric utilities. In 1909, electric drive accounted for slightly less than 25 percent of total capacity for driving machinery; by 1919, electric motors represented over 53 percent of the total horsepower used for this purpose. This major transition was concurrent with changes in the supply of electricity. In 1909, 64 percent of the motor capacity in manufacturing establishments was powered by electricity generated on site; ten years later 57 percent of the capacity was driven by electricity purchased from electric utilities.

Although electric generating capacity in manufacturing continued to increase over this period, electric utilities were expanding so fast that after 1914 their generating capacity exceeded that in all other industrial establishments combined. Of course, utilities were less successful in selling mechanical drive services to firms that already generated their own electricity than to those that were building new facilities. Few manufacturers were willing to write off installed equipment prematurely; a new plant, on the other hand, could be designed around utility power and unit drive.

Many small firms building new facilities could not afford their own electric power plants. Often they rented shaft power along with floor space in large buildings. Utilities made electricity available to these small manufacturers for the first time. In some cases this class of customer was the major source of growth in demand.[58] Thus, the utilities played an important role in increasing the penetration of electric unit drive, and their influence was particularly strong during the second decade of this century.

IN RETROSPECT: THE POWER DISTRIBUTION REVOLUTION

Power production and distribution in manufacturing evolved over a forty- to fifty-year period, between approximately 1890 and 1940. The period immediately following World War I was one of particularly rapid change (Figure 1.3). Steam engines gave way to electric motors and unit drive became the most common method of driving machinery.

Figure 1.3 Some Effects of the Power Distribution Revolution

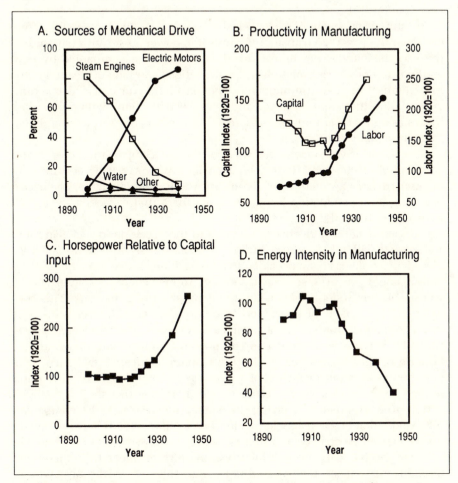

SOURCE: Appendix II: Basic Statistical Data and Estimating Procedures; and Warren D. Devine, Jr., "From Shafts to Wires: Historical Perspective on Electrification," *Journal of Economic History* 43 (1983), Table 3.

Efficiencies in production made possible by unit drive are manifested by significant increases in productivity in manufacturing beginning just after the war and earlier in some industries, such as printing. An important reason for the sharp increase in mechanical drive horsepower relative to capital input is that unit drive required several times as much motor capacity as other forms of electric drive. Finally, unit drive clearly resulted in the use of less energy per unit of output than other methods of driving machinery. Indeed, both energy savings and increased productivity contributed to the dramatic decrease in energy intensity of manufacturing after 1920 (see also Chapter 13).

The shift from steam power to electric power was fundamentally different from the pre-1870 transition from water power to steam. That shift in the way mechanical power was produced was not accompanied by new methods of power transmission and distribution. Adoption of steam did not involve anything like the major changes in factory design and machine organization that went hand in hand with electrification. Rather, manufacturers adopted steam power primarily for reasons of locational and seasonal availability and of cost.[59]

The cost of driving machinery was not the most important factor in the adoption of electric unit drive. This transition was primarily motivated by manufacturers' expectations of significant productivity benefits. Electricity had a value in production by virtue of its form that exceeded savings in costs of mechanical drive.

The special value of electricity was due to the precision in space, in time, and in scale with which energy in this particular form could be transferred. Electricity could easily be provided to any machine located anywhere. Then motors could convert the electrical energy to mechanical energy precisely where the conversion was needed: the drive shaft of a machine. This conversion and transfer of energy could be done on any time schedule — it could be started, stopped, or varied in rate as needed. Finally, electric power could be accurately matched to the power requirements of machines. Because of these attributes of electricity, manufacturers could turn their attention away from problems of power production and distribution and toward improving the overall efficiency of their operations.

Moreover, a fundamental change in viewpoint preceded and accompanied the exploitation of electricity in production. Until the 1890s, most manufacturers viewed electricity in a limited sense: it was simply a good way to transmit mechanical power to factories. In 1891, engineer H. C. Spaulding pointed out that electricity was more than this: it was the best way to distribute power within factories. Two years later, General Electric salesman S. B. Paine demonstrated that electricity could be used to benefit production in other ways as well, and beginning in 1895, a series of discussions led by Professor F. B. Crocker confirmed the innovative view that electricity could serve as a lever in production. The ensuing thirty years saw increasing penetration of electric drive accompanied by numerous innovations in factory design and organization, many of which were possible only because of the precision in space, in time, and in scale with which energy as electricity could be employed.

Notes

1. This description of line shaft drive is due in part to Richard S. Mack, personal communication, Central Washington University, Ellensburg, WA, 1981. Mack is a former employee of a Maine shoe factory that used electric line shaft drive until 1962.

2. Charles Day, "Discussion on the Individual Operation of Machine Tools by Electric Motors," *Journal of the Franklin Institute* 158 (November 1904): 321.

3. Harold C. Passer, *The Electrical Manufacturers, 1875–1900* (Cambridge: Harvard University Press, 1953), 238–240.

4. "New York Notes," *Electrical World* 9 (February 12, 1887): 82.

5. Richard B. DuBoff, "Electric Power in American Manufacturing 1889–1958," Ph.D. diss., University of Pennsylvania (1964), 228.

6. Warren D. Devine, Jr., "The Printing Industry as a Leader in Electrification, 1883–1930," *Printing History* 14 (1985): 27–31.

7. Louis Bell, "Electricity as the Rival of Steam," *Electrical World* 17 (March 14, 1891): 212.

8. Ibid.

9. Passer, *Electrical Manufacturers*, 303–305.

10. F. B. Crocker, V. M. Benedikt, and A. F. Ormsbee, "Electric Power in Factories and Mills," *Transactions of the American Institute of Electrical Engineers* 12 (June 1895): 413–414.

11. Ibid.

12. H. C. Spaulding, "Electric Power Distribution," *Power* 11 (December 1891): 12.

13. "Electrical Transmission of Power at the Edison Works in Schenectady," *Power* 12 (February 1892): 2.

14. "Electric Power Transmission at Columbia, SC," *Electrical World* 23 (May 12, 1894): 656–657; Passer, *Electrical Manufacturers*, 303–305.

15. "Two Motor Driven Factories," *Power* 17 (March 1897): 12–13.

16. Ibid., 13.

17. "Electric Motors at the Ansonia Shops of the Farrel Foundry and Machine Company," *Power* 17 (August 1897).

18. Crocker, Benedikt, Ormsbee, "Electric Power," 420–421.

19. Ibid., 417.

20. Ibid., 415.

21. C. S. Hussey, "Electricity in Mill Work," *Electrical World* 17 (May 9, 1891): 343; "Electric Power for Isolated Factories," *Electrical World* 25 (February 16, 1895): 207; Westinghouse Electric and Manufacturing Company, "Electricity for Machine Driving," Pittsburgh, 1898; A. D. DuBois, "Will It Pay to Electrify the Shops?" *Industrial Engineering and the Engineering Digest* 11 (January 1912): 6–7.

22. Crocker, Benedikt, Ormsbee, "Electric Power," 420–421.

23. G. Richmond, "Electric Power in Factories," *Engineering Magazine* (January 1895); F. B. Crocker, "The Electric Distribution of Power in Workshops," *Journal of the Franklin Institute* 151 (January 1901): 2, 9; G. M. Campbell, "Machine Shop Practice," *American Machinist* (January 25, 1906), 114; F. H. Penney, "Group Drive and Individually Motorized Drive," *Mechanical Engineering* 48 (September 1926): 890.

24. Westinghouse, "Electricity."

25. "Electrical Transmission," *Power* (1892) 2; "Electricity in Industrial Plants," *Manufacturers' Record* 27 (July 19, 1895): 392–393; Crocker, Benedikt, Ormsbee, "Electric Power," 421–422.

26. Penney, "Group Drive," 890.

27. Day, "Machine Tools," 337.

28. Even so, until the 1920s the overall efficiency of driving machinery seldom exceeded about 12 percent — even with electric utility generation.

29. Penney, "Group Drive," 890.

30. DuBois, "Will It Pay," 9.

31. Crocker, Benedikt, Ormsbee, "Electric Power," 414; Crocker, "Electric Distribution," 2; DuBois, "Will It Pay," 8; Penney, "Group Drive," 892.

32. In 1972, for example, energy represented approximately three-quarters of the total cost of driving production machinery (David B. Reister and Warren D. Devine, Jr., "Total Costs of Energy Services," *Energy* 6 [1981]: 305–315).

33. Crocker, Benedikt, Ormsbee, "Electric Power," 427.

34. Crocker, "Electric Distribution," 8.

35. Ibid., 9.

36. Ibid., 6–7.

37. Crocker, Benedikt, Ormsbee, "Electric Power," 427.

38. Day, "Machine Tools," 337.

39. DuBois, "Will It Pay," 4.

40. Ibid.

41. Crocker, Benedikt, Ormsbee, "Electric Power," 429–430; Westinghouse, "Electricity;" Crocker, "Electric Distribution," 4; DuBois, "Will It Pay," 8.

42. Gordon Fox, *Electric Drive Practice* (New York: McGraw-Hill Book Company, 1928), 2, 8, 10; Wilhelm Steil, *Textile Electrification: A Treatise on the Application of Electricity in Textile Factories* (London: George Routledge and Sons, 1933), 111–116.

43. DuBois, "Will It Pay," 5.

44. Warren D. Devine, Jr., "From Shafts to Wires: Historical Perspective on Electrification," *Journal of Economic History* 43 (1983): 366–367.

45. Crocker, Benedikt, Ormsbee, "Electric Power," 428.

46. Steil, *Textile Electrification*, 111–116.

47. Ethan B. Kapstein, "Industrial Power at Scovill Manufacturing Company: A Research Note," Harvard University, 1981, typescript.

48. Crocker, Benedikt, Ormsbee, "Electric Power," 411; Crocker, "Electric Distribution," 23; Day, "Machine Tools," 348.

49. Penney, "Group Drive," 889, 969.

50. William B. Cornell, *Industrial Organization and Management* (New York: Ronald Press Company, 1928), 382–384; Fox, *Electric Drive Practice*, 8, 355.

51. National Academy of Engineering, Committee on Technology and International Economic and Trade Issues, *The Competitive Status of the U.S. Machine Tool Industry* (Washington, D.C.: National Academy Press, 1983), 3.

52. Crocker, "Electric Distribution," 8.

53. Day, "Machine Tools," 333, 338.

54. Ibid., 323.

55. Ibid., 338–339.

56. Fox, *Electric Drive Practice*, 82.

57. R. S. Woodbury, "History of the Milling Machine," in *Studies in the History of Machine Tools* (Cambridge: Technology Press, 1960), 71–72.

58. Raymond C. Miller, *Kilowatts at Work: A History of the Detroit Edison Company* (Detroit: Wayne State University Press, 1957), 159–160.

59. Jeremy Atack, "Fact in Fiction — The Relative Costs of Steam and Water Power: A Simulation Approach," *Explorations in Economic History* 16 (1979): 409–437; Jeremy Atack, F. Bateman, and T. Weiss, "The Regional Diffusion and Adoption of the Steam Engine in American Manufacturing," *The Journal of Economic History* 40 (June 1980): 281–308; Kapstein, "Industrial Power."

Works Cited

Jeremy Atack, "Fact in Fiction — The Relative Costs of Steam and Water Power: A Simulation Approach," *Explorations in Economic History* 16 (1979): 409–437.

Jeremy Atack, F. Bateman, and T. Weiss, "The Regional Diffusion and Adoption of the Steam Engine in American Manufacturing," *The Journal of Economic History* 40 (June 1980): 281–308.

Louis Bell, "Electricity as the Rival of Steam," *Electrical World* 17 (March 14, 1891).

G. M. Campbell, "Machine Shop Practice," *American Machinist* (January 25, 1906).

William B. Cornell, *Industrial Organization and Management* (New York: Ronald Press Company, 1928).

F. B. Crocker, "The Electric Distribution of Power in Workshops," *Journal of the Franklin Institute* 151 (January 1901).

F. B. Crocker, V. M. Benedikt, and A. F. Ormsbee, "Electric Power in Factories and Mills," *Transactions of the American Institute of Electrical Engineers* 12 (June 1895).

Charles Day, "Discussion on the Individual Operation of Machine Tools by Electric Motors," *Journal of the Franklin Institute* 158 (November 1904).

Warren D. Devine, Jr., "From Shafts to Wires: Historical Perspective on Electrification," *Journal of Economic History* 43 (1983).

Warren D. Devine, Jr., "The Printing Industry as a Leader in Electrification, 1883–1930," *Printing History* 14 (1985).

Richard B. DuBoff, "Electric Power in American Manufacturing, 1889–1958," Ph.D. diss., Univ. of Pennsylvania, 1964.

A. D. DuBois, "Will It Pay to Electrify the Shops?", *Industrial Engineering and the Engineering Digest* 11 (January 1912).

"New York Notes," *Electrical World* 9 (February 12, 1887).

"Electric Power Transmission at Columbia, SC," *Electrical World* 23 (May 12, 1894).

"Electric Power for Isolated Factories," *Electrical World* 25 (February 16, 1895).

Gordon Fox, *Electric Drive Practice* (New York: McGraw-Hill Book Company, 1928).

C. S. Hussey, "Electricity in Mill Work," *Electrical World* 17 (May 9, 1891).

Ethan B. Kapstein, "Industrial Power at Scovill Manufacturing Company: A Research Note," Harvard University, 1981, typescript.

Richard S. Mack, personal communication, Central Washington University, Ellensburg, WA, 1981. Mack is a former employee of a Maine shoe factory that used electric line shaft drive until 1962.

"Electricity in Industrial Plants," *Manufacturers' Record* 27 (July 19, 1895).

Raymond C. Miller, *Kilowatts at Work: A History of the Detroit Edison Company* (Detroit: Wayne State University Press, 1957).

National Academy of Engineering, Committee on Technology and International Economic and Trade Issues, *The Competitive Status of the U.S. Machine Tool Industry* (Washington, D.C.: National Academy Press, 1983).

Harold C. Passer, *The Electrical Manufacturers, 1875–1900* (Cambridge: Harvard University Press, 1953).

F. H. Penney, "Group Drive and Individually Motorized Drive," *Mechanical Engineering* 48 (September 1926).

"Electrical Transmission of Power at the Edison Works in Schenectady," *Power* 12 (February 1892).

"Two Motor Driven Factories," *Power* 17 (March 1897).

"Electric Motors at the Ansonia Shops of the Farrel Foundry and Machine Company," *Power* 17 (August 1897).

David B. Reister and Warren D. Devine, Jr., "Total Costs of Energy Services," *Energy* 6 (1891).

G. Richmond, "Electric Power in Factories," *Engineering Magazine* (January 1895).

H. C. Spaulding, "Electric Power Distribution," *Power* 11 (December 1891).

Wilhelm Steil, *Textile Electrification: A Treatise on the Application of Electricity in Textile Factories* (London: George Routledge and Sons, 1933).

Westinghouse Electric and Manufacturing Company, "Electricity for Machine Driving," Pittsburgh, 1898.

R. S. Woodbury, "History of the Milling Machine," in *Studies in the History of Machine Tools* (Cambridge: Technology Press, 1960).

2 | *Electricity in Information Management: The Evolution of Electronic Control*

Warren D. Devine, Jr.

The first of the two sections of this chapter recounts the beginnings of electronic control in manufacturing by looking primarily at machine tools. We focus on machine tools not only because of their importance in the evolution of electronic control but also because of their ubiquity in production: almost all manufactured goods are produced with machine tools or with machines made by using these tools.

Our account begins in the 1920s with the use of various mechanical and hydraulic devices as the first automatic controls. Such automatically controlled machines were inflexible: only a limited and fixed sequence of control movements could be carried out without human intervention. It is of major significance to our story that virtually all progress beyond this stage can be attributed to the application of electricity and electronics. Beginning in the 1930s, electric devices made it increasingly easier to vary sequences of machine movements under automatic control. By the 1960s, machines were in use whose sequences of operations could be altered simply by making changes to a paper or magnetic tape. With the application of electronic digital computers in the early 1970s, machine tools were beginning to become truly flexible instruments of production.

The second section of the chapter outlines the impact of microelectronics on manufacturing production, a stage of development that is still unfolding and that strongly parallels the earlier history of electricity and the electric motor described in Chapter 1.

Until the 1970s, electronic control and information technology affected the operation of *individual* machines and processes rather than the interactions between machines or the organization of production activities. By this point, certain problems that accompanied centralized computer control had slowed progress toward automatic production by *groups* of machines as well as progress toward any important organizational changes. With the advent of the microcomputer, however, each machine could be endowed with a great amount of information-handling capacity. Significantly, this endowment facilitated the electronic linking of computer-controlled machines and processes. Today certain office activities that support production are being computerized and linked with one another. But it is the electronic integration of production *and* these supporting activities — made possible by computer technologies — that has the greatest implication for organizational change in contemporary manufacturing.

ELECTRICITY AND AUTOMATIC CONTROL: TECHNOLOGICAL PROGRESS PRIOR TO THE MICROCOMPUTER

Mechanical and Hydraulic Methods of Control

Late in the nineteenth century, the limited amount of electric drive in American industry was provided by direct-current (dc) motors. The prevailing type of dc motor was the shunt-wound machine, which could easily vary in speed over a range of three or four to one by changing a rheostat. With a gear box, an even wider range of speeds could be obtained.

The alternating current (ac) polyphase induction motor was first marketed in 1892. For the same output, ac motors were superior to dc motors in a number of respects: they were smaller, lighter, simpler, did not spark, required very little attention, and were significantly cheaper. But ac motors had one principal drawback: their speed could not be varied without seriously impairing performance. Frequency, not voltage, governs the speed of an induction motor, and it was not practical to provide variable-frequency current. For a time, whether ac or dc motors were employed depended on which current was available and whether it was necessary or desirable to vary the speed of machines. But after the turn of the century, utilities increasingly supplied ac power. Now compatibility with the rapidly growing utility system was another of the ac motor's advantages.[1] Where necessary, an adequate range of speeds could sometimes be provided without excessive loss of power by means of geared transmission systems.

In addition to speed changing, machine tool operation required machinists to manually perform a number of control movements, such as feeding the workpiece into the tool or moving the tool into or across the workpiece. Milling machine operators, for example, hand-fed the workpiece — mounted on a sliding table — against the cutting wheel, moved the work rapidly between cuts, and quickly returned the table to its starting position. Prior to World War I, attempts were made to automate these kinds of movements using geared transmission systems.[2] This initial work in automatic machine tool control was based upon experience using gears to change the speed of ac motor-driven machines.

After World War I, designers tried to use hydraulic mechanisms to vary the speed of machine tools.[3] Hydraulics were not actually new — having been employed in other industrial applications — but they had to be adapted to the specific characteristics of each kind of tool. By the late 1920s, simple and efficient units of comparatively small size had been developed, and their cost was competitive with that of changing speed via gears. Tool manufacturers also found that some manual control movements (including some that had been automated via geared systems) could be performed efficiently and automatically by hydraulic mechanisms. For example, in 1927 the Cincinnati Milling Machine Company produced the first standard milling machine that utilized a hydraulic system for automatic movement of the machine's sliding table.[4]

The principal advantages of hydraulic drive and control methods over geared systems were that hydraulics offered a high degree of flexibility in speeds, provided a cushioning effect that reduced shock to the machine when entering work or reversing, and withstood extreme peak loads. Moreover, it was possible to adapt hydraulic equipment to carry out a fixed sequence of control movements that ensured each motion would begin only after the preceding one was completed. These characteristics were especially valuable in rapid, high-volume production.[5]

Geared and mechanical-hydraulic machine tool control systems allowed machinists to set in motion a predetermined (or "programmed") sequence of movements. These sequences, however, were limited in complexity; irregularly shaped workpieces, for example, could not be produced automatically. More significantly, the program was fixed and inflexible; it was incorporated into the very design of the machine's power transmission system.

The earliest and most common device that enabled machines to produce irregular shapes automatically was the cam. A cam is a solid piece of material of virtually any shape that mechanically adjusts the position of a lever or other machine element as it rotates. Thus a cam enables a machine to automatically carry out a predetermined sequence of movements or events; the sequence is fixed by the shape of the cam.[6]

Cam-following tools were the first programmable automatic machines in that the sequence of movements to be followed was not part of the design of the machine. All the information required to specify the ultimate shape of the workpiece was stored in the cam. To produce a differently shaped piece, a new master cam could be built.

However, cam control was successfully applied only to woodcutting tools. An important weakness of this technique was that all the force required to position the cutting tool had to be furnished by the cam itself. It proved difficult to build a cam and a cam-following mechanism strong enough to accurately transfer motion to metal cutters; the cam wore out quickly. Very little progress around this difficulty was made until the advent of electromechanical devices for machine tool control.[7]

Electromechanical Control: Contour Following

World War I highlighted the need for fully automatic steering systems for naval vessels and for automatic control of the position of heavy loads, such as shipboard guns. In response, engineer Nicholas Minorsky developed for the U.S. Navy what today would be called a "servomechanism."[8] A servomechanism is a system that automatically controls the position of some device or machine element. The term implies the use of "feedback," or comparison of the desired position of the element with its measured position, and amplification of the difference between these positions, or error (see Figure 2.1). The system is automatic: its goal is to reduce the error signal to zero, when position adjustment ceases.[9]

During the 1920s a few machine tool builders began to experiment with servomechanisms. In the first application of record, engineer John C. Shaw working in the shop of Joseph Keller developed an electrical sensing device or "tracer" that required only very light contact with the master cam of a cam-following machine tool. On this machine, known as the Keller Duplicating System, the tracer's deflections were used as inputs to a servomechanism that moved a sliding table holding the workpiece.

In the 1930s tracers and servomechanisms with electrical and electronic components were incorporated into three milling machines that were commercial successes: Cincinnati's Die-Sinker, Hydro-Tel, and Profiler. On these "copy-milling" machines, the tracer or stylus exerted only fingertip pressure on the master cam or model of the part to be produced. This eliminated the need to fabricate the durable and costly models that were required for cam control. Significantly, it meant that hard metal parts could now be produced by automatic machines that were programmable. Although actual cutting time was not necessarily less than that of conventional manual machine tools, overall production time was often reduced by more than 50 percent. This was because the operator no longer needed to stop the machine frequently to check the workpiece for dimensional accuracy, take trial cuts, check again, or refer repeatedly to drawings.[10]

Figure 2.1   A Servomechanism

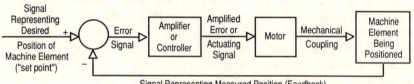

Contour-following or tracer control was extended from milling machines to lathes and then to other machine tools. Eventually, the tracing principles were taken several steps further. On Pratt and Whitney's Veltrace machine, developed in the late 1940s, the tracer did not actually touch the model, allowing models that were very soft or fragile to be used again and again. Cincinnati's Line Tracer Hydro-Tel of the early 1950s could trace directly from a drawing by using a cathode ray light and phototube to translate the drawing into commands to the servomechanisms.[11]

Tracer control greatly enhanced a machine's ability to store and use information. With this system, the information source was not a metal cam, but a model of the part to be machined made of an easily formed material, such as wood or plaster of paris. By permitting the use of soft, easily formed models, tracer control reduced the cost of storing information and providing it to the cutting tool, that is, of *reprogramming* the machine. Lowering the cost of reprogramming an automatic machine means that it can be more readily used to make an assortment of products; in other words, it becomes more flexible. It is significant that the initial use of electromechanical feedback control devices on machine tools was responsible for this increase in flexibility.

Numerical Control: Mathematical Modeling

By the late 1930s servomechanisms were being used in several industrial settings other than metal working, the most notable being steel strip mills and paper mills. In these applications, the speed of roller-drive motors was automatically controlled so as to maintain proper tension in the material being processed. A little later, the military requirements of World War II led to particularly rapid progress in the emerging field of control engineering. When the war ended, a host of technological resources became available to American manufacturers. These included proven servomechanism components, such as electronic controllers;[12] analytical techniques and devices, perhaps the most important being the analog computer;[13] and engineers with experience in control systems.[14] A chronology outlining the evolution of control engineering and showing supporting developments in electronics and computers is given in Figure 2.2.

In 1948, the goal of easily varying the sequence of movements of an automatic machine tool — as contrasted with merely setting in motion a fixed sequence — was being pursued in two principal ways: tracer control and numerical control. The first, as we have seen, required a physical model of the object to be fabricated (or at least a complete drawing of it, as with the Cincinnati Line-Tracer Hydro-Tel). The second required not an image of the finished object or part, but only an abstraction of it: a mathematical model and instructions to the machine.

Figure 2.2 Chronology: Evolution of Control Engineering and Supporting Developments in Electronics and Computers

Control Engineering

Automatic control is a specialty
of mechanical engineering

1900	1910	1920	1930	1940	1950	1960	1970	1980

1918-1922 First work on servomechanisms

1920s Servomechanisms for naval applications

late 1920s Principal of feedback used in telephony

late 1930s Electric feedback controls for steel and paper mill drives

1940 MIT Servomechanisms Laboratory established. Analog computers used to design controllers

1940s Servomechanisms extensively used in weapons

1950 Electronic controllers

1955 Transistorized electronic controllers. Miniaturization begins

1960 Digital electronic control; mainframe computers

1968 Digital real-time control in space program

1972 Programmable logic controller

1975 Digital electronic control; microcomputers

1980 Distributed/hierarchical control

Electronics and Computers

1906 Vacuum tube amplifier

1927 Negative feedback amplifier

1930 Mechanical differential analyzer (analog computer)

1933 Stabilization of analog circuits in instrument

1942 Electronic differential analyzer (analog computer)

1945 Electronic digital computer (ENIAC)

1947 Point-contact transistor

1951 Commercial digital computer with internally stored program (UNIVAC)

1953 IBM 701

1958 Transistorized digital computer (Philco 2000, IBM 709)

1959 Integrated microelectronic circuit

1959 Fully transistorized minicomputer (PDP-1)

1962 Integrated circuits manufactured in large quantities

1967 Integrated circuit digital computer. Very large-scale microelectronic integration

1971 Microprocessor

1974 Microcomputer

1975 Microprocessor bundled with digital process instrumentation

1975 Personal computer

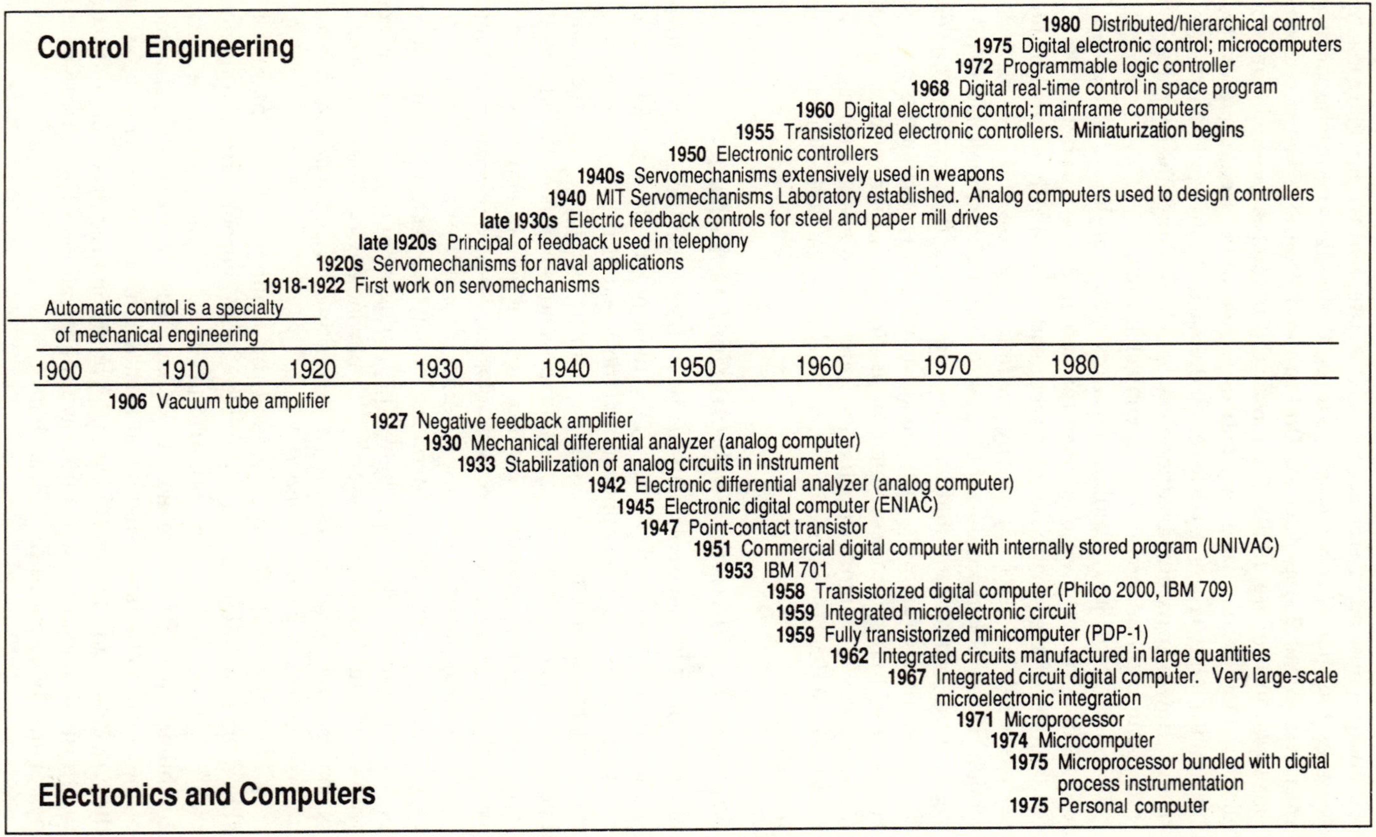

At this time the U.S. Air Force needed new techniques for building a generation of aircraft whose parts were designed to be machined rather than riveted. They were to be made of high-duty alloys of steel or aluminum and were of a form and specification that required the performance of a complex sequence of machine operations to extremely accurate standards. Application of production line techniques to the machining of such parts in small quantities was inappropriate. On the other hand, the use of general-purpose, manually controlled machine tools for such work would involve unwarranted expenditures on jigs and fixtures and an equally great expenditure of time by many skilled machinists. In an attempt to fill the gap between these extremes in production technique and organization, a project was initiated that was to take full advantage of the newly acquired expertise in electromechanical control. A contract was awarded to the Massachusetts Institute of Technology, whose engineers demonstrated the first numerically controlled machine — a milling machine — in 1953.[15]

The pattern to be milled was represented by binary digits and stored by punching the digits into a paper tape. The tape was "read" by metal "fingers" that reported the presence or absence of holes via electric pulses. These pulses were converted to a continuous electric signal and used as input to the servomechanism that moved the tables of the milling machine.[16] Later the dimensions of the part and instructions could also be provided to the machine on magnetic tape.[17]

Tape control represented a further step in the reduction of the cost of information storage and use in the control of machines. The information cost was less than that associated with tracer control because a manually produced, wear-prone physical model was replaced by a mathematical model, and because the mathematical model could be easily changed (i.e., the machine reprogrammed). Indeed, one respected observer has asserted, "Tape control . . . undoubtedly was the greatest advance in machine-control history. It enabled a very long and complex program to be imposed on a machine, yet it could easily be altered or replaced."[18]

By 1955 there were still less than a dozen numerically controlled machine tools being used in production, and most of these on an experimental basis. Moreover, only a very small fraction of new machine tools incorporated any form of electric feedback control. The new automatic control technology initially had little impact on fabrication in American industry.[19]

Flexible Automation

By 1967 numerical control had become thoroughly established. Machines that were controlled numerically included lathes, multipurpose drilling machines, jig borers, tapping machines, and three-dimensional contour millers. Many of their control systems employed fully transistorized circuits.

In addition, to eliminate the need for shifting a workpiece from one specialized machine to another, different tools — capable of performing a variety of operations — were incorporated into a single multispindle, multi-function unit called a "machining center." For example, the Sunstrand Omnimil was designed for straight and contour milling, boring, and thread-ing; the Hughes MT-3 Machining Center permitted use of forty-five differ-ent tools in one setup, with tool changes made automatically in three seconds. As well as selecting and positioning tools, the tape and electronic feedback control systems of such centers adjusted cutter speeds, table feeds, and activated coolant flows.[20] Altogether, numerically controlled machine tools used in production numbered over 10,000 and accounted for perhaps 20 percent of the annual value of output of the machine tool industry.[21] A chronology of the evolution of automatic machine tool control and process control is provided in Figure 2.3.

Numerically controlled machinery had a number of advantages over con-ventional manually controlled machinery. The time required to get a newly designed part into production — the machine setup time — was sometimes as much as 65 percent to 75 percent less with numerical control. In addi-tion, more uniform quality could be achieved with numerical control because all the pieces in a lot were machined using the same tape, under the same conditions. For example, the weight of a certain aircraft wing panel varied between 950 and 990 pounds when produced conventionally; numer-ically controlled machining produced panels weighing 954 to 959 pounds.[22] Finally, errors that sometimes accompanied manual machining were elimi-nated, and, as a result, assembly was more efficient and inspection costs were reduced.

However, the most significant aspect of program control via tape tran-scended any advantage of this technique in producing a particular part. In his 1967 book on numerical control, R. M. Dyke titled the opening chapter "Numerical Control — Flexible Automation."[23] He gave the main reasons for adopting numerically controlled machine tools: (1) numerical control pro-vides flexible automation because it is adaptable to many different job requirements; and (2) numerical control makes automation of short and medium production runs economical.[24] Because these aspects of numeri-cally controlled machine tools are of far-reaching significance, they are examined in greater detail below.

Machine Flexibility

Fabrication with numerically controlled machines was extremely flexible. To make changes in the shape or finish of the part being made, or to change from making one type of part to another, it was only necessary to change the tape and perform a limited number of setup operations. Changes in a punched tape program (reprogramming) were readily made by repunching the tape, compared with the costly changes required when using conven-

Figure 2.3 Chronology: Evolution of Automatic Machine Tool Control and Automatic Process Control

Automatic Machine Tool Control

Timeline: 1900 1910 1920 1930 1940 1950 1960 1970 1980 1990

Mechanical and hydraulic control
(geared and hydraulic systems, cam-following)

Manual control

Electromechanical control
(servomechanisms, contour-following)

1953 Numerical control

1967 Machining center

1970 Computerized numerical control (centralized mainframe or minicomputer)

1973 Programmable logic controllers for machine tools

mid-1970s Computerized numerical control (decentralized microcomputer)

1980s Clusters of automatic machine tools under distributed/hierarchical control

Automatic Process Control

1910 Continuous processes displace batch processes

1930 Pneumatic controller (beginning of automatic control)

1940 PID pneumatic controller (individual fluid streams controlled automatically)

1940 Pneumatic transmitters for remote display of instrument readings (beginning of trend toward centralized control rooms)

1950 Electronic controllers displace pneumatic controllers on individual fluid streams; electronic transmitters for control room display

1955 Digital process instrumentation

1957 Digital computer in control room to monitor and record plant performance

1959 Digital computer used to determine and adjust set points of electronic controllers

early 1960s Fluid streams under centralized computer control

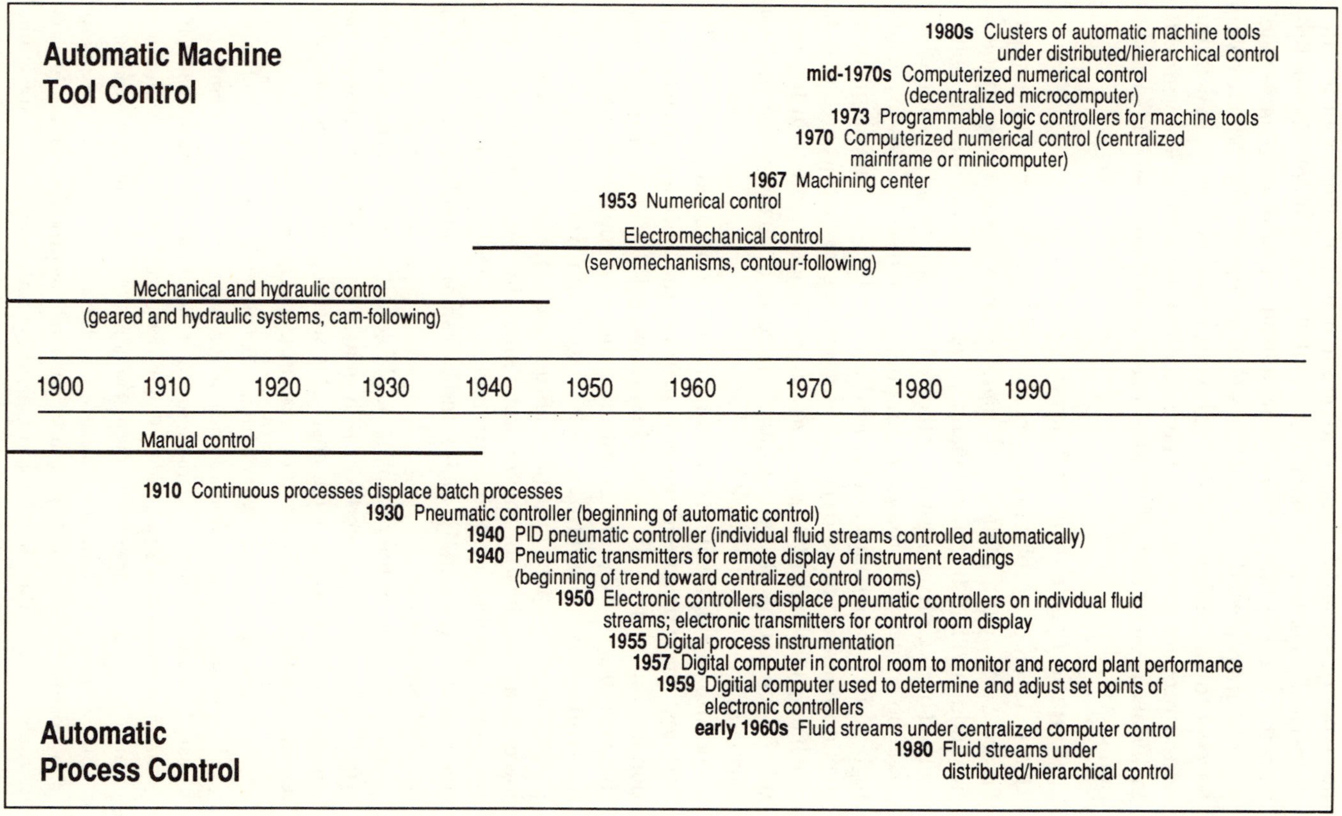

1980 Fluid streams under distributed/hierarchical control

tional tooling. In order to repeat a production run, the tape was simply taken out of storage and installed on the machine. Because of this flexibility, numerically controlled machinery usually had a much higher utilization rate than conventional manual or special-purpose automatic machinery. Some users reported that machines actually operated productively up to 85 percent to 95 percent of a shift. Thus throughput increased with the adoption of numerical control, while actual cutting speeds (determined by the materials involved and other job-specific conditions) generally remained the same.

Automatic Production for Small Lots

Milestones in the evolution of production organization are summarized in Figure 2.4.[25] By the 1960s, much high-volume production (as in the automotive industry) had become dependent on special-purpose automatic equipment. The high fixed cost of dedicated machinery was justified because it could be spread over the great number of identical pieces produced. Such high-capital costs could not be justified for low- and medium-volume production runs, up to, say, several hundred pieces. Yet according to a major machine tool builder, the average size of production lots in the mid-1960s was fifteen to fifty pieces in about 75 percent of all metal-working manufacturing in the United States!

Because of the flexibility of numerically controlled machinery, its high first cost could be spread not only over the pieces in a given (small) production lot, but over all the pieces in a number of different production lots — that is, over a broad scope of output. A small lot could be produced automatically at a cost competitive with that which would result if a lot were produced with special-purpose automatic machines in high volume, and lower than the cost of producing a small lot by conventional, manually controlled machinery. From the mid-1960s, then, numerically controlled machinery increasingly facilitated automatic production of small- and medium-sized lots.

To improve upon the flexibility available with tape-controlled machines, it would be necessary to endow the system with greater information-handling capacity and at a reasonable cost. This was regarded as the essential problem in automatic machine tool control as early as 1952:

> A rough measure of the cost of automatic control is the amount and nature of the information the automatic machine must handle. To perform a complicated operation such as manufacturing a metal part, we must *build into the machine* [italics added] a great deal of information-handling capacity, for it has to carry out a whole complex of instructions. This equipment is expensive. . . .
> But suppose we want an automatic machine that will make not one product or part, but a number of different kinds of products, and only a few of each. . . . Now the machine must handle a different set of instructions for

Figure 2.4 Milestones in the Evolution of Production Organization

1900	1910	1920	1930	1940	1950	1960	1970	1980	1990

1900 Machines arranged along line shafts

1910 Machines gouped by operation or product

1913 Flow-line assembly (Ford Motor Company)

1920 Assembly line common

1925 Large, metal-working transfer machine

1930 Transfer machine for engine manufacture

1941 80-station transfer machine. Inflexible

1948 Word "automation" first used. Transfer machine common

1957 Greater mechanical integration (link lines and centralized control stations). Limited flexibility

Improved flexibility

1969 Highly developed automatic transfer machines.

1980s Computerized materials handling devices; flexible manufacturing systems

1985 Modular assembly (automatic guided vehicles)

1988 Computer-integrated manufacturing

each product. In other words, it must be able to deal with more information. ... This is the essential problem in automatizing machine tools.[26]

The year 1952 also marks the dawn of the commercial digital computer. As electronic digital computers underlie subsequent progress in automatic control and information management, it is appropriate at this point to briefly recount their emergence and first applications in production.

Computers in Control

The first large-scale automatic digital calculator was the Mark I, completed at Harvard University by Howard Aiken in 1943. This computing machine employed motor-driven relays for logical operation and storage, and received instructions and data on punched paper tape. In 1945 a device called ENIAC (Electronic Numerical Integrator and Computer) was built at the University of Pennsylvania for the U.S. Army. This first electronic digital computer used nearly 18,000 vacuum tubes and consequently occupied over 3,000 cubic feet and weighed 30 tons. The ENIAC's instructions were changed by moving external electric cables and switches.

The first commercial electronic digital computers, starting with the UNIVAC in 1951, incorporated a significant improvement: they could store instructions or programs internally. Moreover, internal programs known as "compilers" became available that could automatically translate programs written in algebraic notation into instructions usable by the machine.[27] Internal storage and compilers made digital computers easy to reprogram. Indeed, according to a review of the evolution of computers done at the National Bureau of Standards a quarter-century later, this reprogrammability was "probably the most distinctive feature of the new computer of the 1950s and the single characteristic which has led to the so-called ... computer revolution of the latter half of the 20th century."[28]

Stored-program digital computers provided a large market for the transistor — a small, low-power amplifier invented in 1947 as a replacement for the large, power-hungry vacuum tube. The synergy between a new component and a new application resulted in the rapid growth of both. Beginning in 1959, computers with vacuum tubes were displaced by computers with transistors; the new machines were smaller, more reliable, and — most importantly — faster. Speed is critical in the application of digital computers to control. When functioning as a controller, a digital computer calculates the required control action using information on the current state of the system. Thus the computer can be useful as a control tool only if the time required to perform the calculations is short compared with the period in which significant changes take place in the system being controlled. Until the late 1950s and early 1960s computers were not always fast enough.

The use of digital computers as industrial controllers occurred earlier in materials processing than in fabrication. (Figure 2.3 provides a chronology of the developments leading to digital computer control in the process industries.) A Monsanto ammonia plant and a B. F. Goodrich vinyl plastic plant were placed under computer control in 1960.[29] Chemical processing installations such as these have hundreds of fluid streams and thus hundreds of control loops that are — in principal — the same as the servomechanism shown in Figure 2.1. These feedback loops control fluid flow rates (by adjusting the positions of valves, for example), pressures, temperatures, and liquid levels.[30] The operation is complicated because all of these variables are interrelated according to the physical laws that govern the process. The large mainframe computer of the 1960s seemed uniquely capable of replacing — with a single unit — all the electronic controllers used in the individual loops.[31] More important, it appeared that such a computer could also be used to optimize the performance of the overall system by coordinating the operation of those loops.

Nearly 300 computer-controlled chemical processes were in operation in the late 1960s, and there were reports that the new digital control systems had increased productivity at some of these sites by 10 percent to 20 percent.[32] But large control computers were very expensive and thus suitable only for large plants. In these installations, thousands of wires transmitted signals to a central computer, with problems of signal scaling, degradation, and interference; hundreds of output signals actuated control devices, initiated alarms, and drove data displays.[33] These systems were highly complex. Moreover, since it was possible for the malfunction of a single component to bring the computer to a halt or cause improper operation, plant managers insisted on having costly backup control systems.[34] By the early 1970s it was apparent that complexity, vulnerability, and resulting high costs had slowed any trend toward centralized computer control in the process industries, an occurrence that some observers characterized as a basic "failure of centrality."[35]

In the late 1960s and early 1970s the digital computer was also being applied in the fabricating industries, most conspicuously to numerically controlled machine tools. With conventional numerical control, a punched paper tape was decoded by a mechanical or optical reader attached to the machine itself. This was unnecessary with computerized numerical control. Programs governing the fabrication of every part a machine made could be stored in the computer's memory and could be altered much more easily than punched paper tape. The computer then used one of the programs to control the machine.[36]

The principal advantage of the digital computer was that it provided the automatic machine with much greater information storage and easier reprogrammability, which increased its flexibility.[37] But computers were expensive; their cost could seldom be justified if applied to a single numerically controlled machine. Thus, a single computer was often used to control a number of machines — sometimes even as many as ten to fifteen! This raised in the fabricating industries the same problems confronted in the process industries: the complexity and vulnerability of multiple operations under centralized computer control. Sales of such machine tool systems failed to live up to expectations.[38]

These problems set the stage for the microcomputer, which would allow computer control to take place at the machines themselves. More significantly, microcomputers would allow computer-controlled machines to be linked with one another and with the various activities that support production — with important implications for organizational change.

ELECTRONIC INFORMATION MANAGEMENT: ORGANIZATIONAL CHANGE IN CONTEMPORARY MANUFACTURING

Computer-Controlled Production Machinery

By the late 1960s most of the computers used in American industry were transistorized. Transistors had led to steady and appreciable reductions in the cost of computing on large mainframe machines and had facilitated the introduction of smaller, more limited-purpose minicomputers. But the transistor was a transitional technology — the last stage in a concept of electronics built around circuits consisting of individual components.

In 1959 Robert Noyce and his co-workers at Fairchild Semiconductor developed a process for making and electrically interconnecting transistors and other circuit elements on a single piece of semiconducting material, such as pure silicon. The era of microelectronics had begun. By the mid-1970s an individual integrated circuit — on a chip less than one-quarter inch square — could contain more electronic components than the most complex piece of electronic equipment that could be built in 1950.[39]

Noyce's Intel Corporation introduced the microprocessor, or "computer on a chip," in 1971. The microprocessor differed significantly in design from other integrated circuits in that it was programmable in essentially the same way that very much larger computers were programmable.[40] Intel's chips became the central processing units of the first microcomputers. By the mid-1970s microcomputers were available that had more computing capacity than the the first large digital computer, ENIAC; they occupied 1/30,000 the volume, were twenty times faster, and cost 1/10,000 as much. Thus began the dispersal of computers to the very sources of the information they collect and to the very processes or machines they control.[41]

For years, conventional manual and semiautomatic machine tools had utilized simple electrical devices fabricated from many types of contacts, relays, and switches to control functions that were essentially on-off in behavior. In these systems — sometimes called discrete controllers — the electromechanical hardware was interconnected with wires (or "hard wired"), and this wiring determined the controller's logical behavior. In the early 1970s discrete control systems were developed that incorporated integrated microelectronic circuits instead of conventional hardware. Rather than rewiring, the connections between circuits in these devices could be readily changed electronically, modifying the logical behavior of the controller. Thus these devices were called "programmable logic controllers."

This type of controller, like relays and switches, was used to control functions that had only two states: on or off. Just as sequences of relays and switches could control a series of events in a machine, so also could the new controller. The microelectronic device, however, was much more flexible: it could be reprogrammed by machinists who possessed little or no computer-programming skill. As this made it easier to vary the sequence of operations of the controlled machine, the machine, too, became more flexible.[42]

In 1973 General Automation, Inc., marketed a numerically controlled lathe that took full advantage of the new electronic technology. On this machine tool, more than 90 percent of the electromechanical devices (used until then to transform taped instructions into signals that drove the machine's servomechanisms) were replaced by integrated microelectronic circuitry. A microprocessor-based controller added relatively little to the cost of a machine tool and helped increase its sales appreciably.

Then came more complete integration of computer and machine. The use of a small computer to control one or perhaps as many as four machines offered the programming advantages of control by a larger central computer, yet was cheaper, less complex, and more reliable. In addition, since such a system included one or only a few machines (as opposed to as many as ten to fifteen under central computer control), a malfunction of the computer was much less costly.[43]

By 1982 electromechanical tape readers had been largely supplanted by microcomputers mounted on each machine. A numerically controlled machine tool endowed with a microcomputer possessed capabilities that had not been available with conventional numerical control. These included, for example, interpolation; switchable inch-centimeter readout and programming; a wide range of tool offsets, spindle speeds, and feed rates; and enhanced remote control. These features further enhanced the machine's flexibility.[44] In 1984 around 5,000 metal cutting machines under microcomputer control were shipped to end users. Although this number was only about 5 percent of all metal cutting machines purchased that year, their value was high, representing over 40 percent of the value of all metal-cutting machines shipped in that year.[45]

During the 1980s computerized numerical control spread beyond metalworking to a wide variety of fabricating operations. For example, one application outside the sectors normally associated with machine tools is a computer-controlled cloth-cutting machine. A certain company employed such a machine in order to make sails of the consistent quality and in the high volumes required for popular racing classes. The reprogrammability of the machine proved its worth when recession hit the sailboat market. The company was able to move into new business areas, such as cutting fibers for airplane composite structures, basing the new line of business on the same machine.[46] Flexibility was the key to the company's scope of supply and perhaps to its very survival.

Linking Production Units

Distributed/Hierarchical Control

The preceding section noted some of the improvements in individual automatic machines brought about by microelectronic circuitry and the microcomputer. While important, these improvements have not been as dramatic as those made possible in controlling large clusters of machines, very complex processes, or entire plants. Here, the emphasis on automatic control changed from one of centralization to one of decentralization. This new approach, first applied in the process industries, is based on distributed data acquisition and information processing, and decentralized control. Implementation requires a hierarchy of computers, each performing tasks appropriate to its position in the hierarchy.[47]

In the most modern chemical plants, for example, each control loop is under the direction of a microcomputer in close proximity to it. The computer acquires data from sensors, conditions the data, calculates the required control action, and generates an actuating signal. In addition, because the operation of each loop must be coordinated with that of others, the microcomputers exchange information via a supervisory computer. This may be a larger computer — a minicomputer — at a higher level in the hierarchy. Such a control philosophy has been called "distributed intelligence," or more precisely, "distributed/hierarchical control."[48]

The advantages of distributed computing over central computing are many. Control equipment and maintenance costs are lower; the control system is more flexible — more easily changed or expanded; interference in transmitted signals is reduced; but perhaps most important, distributed control is more reliable than central control. With a single large computer, one failure could stop the entire facility; with a distributed system, the immediate impact of a malfunction is limited to the area in which it occurs.[49]

In the early 1980s distributed/hierarchical control was applied in the fabricating industries. In a typical system, several microcomputer-controlled machine tools are linked by a minicomputer, and several minicomputers are tied in turn to a large mainframe computer. The programs for the manufacture of every part the firm makes are stored in a central data base, and they are transferred automatically from the mainframe computer to appropriate machines in the network. In addition, information about the status of each machine, the volume of its production, and the quality of the finished parts flows back to the central computer from the distributed ones. As many as 100 machine tools have been connected in such a hierarchy.[50] This distributed/hierarchical approach to control provides the basis for the flexible production systems discussed below.

Flexible Manufacturing Systems

Clusters of different kinds of machine tools under distributed/hierarchical computer control are frequently formed that are capable of performing most or all of the operations necessary to produce not just a single part, but a whole family of parts. In many of these clusters the workpieces are moved from one machine to another by manual means. But computer technology is enhancing the *mechanical* integration of clustered machines as well as making possible their *electronic* integration. If industrial robots or other computerized materials-handling devices are employed to transfer workpieces between the machines of a cluster, the entire ensemble becomes what is called a "flexible manufacturing system."[51]

A typical flexible manufacturing system might consist of five to ten machines under distributed/hierarchical control, a belt conveyor, and several robots.[52] The central computer schedules work and flow of materials among the machines and downloads programs to microcomputers at individual machines. Once a pallet of workpieces is set in place, the individual pieces proceed automatically from tool to tool, where they are machined in the proper sequence. Microcomputers control the machining and signal the central computer when a task is finished. Computerized materials-handling devices may also be used to load and unload the system and to store and retrieve parts.[53]

When used to make many identical parts, such a system is unlikely to be as fast as special-purpose automatic machinery. However, production of a different part requires only that different programs be loaded. Conceptually, a flexible manufacturing system can have several part styles in production at once, can route production around a malfunctioning machine, can include automatic inspection equipment, and can adjust its programs in response to feedback based upon a statistical profile of the parts in process.[54] In such an ensemble, production planning, materials handling, and fabrication are mechanically as well as electronically integrated into a single system.

Assembly is also becoming more flexible with the application of electronic and computer technology. In a growing number of plants, teams of people assemble the components of a product brought to them on battery-powered, unmanned platforms. These platforms — called automatic guided vehicles — move along energized guide paths buried in the factory floor and stop at the various work teams. The vehicles can be virtually any size and shape, and they can carry a wide variety of tools in addition to the components to be assembled.

In 1974, a Swedish automobile assembly plant was the first to employ automatically guided vehicles. The system was marketed in North America in 1982 and adopted by the General Motors Corporation for four large assembly plants beginning in 1985. Reportedly, advances in computer technology were the basis for this manufacturer's confidence in a system previously tested only in Europe. Automatically guided vehicle systems are under distributed/hierarchical control. On-board microcomputers coordinate and monitor all operations on the carrier and automatically activate certain features, such as hydraulic workpiece-positioning devices. Minicomputers located on the factory floor direct the vehicles to work teams and process communications between them and the central, supervisory computer.[55]

In plants that are built around such carriers, flow-line assembly is replaced by modular assembly. This constitutes a major change in the organization of production. The physical layout and constant speed of the conventional assembly line require the breaking down of workers' jobs into simple, repetitive tasks, such as the installation of a single small part and a few screws. With modular assembly, on the other hand, teams of workers perform logical blocks of tasks, such as the installation of all the accessories on an engine. Stopping an assembly line to adjust a part that is not installed correctly would idle hundreds of workers; commonly, an incorrectly fitted part is "tagged," the expectation being that it will be repaired at the end of the line. With modular assembly, a carrier cannot move from a cell until released by the work team; the group is responsible for fixing malfitted parts then and there. Manufacturers expect this new mode of organization and responsibility to bring significant improvements in product quality.

In an assembly plant designed around automatically guided vehicles, production planning, parts movement, and assembly are mechanically and electronically integrated into a single system. Industrial engineers say that such a plant can be an extremely flexible one. The computer system can keep track of hundreds of carriers and decide on a minute-by-minute basis what team to assign them to, based on variations in the model mix. Thus it is easier to produce quite different products in the same plant — cars and trucks, for example. This would not be possible with flow-line assembly since lines are usually tooled for just one product. In addition, in the new modular plants, there is no tearing out of assembly lines and installation of new ones with each model change; new models can be introduced more quickly and at lower cost.[56]

Computerizing Activities That Support Production

Mechanization and automation of production had proceeded quite far before computers were introduced. The many office activities that support production, however, remained largely unmechanized until the advent of computer-related technologies.

There are three broad areas of activity in the offices of most manufacturing firms: professional tasks (e.g., design, engineering, production management, product distribution), clerical tasks (e.g., word processing, filing), and communications. Early computer applications were largely limited to a few professional tasks: either complex scientific or engineering calculations or repetitive and voluminous bookkeeping computations were done on large central computers. Today many professional tasks are done at personal electronic workstations. Examples include computer-aided design (CAD) stations, desktop computers for engineering analysis, and terminals for updating customer accounts.

Computers are also permitting the mechanization and automation of the more labor-intensive clerical tasks such as word processing and filing. Word processing technology allows a worker to key or enter material only once and to store it for subsequent changes (some of which can be carried out automatically) and for printing. Even the machine-entry process itself is a candidate for mechanization via several emerging electronic technologies. Electronic filing enhances the ability to locate and retrieve information, and it reduces the space required for information storage.

The use of computers to enhance communications, the third area of activity in offices, has also begun. Some computerized office activities have been linked and integrated electronically, just as computerized production machinery has been linked. The result — sometimes called "office automation" — can yield benefits not available when office activities are carried out independently.

With office automation, the electronic workstation is no longer just a specialized piece of equipment for a specific task. It includes word processing capability and other utilities such as spreadsheets, calendars, and directories. Significantly, it also includes communications capabilities. With computer-related office communication systems, text can be instantaneously transmitted to terminals on other workers' desks or to workstations; spoken messages can be recorded and automatically transmitted by telephone; documents can be filed and retrieved anywhere at any time without recourse to a voluminous paper filing system. In sum, office automation integrates aids for all of a worker's tasks; more important, it permits easy and rapid sharing of information among workers.

Electronic Integration of Production and Support Activities

We have seen that computer-controlled production machinery can be linked and coordinated electronically and that many of the office activities that support production can be mechanized and linked via electronic computers as well. But the impact of electronic information technologies on the production system as a whole is potentially even greater. In reality, of course, *all* the components of production organizations — those located on the factory floor as well as those located in offices — are interdependent. They are linked and coordinated by flows of information that direct and monitor their activities much as the human nervous system directs and monitors the body. A good deal of this information traffic is routine; yet only a small part has so far been entrusted to computer technologies. A major challenge — and major opportunity — before manufacturers is computerizing the processing and transmitting of information between components of organizations, as well as the routine decision making that is based on that information.

There are several examples of the electronic integration of the various components of production enterprises. One, which stresses the sharing of a common data base, is known as "computer-integrated manufacturing." Beginning with the design process, the following paragraphs outline some of the many uses for the data describing a particular part.

The technology of computer-aided design began as a drafting aid in the late 1960s. A designer can sketch a part with a light pen on a special screen and enter important dimensions at a keyboard; the computer system converts this information into a precise engineering drawing. Frequently used details can be stored for recall much as the manual draftsman uses a template. Drawings can be displayed on the screen at any angle and at any scale. Cross sections are easily generated. A digital plotter can produce paper drawings as needed, thus automating the laborious drafting process. If access to video monitors replaces the use of paper drawings, the problem of working with an out-of-date version of the design can be almost eliminated.[57]

Few computer-aided design systems stop here. The geometry data of the part are also needed in structural, dynamic, or other engineering analyses. With a manual system, an engineer or technician would take measurements from a paper drawing and key them into a computer; the computer-aided system eliminates the data transfer steps, enabling the engineer to concentrate on the analyses. Furthermore, by using animation, the designer can "move" parts on his screen and "assemble" several parts to check clearances. Many models and prototypes can be eliminated.

Once the design of a part is completed, manufacturing engineers must determine how to make it. Computer aids can help in two ways, again drawing on the geometry data stored in the system. First, by maintaining a file of all parts being made, manufacturers frequently find that the new design is not really new. Other parts from previous designs may be suitable for the new purpose. Second, computer software can help the manufacturing engineer determine how best to make the part and can automatically generate the numeric programs to control the machines that will form it.

By this point, the data file on the part includes its geometry, its material, its relationship to the rest of the assembly, and instructions for its manufacture. The data file will also include the date on which delivery of the assembly is needed. Armed with similar information about all parts in process, computerized resource planning tools can schedule the ordering of materials and can schedule the work through the factory. At the appropriate time, materials-handling equipment will deliver the workpiece to a machine tool, the tool's control program will be loaded by a master computer, and the part will be processed. It will then be delivered for further processing or assembly. Throughout this period from design through delivery, computers can also monitor progress and gather data for cost accounting.

Two points about computer-integrated manufacturing deserve emphasis. First, the primary focus is the linking and coordinating of design, engineering, production management, fabrication, and even product distribution and field service via an electronic network of commonly available information. Second, a major benefit of this electronic integration is the increase in speed and flexibility with which the organization can react to the competitive environment.

PROSPECT: A CONTEMPORARY ORGANIZATIONAL REVOLUTION?

The information revolution is still young and its course and potential are difficult to predict. Some of its implications become clearer, however, with increased understanding of its past.

In this chapter we have seen that from their first application to machine tools, electronic control technologies freed production from constraints, making it more flexible. This process is continuing and broadening today. By facilitating the integration of the information resources of an entire production enterprise, electronic computer-related technologies appear to permit, even encourage, a rethinking and rearranging of the ways that production activities are organized. Freed from the constraints inherent in time-honored structures, factories, offices, and entire manufacturing enterprises can be integrated and organized in more efficient ways. These incipient changes may prove to be very much analogous to the revolutionary changes that accompanied the broad adoption of electricity in production three-quarters of a century ago.

Enterprise-wide electronic integration has just begun. The field is still in its infancy, and linking components together is complicated by insufficient standardization. Establishing adequate communication between computer-controlled devices from different vendors is often difficult, if not impossible. Moreover, it is frequently unclear what information is actually needed to effectively manage a complex, flexible manufacturing enterprise. Since computer technologies allow the collection and transmission of enormous amounts of data, everything imaginable and measurable is sometimes compiled, providing managers with much superfluous information.[58]

Thus most organizations have integrated only some of their activities electronically and are far from realizing their full potential. At some factories, the initial emphasis has been on scheduling and production, while design and planning dominate at others. Some offices have stressed communication among workers, while others have emphasized tools for the workers. In almost all cases, major functions have yet to be included in the electronic network.

If the pace of electronic integration is slow, it may also be due to the sizable costs that are involved. Investments in computer-integrated manufacturing have a longer time horizon than do conventional equipment purchases. And many of the benefits — such as higher product quality or enhanced ability to respond to future surprises — are inherently difficult to quantify. Moreover, computer integration requires great technical skill and must be custom tailored for each organization; diffusion throughout manufacturing may be slow because of limited expertise. Thus the pioneers of electronic integration may be acting on faith, just as many of the pioneers of electrified mechanical drive did. Computer technologies are almost certain to bring further organizational change and productivity gains to manufacturing, but it is not clear whether that time is imminent or some years in the future.

Notes

1. Warren D. Devine, Jr., *An Historical Perspective on the Value of Electricity in American Manufacturing*, ORAU/IEA-82-8(M) (Oak Ridge, TN: Oak Ridge Associated Universities, Institute for Energy Analysis, September 1982), 41.

2. R. S. Woodbury, "History of the Milling Machine," *Studies in the History of Machine Tools* (Cambridge, MA: Technology Press, 1960), 91–95.

3. Hydraulic mechanisms are operated by liquid (usually a light oil) flowing under high pressure in a circuit of pipes. On machine tools, pressure is maintained with a pump driven by a constant-speed electric motor. Variable-speed translational or rotational motion can be obtained by routing the liquid to cylinders with pistons or to hydraulic motors, respectively.

4. Woodbury, "Milling Machine," 95.

5. S. Einstein, "Machine-Tool Milestones, Past and Future," *Mechanical Engineering* 52 (November 1930): 960–961; J. P. Ferris and E. Wiedmann, "Progress in the Use of Hydraulic Equipment on Production Machinery," *Mechanical Engineering* 54 (July 1932): 477–482; H. D. Wagoner, *The U.S. Machine Tool Industry from 1900 to 1950* (Cambridge: MIT Press, 1968), 16–18.

6. Cam-controlled devices have long been used on tools. As early as 1818 Thomas Blanchard, at the Springfield, Massachusetts Armory, invented a "copying lathe" to produce automatically the irregular shapes of military gunstocks. The desired or master shape was placed in a tracing cradle; a number of cams "profiled" the

master, adjusting cutting tools so as to shape a similar gunstock from a rough block of wood. This was the first of a class of tools known as "cam following" tools. Later, during the Civil War, the cam-controlled automatic screw machine was developed. Prior to the turn of the century, the turret lathe was made automatic by cam control. This is a lathe on which a number of different tools are permanently mounted on a rotatable base or turret so that they can be brought into use when needed (Woodbury, "Milling Machine," 97–98; J. R. Bright, "The Development of Automation," in *Technology in Western Civilization* II, eds. Melvin Krantzberg and Carroll W. Pursell [New York: Oxford University Press, 1967], 644).

7. William Pease, "An Automatic Machine Tool," *Scientific American* (September 1952), 56; Woodbury, "Milling Machine," 97–98; T. C. Rolt, *Short History of Machine Tools* (Cambridge: MIT Press, 1965), 241.

8. S. Bennett, "Nicholas Minorski and the Automatic Steering of Ships," *IEEE Control Systems Magazine* 4 (November 1984): 10–15.

9. Harold Chestnut and Robert W. Mayer, *Servomechanisms and Regulating System Design*, 2nd ed. (New York: Wiley, 1959), 3; Alan Andrews, *ABCs of Synchros and Servos* (Indianapolis, IN: H. W. Sams, 1962), 7–10.

10. Rolt, *Machine Tools*, 241; Woodbury, "Milling Machine," 98–100.

11. Woodbury, "Milling Machine," 100.

12. In the late 1940s and early 1950s an electronic controller consisted of vacuum tube amplifiers and other electric circuit elements that together could amplify or otherwise transform the error signal in a servomechanism into an appropriate actuating signal (see Figure 2.1).

13. Many control problems — whether they arise in a mechanical, hydraulic, thermal, or electrical system — can be described using the same mathematical expressions. This means that if it is difficult or impossible to experiment with a system while devising a method of controlling it, it is always possible to do so by analogy, using a more manageable system. The electronic analog computer is an especially manageable analogic system that displays — as a continuous, time-varying voltage — the predicted performance of whatever system it is "wired" to represent. This type of computer proved to be a valuable tool for designing servomechanisms.

14. Chestnut and Mayer, *Servomechanisms*, 14–16; Otto Mayr, *Origins of Feedback Control* (Cambridge: MIT Press, 1970), 132; and Otto Mayr, *Feedback Mechanisms in the Historical Collections of the National Museum of History and Technology* (Washington, D.C.: Smithsonian Institution Press, 1971), 126–127; S. W. Herwald, "Recollections on the Early Development of Servomechanism and Control Systems," *IEEE Control Systems Magazine* 4 (November 1984): 29–30.

15. R. F. Paul, "The Early Stages of Robotics," *IEEE Control Systems Magazine* 5 (February 1985): 27–28; Rolt, *Machine Tools*, 238–242.

16. Tape control did not originate with the Air Force-MIT project. A loom invented for the silk industry by French weaver Charles Jacquard in 1801 was the first practical industrial machine to be controlled by information stored in this way. The pattern of the Jacquard loom's weave was made to correspond to the pattern of holes punched in a series of cards (Thomas G. Gunn, "The Mechanization of Design and Manufacturing," *Scientific American* [September 1982], 117). The monotype — a two-machine system for casting lead type for printing — was developed during the 1890s. The first machine produced a punched paper tape coded for the type desired; the second used the tape to cast type from molds (Warren D. Devine, Jr., *Technological Change and Electrification in the Printing Industry, 1880–1930*, ORAU/ IEA-84-8 [Oak Ridge, TN: Oak Ridge Associated Universities, Institute for Energy Analysis, January 1985]). Perforated paper rolls were used to actuate player pianos

beginning in 1904; similar sheets controlled a lathe built in 1948 by the Arma Corporation (Bright, "Development of Automation.").

17. Pease, "Automatic Machine Tool," 57–59; Woodbury, "Milling Machine," 100–102.

18. Bright, "Development of Automation," 645.

19. J. R. Bright, *Automation and Management* (Cambridge: Harvard University Press, 1958), 19.

20. M. Dyke, *Numerical Control* (Englewood Cliffs, NJ: Prentice Hall, 1967).

21. Bright, "Development of Automation," 645.

22. E. Brent Sigmon, *Electronic Information Management and Productivity*, ORAU/ IEA-85-3(M) (Oak Ridge, TN: Oak Ridge Associated Universities, Institute for Energy Analysis, April 1985), 11–12.

23. The word "automation" first appeared in 1948 referring to the automatic handling of parts in process in the metal-working industries. By 1954 to most industrialists the word implied continuous automatic production made possible by automatic control. To nontechnical people, however, "automation" had become synonymous with almost any kind of technical change (Bright, "Development of Automation," 635–641). Our use of the term here is that of the industrialist for whom Dyke was writing.

24. Dyke, *Numerical Control*, 13.

25. As described in Chapter 1, the spread of electric unit drive after World War I made it possible to locate production machines according to criteria other than those dictated by a factory's power distribution system. For example, machines could be grouped by operation (i.e., milling, grinding, etc.), or the various kinds of machines needed to produce a particular product could be clustered together or placed relative to one another in accordance with the required sequence of operations. In some industries workpieces came to be moved from operation to operation by conveyors, thus mechanically linking a series of separate machines into a production line and dramatically increasing the flow of work. After World War II, "transfer machines" became common in high-volume metal-working industries such as automobile manufacture. A transfer machine replaces separate machine tools with a series of tools mounted on a common base; workpieces are automatically — and sometimes very rapidly — transferred from one automatic work station to the next. Although these and other techniques brought significant economies, they reduced flexibility in the production process and increased uniformity in the design of products (Bright, *Automation and Management*, 100–102; C. R. Hine, *Machine Tools for Engineers*, 2nd ed. [New York: McGraw-Hill, 1959], 401–410; E. D. Lloyd, *Transfer and Unit Machines* [Brighton, England: The Machinery Publishing Company, 1969], 9–22).

26. Pease, "Automation Machine Tool," 107.

27. A. Augarten, *Bit by Bit: An Illustrated History of Computers* (New York: Ticknor and Fields, 1984); Robert Casey and Robert Friedel, *A Century of Electrical Progress* (New York: Institute of Electrical and Electronics Engineers, 1984); H. B. O. Davis, *Electrical and Electronic Technologies: A Chronology of Events and Inventors from 1940 to 1980* (Metuchen, NJ: Scarecrow Press, 1985).

28. R. M. Davis, "Evolution of Computers and Computing," *Science* 195 (March 18, 1977): 1096.

29. Bright, "Development of Automation," 645; S. Kahne, I. Lefkowitz, and C. Rose, "Automatic Control by Distributed Intelligence," *Scientific American* (June 1979), 85.

30. L. B. Evans, "Impact of the Electronics Revolution on Industrial Process Control," *Science* 195 (March 18, 1977): 1146.

31. The input to an electronic controller might be, for example, a voltage representing the difference between the desired and measured flow of a liquid in a pipe, similar to the position error signal in the servomechanism of Figure 2.1. If the magnitude of this continuous electric signal is sampled at short intervals and assigned a numerical value, it becomes useful to a digital computer. The computer can be programmed to analyze such digital error signals and generate actuating signals, just as electronic controllers do. (Mayr, *Feedback Mechanisms*, 126; Kahne, Lefkowitz, Rose, "Automatic Control," 80).

32. A. H. Hix, "Status of Process Control Computers in the Chemical Industry," in *Proceedings of the IEEE*, Special Issue on Computers in Industrial Process Control (January 1970), 4.

33. Ibid.; D. L. Morrison et al., "Advances in Process Control," *Science* 215 (February 12, 1982): 814.

34. Douglas A. Cassell, *Microcomputers and Modern Control Engineering* (Reston, VA: Reston Publishing Company, 1983), 140–141; Kahne, Lefkowitz, Rose, "Automatic Control," 85.

35. "Over the past 25 years engineers have invented a variety of methods for analyzing systems and for designing control strategies. . . . All of these rest on the common presupposition of centrality: all the information available about the system, and the calculations based upon this information, are centralized, that is, take place at a single location. (But) when considering very large systems, the presupposition of centrality fails to hold, either due to the lack of centralized information or to the lack of centralized computing capability." (N. R. Sandell, Jr. et al., "Survey of Decentralized Control Methods for Large Scale Systems," *IEEE Transactions on Automatic Control* AC-23, Special Issue on Large-Scale Systems and Decentralized Control [April 1978], 108.)

36. Institute of Electrical and Electronics Engineers, *IEEE Spectrum*, Special Issue on Technology '74 (January 1974), 48–49.

37. C. A. Hudson, "Computers in Manufacturing," *Science* 215 (February 12, 1982): 820.

38. *IEEE Spectrum* (January 1974), 49.

39. R. N. Noyce, "Microelectronics," *Scientific American* (September 1977), 63–65; R. Friedel, "Sic Transit Transistor," *American Heritage of Invention and Technology* (Summer 1986), 34–40.

40. Cassell, *Microcomputers*, 198–199.

41. Noyce, "Microelectronics," 63, 65.

42. *IEEE Spectrum* (January 1974), 44–45; Cassell, *Microcomputers*, 103, 104, 109.

43. *IEEE Spectrum* (January 1974), 48, 52; B. M. Oliver, "The Role of Microelectronics in Instrumentation and Control," *Scientific American* (September 1977), 188.

44. Gunn, "Design and Manufacturing," 125; Alan Cane, "Manufacturing Automation: A Revolution on the Factory Floor," *Financial Times* (July 16, 1982), III:1, 7.

45. Sigmon, *Electronic Information Management*, 12.

46. Howard Rudnitsky, "Cutting It Close," *Forbes* (April 30, 1984), 78–79.

47. Kahne, Lefkowitz, Rose, "Automatic Control"; Morrison et al., "Process Control."

48. Cassell, *Microcomputers*, 141–150.

49. Morrison et al., "Process Controls," 814, 818; Cassell, *Microcomputers*, 150–151; A. Z. Spector, "Computer Software for Process Control," *Scientific American* (September 1984), 184.

50. Gunn, "Design and Manufacturing," 126.

51. Ibid.

52. Tom Hughes and Donald E. Hegland, "Flexible Manufacturing — The Way to the Winner's Circle," *Production Engineering* (September 1983), 54–63.

53. Gunn, "Design and Manufacturing," 126; Arthur J. Roch, Jr., "Flexible Machining in an Integrated System," in *Design and Analysis of Integrated Manufacturing Systems*, ed. W. Dale Compton (Washington, D.C.: National Academy Press, 1988), 38–41.

54. The higher the level of integration among machines, the greater the need for some form of automatic inspection of products. A worker operating a machine tool manually can note a defect and stop work immediately, but a machine running autonomously could, through a mechanical failure or a programming error, ruin an entire batch of parts. Information from sensory devices on the machines can be utilized to accept or reject individual parts. Moreover, the information can also serve to build a statistical data base (Gunn, "Design and Manufacturing," 127). Statistical feedback makes it possible for the computer that controls each machine tool to distinguish between random defects and systematic defects and to adjust the tool accordingly during operation. Thus, microcomputers have provided the discrete parts manufacturer with the means of obtaining useful feedback from his processes — and hence continuous closed-loop control over them — by methods analogous to those used in bulk process manufacturing.

55. "The Volvo AGVs Have Landed," *Via Volvo* (Winter 1985), 19–21, 24.

56. John Holusha, "A New Way to Build Cars," *The New York Times* (March 13, 1986), 30.

57. Gunn, "Design and Manufacturing"; Donald E. Hegland, "Out in Front with CAD/CAM at Lockheed-Georgia," *Production Engineering* (November 1981), 45–48; Donald E. Hegland, "Putting the Automated Factory Together," *Production Engineering* (June 1982), 56–85; "The Automated Factory: A Progress Report," *Production Engineering* (June 1983), 40–46.

58. Lawrence C. Seifert, "Design and Analysis of Integrated Electronics Manufacturing Systems," in Compton, *Integrated Manufacturing Systems*, 20; and William B. Rouse, "The Human Role in Advanced Manufacturing Systems," in Compton, *Integrated Manufacturing Systems*, 150–151, 161.

Works Cited

Alan Andrews, *ABCs of Synchros and Servos* (Indianapolis: H. W. Sams, 1962).

A. Augarten, *Bit by Bit: An Illustrated History of Computers* (New York: Ticknor and Fields, 1984).

S. Bennett, "Nicholas Minorski and the Automatic Steering of Ships," *IEEE Control Systems Magazine* 4 (November 1984): 10–15.

J. R. Bright, *Automation and Management* (Cambridge: Harvard University Press, 1958).

J. R. Bright, "The Development of Automation," in *Technology in Western Civilization II*, eds. Melvin Krantzberg and Carroll W. Pursell (New York: Oxford University Press, 1967).

Alan Cane, "Manufacturing Automation: A Revolution on the Factory Floor," *Financial Times* (July 16, 1982), III:1, 7.

Robert Casey and Robert Friedel, *A Century of Electrical Progress* (New York: Institute of Electrical and Electronics Engineers, 1984).

Douglas A. Cassell, *Microcomputers and Modern Control Engineering* (Reston, VA: Reston Publishing Company, 1983).

Harold Chestnut and Robert W. Mayer, *Servomechanisms and Regulating System Design*, 2nd ed. (New York: Wiley, 1959).

W. Dale Compton, ed., *Design and Analysis of Integrated Manufacturing Systems* (Washington, D.C.: National Academy Press, 1988).

H. B. O. Davis, *Electrical and Electronic Technologies: A Chronology of Events and Inventors from 1940 to 1980* (Metuchen, NJ: Scarecrow Press, 1985).

R. M. Davis, "Evolution of Computers and Computing," *Science* 195 (March 18, 1977): 1096–1102.

Warren D. Devine, Jr., *An Historical Perspective on the Value of Electricity in American Manufacturing*, ORAU/IEA-82-8(M) (Oak Ridge, TN: Oak Ridge Associated Universities, Institute for Energy Analysis, September 1982).

Warren D. Devine, Jr., *Technological Change and Electrification in the Printing Industry, 1880–1930*, ORAU/IEA-84-8(M) (Oak Ridge, TN: Oak Ridge Associated Universities, Institute for Energy Analysis, January 1985).

R. M. Dyke, *Numerical Control* (Englewood Cliffs, NJ: Prentice Hall, 1967).

S. Einstein, "Machine-Tool Milestones, Past and Future," *Mechanical Engineering*, 52 (November 1930): 959–962.

L. B. Evans, "Impact of the Electronics Revolution on Industrial Process Control," *Science* 195 (March 18, 1977): 1146–1151.

J. P. Ferris and E. Wiedmann, "Progress in the Use of Hydraulic Equipment on Production Machinery," *Mechanical Engineering* 54 (July 1932): 477–482.

R. Friedel, "Sic Transit Transistor," *American Heritage of Invention and Technology* (Summer 1986), 34–40.

Thomas G. Gunn, "The Mechanization of Design and Manufacturing," *Scientific American* (September 1982), 114–130.

Donald E. Hegland, "Out in Front with CAD/CAM at Lockheed-Georgia," *Production Engineering* (November 1981), 45–48.

Donald E. Hegland, "Putting the Automated Factory Together," *Production Engineering* (June 1982), 56–85.

Donald E. Hegland, "The Automated Factory: A Progress Report," *Production Engineering* (June 1983), 40–46.

S. W. Herwald, "Recollections on the Early Development of Servomechanism and Control Systems," *IEEE Control Systems Magazine* 4 (November 1984): 29–32.

C. R. Hine, *Machine Tools for Engineers*, 2nd ed. (New York: McGraw-Hill, 1959).

A. H. Hix, "Status of Process Control Computers in the Chemical Industry," in *Proceedings of the IEEE*, Special Issue on Computers in Industrial Process Control (January 1970), 4–10.

John Holusha, "A New Way to Build Cars," *New York Times* (March 13, 1986), 30.

C. A. Hudson, "Computers in Manufacturing," *Science* 215 (February 12, 1982): 818–825.

Tom Hughes and Donald E. Hegland, "Flexible Manufacturing — The Way to the Winner's Circle," *Production Engineering* (September 1983), 54–63.

Institute of Electrical and Electronics Engineers, *IEEE Spectrum*, Special Issue on Technology '74 (January 1974).

S. Kahne, I. Lefkowitz, and C. Rose, "Automatic Control by Distributed Intelligence," *Scientific American* (June 1979), 78–90.

E. D. Lloyd, *Transfer and Unit Machines* (Brighton, England: The Machinery Publishing Company 1969).

Otto Mayr, *Feedback Mechanisms in the Historical Collections of the National Museum of History and Technology* (Washington, D.C.: Smithsonian Institution Press, 1971).

Otto Mayr, *Origins of Feedback Control* (Cambridge: MIT Press, 1970).

D. L. Morrison, Richard H. Snow, and John P. Lamoureux, "Advances in Process Control," *Science* 215 (February 12, 1982): 813–818.

R. N. Noyce, "Microelectronics," *Scientific American* (September 1977), 62–69.

B. M. Oliver, "The Role of Microelectronics in Instrumentation and Control," *Scientific American* (September 1977), 180–190.

R. F. Paul, "The Early Stages of Robotics," *IEEE Control Systems Magazine* 5 (February 1985): 27–31.

William Pease, "An Automatic Machine Tool," *Scientific American* (September 1952), 101–115.

Arthur J. Roch, Jr., "Flexible Machining in an Integrated System," in *Design and Analysis of Integrated Manufacturing Systems*, ed. W. Dale Compton (Washington, D.C.: National Academy Press, 1988).

T. C. Rolt, *Short History of Machine Tools* (Cambridge: MIT Press, 1965).

William B. Rouse, "The Human Role in Advanced Manufacturing Systems," in *Design and Analysis of Integrated Manufacturing Systems*, ed. W. Dale Compton (Washington, D.C.: National Academy Press, 1988).

Howard Rudnitsky, "Cutting It Close," *Forbes* (April 30, 1984), 78–79.

Nils R. Sandell, Jr., Pravin Varaiya, Michael Athans, and Michael G. Safonov, "Survey of Decentralized Control Methods for Large Scale Systems," *IEEE Transactions on Automatic Control* AC-23, Special Issue on Large-Scale Systems and Decentralized Control (April 1978), 108–113.

Lawrence C. Seifert, "Design and Analysis of Integrated Electronics Manufacturing Systems," in *Design and Analysis of Integrated Manufacturing Systems*, ed. W. Dale Compton (Washington, D.C.: National Academy Press, 1988).

E. Brent Sigmon, *Electronic Information Management and Productivity*, ORAU/IEA-85-3(M) (Oak Ridge, TN: Oak Ridge Associated Universities, Institute for Energy Analysis, April 1985).

A. Z. Spector, "Computer Software for Process Control," *Scientific American* (September 1984), 174–186.

"The Volvo AGVs Have Landed," *Via Volvo* (Winter 1985), 19–21, 24.

H. D. Wagoner, *The U.S. Machine Tool Industry from 1900 to 1950* (Cambridge: MIT Press, 1968).

R. S. Woodbury, "History of the Milling Machine," *Studies in the History of Machine Tools* (Cambridge, MA: Technology Press, 1960).

Part II
Electricity in Manufacturing:
Its Role in Materials Processing

Introduction

The next five chapters focus on the use of electricity to transform materials from one physical or chemical state to another. We deal primarily with that segment of manufacturing in which the materials processed come as bulk solids or fluids or slurries, rather than as discrete pieces to be worked individually. Such bulk materials processing is the type of production that predominates in the primary metals, glass, pulp and paper, petroleum refining, and chemicals industries.

Like the preceding part, this part of the book distinguishes between two periods of history. Chapter 3 describes several very early developments in electric materials processing, when new commercial products and new industries were made possible by uniquely electrical production techniques. Chapters 4 through 7 deal with more recent times, when both new and evolving applications of electricity are facilitating important improvements in production techniques for a wide variety of basic industrial commodities.

The electric technologies described in Chapter 3 were, for the most part, developed during the 1880s and 1890s — shortly after the first commercial production of electricity, in the latter part of what has been called the "heroic age" of American invention. This was a time particularly favorable to the individual inventor-entrepreneur because the business of inventing and commercializing new products and the techniques of making them had not yet come to be dominated by large institutions. Working largely alone, or with a small number of associates, individuals like Charles Hall, Hamilton Castner, and Edward Acheson developed practical chemical processes uniquely dependent on electricity for manufacturing a variety of chemicals and metals. We will refer to these techniques as "electrochemical processes" or, simply, as "electroprocesses."

With certain electroprocesses, substances such as aluminum, chlorine, and silicon carbide could be produced; these materials had not been previously available commercially and were to become of great importance to society. Other electroprocesses yielded materials that were already available, such as caustic soda and phosphorus, but electric methods of making these products had many advantages over chemical methods. These advantages included the high purity of the output, plus the ability to use plentiful raw materials as input and to make additional valuable products in the same plant. As the demand for electrochemicals increased, major new industries arose.

As important as these new materials and new industries became, they did not represent the only impacts of early electroprocessing on American manufacturing. During the nineteenth and early twentieth centuries, most bulk materials processing was carried out in batches, that is, discontinuously with respect to time. Continuous processing — the mode of operation characteristic of the electroprocess industries — was preferable because it eliminated the time and labor spent in emptying, cleaning, and refilling reaction vessels between batch cycles and in waiting for reaction conditions to become reestablished. The continuous processing techniques pioneered by the electroprocess industries have now become standard throughout bulk materials manufacturing.

The electric technologies discussed in Chapters 4 through 7 are those that have become important in recent years. These technologies have not provided the basis for entirely new products and major industries, nor are they employing an energy form that is itself a newcomer to the industrial scene, as was the case with electricity in the late nineteenth century. As a result, their histories are not as dramatic, nor do their pioneers seem as heroic as those of the earlier period. Nevertheless, current applications of electricity in the processing industries have made possible major improvements in the production of many basic industrial commodities.

Chapter 4 provides background on how these recent developments reflect such forces as the changing relative costs of electricity and fuels, the need for pollution abatement, and the rising costs of raw materials. Such factors have had a mutually reinforcing effect in driving the movement toward increased use of electricity in industries that had previously been largely dependent on fuel-based technologies.

No industries better illustrate the recent inroads that electricity has made than those engaged in high-temperature materials processing (Chapter 5). When electricity is used as a source of heat, it is most competitive for those processes in which the temperatures required for the materials transformations are high — that is, thousands of degrees Fahrenheit. Steelmaking and glassmaking require such temperatures, and the use of electric techniques instead of fuel-based techniques for thermal processing has proceeded the farthest in these industries. In particular, the long-established steel industry, in many ways the very symbol of the industrial age, may be in a transition that will eventually be judged to be as dramatic as the rise of the pioneering electroprocessing industries nearly a century ago.

Electrothermal processing is leading to revolutionary changes in the entire system of steel manufacturing in the United States. Because the ferrous raw material for the electric arc furnace is scrap, not primarily pig iron, such furnaces can dispense with the coke ovens and blast furnaces that have been typical of the fuel-based open-hearth and basic oxygen processes. As a consequence, these new mills are not held hostage by the logistics of coal, iron ore, and limestone supply, or to the economies of scale dictated by conventional steelmaking. Built with relatively modest investment (capital costs are far lower than those of conventional steel facilities), electric "minimills" can draw on local resources of scrap and can produce for local markets. Thus, in contrast to fuel-based steelmaking, the electric steel production system, in addition to its more favorable cost and efficiency characteristics, is a geographically decentralized one.

To date, changes in glassmaking are less far reaching. Glassmaking furnaces can be fired by natural gas or heated via electricity, or both gas and electricity can be used in the same installation in various proportions. This latter arrangement — in which gas-fired glass production is substantially augmented by the addition of electric heat — has so far been the principal way electricity is used in furnace operations. The relative prices of gas and electricity have, therefore, been important in determining glassmakers' energy choices. Indeed, since the 1960s the price of electricity for glass manufacturing relative to that of gas has dropped by half, and the fraction of total energy purchased by the industry in the form of electricity has doubled. But all-encompassing systemic changes, as in the steel industry, have not yet begun in glassmaking.

Unlike steelmaking and glassmaking, where high-temperature thermal processes are employed, most transformations in the paper, chemicals, and petroleum-refining industries require comparatively low temperatures (Chapter 6). These temperatures can be achieved most economically using steam or heated gas from the combustion of fuels. Even so, aided by falling electricity prices compared to those for fuels, the low-temperature processing industries have, over the past two decades, increased their use of electricity relative to their use of fuels.

A significant component of their growing use of electricity has been the substitution of electromechanical processing — the use of mechanical energy produced by electric motors — for thermal or chemical processing. Important examples are mechanical fiberization in the pulp and paper industry, and the separation of components of liquid and gaseous mixtures in the chemical industry. In petroleum refining, the production of higher quality products from lower quality feedstocks has been achieved primarily through additional processing, and the various refining processes involved have used increasing amounts of electromechanical energy.

Electromechanical energy also displaces thermal processing indirectly through the use of oxygen. Oxygen, separated from air electromechanically, is used in enhancing or replacing a variety of conventional manufacturing processes. For example, when oxygen is used to enrich air for fuel combustion in glass melting, cement calcining, copper smelting, and catalyst regenerating in petroleum refining, combustion temperatures are raised and throughput increased. Oxygen is also the key to new high-yield, energy-efficient processes that replace conventional thermal processes for reforming low-grade fuels such as coal and petroleum residuum in the production of the synthesis gases that are the basic building blocks for the petroleum refining and organic chemicals industries.

Chapter 7 moves beyond the inroads being made by electric-based technologies in high- and low-temperature thermal processes. It profiles an emerging group of electrotechnologies that are founded on electricity's ability to produce desired effects in materials in ways that are not otherwise possible.

One such technology that is widely applied is dielectric heating, in which the energy contained in electromagnetic waves is absorbed within the material being heated, not conducted to the interior from the surface as occurs with fuel-fired heating. Although dielectric heating technology can cover a broad range of temperatures, the major applications thus far have been in processing nonconducting materials that cannot withstand other methods of heating to high temperatures without damage. These applications include drying paper during its manufacture, vulcanizing rubber, and drying certain foods.

This new wave of electrical processes — dielectric heating being just one example — is very promising. The technologies resemble the early electrolytic and electrothermal applications covered in Chapter 3 in that the effects achieved are unique to energy in the form of electricity. Together with the electroprocessing technologies already established over the past two decades, they point the way toward the continuing electrification of the nation's materials processing industries.

3 | Early Developments in Electroprocessing: New Products, New Industries

Warren D. Devine, Jr.

Commercial production of electricity began in 1882. During the ensuing quarter century, practical chemical processes uniquely dependent on electricity were developed for manufacturing a variety of chemicals and metals. By the 1920s, major new industries centering on these electrochemical processes had grown and matured.

Electrochemical processes (or, simply, "electroprocesses") are usually divided into two major categories: electrolytic processes and electrothermal processes. In electrolytic processes, certain specific chemical changes are brought about via electricity. In electrothermal processes, electricity is used to produce heat, which in turn brings about or makes possible a desired physical or chemical change.[1]

The electroprocess industries are usually classified according to whether their principal products are chemicals or metals. Most firms employing electrolytic or electrothermal processes are assigned to Standard Industrial Classification (SIC) 28, "Chemicals and Allied Products," or to SIC 33, "Primary Metals Industries."[2] These electroprocess firms account for an important share of the economic activity of SICs 28 and 33.[3]

INNOVATIVE PROCESSES USING ELECTRICITY

This chapter considers a number of technical and entrepreneurial achievements that were made possible by electricity. Brief accounts are given of the rise of wholly new industries: aluminum, alkalies and chlorine, manufactured abrasives, and phosphorus. Common themes in the development and growth of these industries are noted, particularly the spread of continuous processing and its relationship to both Niagara power and the use of individual electric motors for driving machinery. The chapter concludes by pointing out a number of striking parallels between the ways electric motors have affected production in the fabricating industries and the ways certain electroprocesses have affected production in the bulk processing industries.

Aluminum

Although aluminum is the most abundant metal in the earth's crust, it proved difficult to reduce to a pure metallic state. Indeed, until the mid-nineteenth century only enough pure aluminum had been produced to allow chemical study of its properties. In 1854, Henri St. Claire Deville obtained support for his experiments from the French government and by 1874 had developed a batch process for the reduction of aluminum chloride with sodium in a fused bath.

Expensive metallic sodium, however, represented about 50 percent of the cost, and for many years aluminum remained a semiprecious metal. Uses were restricted to jewelry and a few special purposes, including a 100-ounce cap for the top of the Washington Monument. By 1886 the price of aluminum by the Deville process had been reduced from around $100 per pound to from $8 to $12 per pound due chiefly to the use of cryolite (sodium-aluminum fluoride) as a flux in the bath, to the reduced cost of sodium, and to improvements in equipment.[4]

Large-scale production of aluminum was not achieved until the advent of the Hall-Heroult process. In 1886, Charles M. Hall produced the first aluminum in the manner employed today: electrolysis of aluminum oxide (Al_2O_3), or alumina, dissolved in a molten bath of cryolite. That same year Paul L. T. Heroult was granted a French patent for a process similar to that of Hall. Heroult could not obtain a United States patent, however, as Charles Hall and his sister Julia were able to demonstrate that the Hall process had been practiced prior to the date of Heroult's French patent.

The electrolytic cell used by Hall was a clay crucible lined with carbon and heated externally with a gasoline burner; current for the carbon-rod electrodes was provided by a battery. Heroult's apparatus was more akin to an electric resistance furnace. Heat was provided internally, by the resistance of the material in the cell to an electric current; the current was supplied from a generator turned by a steam engine.[5]

Charles Hall and Alfred Hunt formed the Pittsburgh Reduction Company in 1888. The first plant — in Pittsburgh, Pennsylvania — employed two electrolytic cells or "pots" with current supplied by Westinghouse generators turned by coal-fired steam engines. Within a year aluminum production was 1,300 pounds per month, and the metal was offered for sale at $2 to $3 per pound. In 1891 the company built a new plant in New Kensington, Pennsylvania, so as to take advantage of local supplies of coal and natural gas and reduce somewhat their high electricity costs. Numerous improvements in the electrolytic process were also made, the most significant being the change from external heating of the pots to internal heating — the method employed by Heroult. The company's energy and materials costs per pound of aluminum decreased, and by 1894 the price of the metal was between 50¢ and 75¢ per pound.[6]

Hall realized that the price of aluminum could be lowered still further, not only by using cheaper electricity, but also by expanding operations to take advantage of economies of scale in plant and equipment. In 1895, the Niagara Power Company completed the first large hydroelectric development in the United States at Niagara Falls, New York. In August of that year, the Pittsburgh Reduction Company became Niagara's first industrial customer, using 1,300 kW. The next year a second plant was built, with twice the power demand of the first. Total output was over 100,000 pounds of aluminum per month, and the price of the metal subsequently fell to between 20¢ and 30¢ per pound.[7]

When Hall and his associates first offered aluminum at a price approximately one-half that of the chemically reduced metal, they quickly captured the existing market: fabrication of a few luxury and novelty items. Electrolytic aluminum was essentially a new commodity for which no mass market existed. Indeed, almost every new application had to be won in contest with long-established metals — copper, tin, brass, zinc, and iron — whose characteristics and suitability for specific purposes were well known. The Pittsburgh Reduction Company was faced with accurately determining the characteristics of aluminum under various user conditions, developing suitable applications, and conducting a campaign of education and demonstration.

The steel industry proved to be the first large and expanding market. It was found that small amounts of aluminum added to molten steel resulted in more complete deoxidation, thus improving steel quality. During the early 1890s, this market used nearly half of the output of the Pittsburgh Reduction Company. By the mid-1890s, the advantages of aluminum cooking utensils had been demonstrated. This market soon absorbed more aluminum than the steel industry and remained predominant until the turn of the century.

New markets began to appear every year, such as lithographic printing plates, bicycle parts, cameras, paint, and foil. Before World War I aluminum was used to a considerable extent in electric transmission cables, and — in the form of copper alloys — for certain automobile parts that did not require great strength. Thus the aluminum industry grew steadily, based on new and expanding markets that were created largely by the efforts of the industry itself. Less than twenty years after its founding, the Pittsburgh Reduction Company — now the Aluminum Company of America (Alcoa) — was producing over 1.5 million pounds of aluminum per month.[8]

Electrolysis of alumina has always been a continuous process — and only minor changes have been made in it since the 1890s. Each reduction cell is simply a carbon-lined steel pot, with carbon anodes penetrating the contained bath from above. Electric current enters the pot through the anode, passes into the bath and through a layer of molten aluminum at the bottom of the pot, and exits via the carbon lining. The anodes are lowered further into the bath as they are oxidized (or are raised as the level of molten aluminum rises) so as to maintain optimum spacing between them and the molten aluminum layer. Alumina is added to the bath periodically, and metal is tapped from the bottom.

Pots are operated at design conditions for extended periods of time. Operation at reduced production rates and shutdowns are avoided because the necessary thermal equilibrium is lost with consequent reduced production efficiency even during the return to normal operation. Interruptions in power supply are very costly. Equipment degradation occurs if a power outage exceeds three hours in duration. The bath solidifies after four hours and, if the pot is to be salvaged, the charge must be removed mechanically.[9]

Alkalies and Chlorine

Caustic soda (NaOH, or sodium hydroxide) has always been one of the half-dozen most important chemicals used in the United States, both in terms of quantity consumed annually and in diversity of application. Until the 1890s, caustic soda was made chemically using lime, frequently in plants that employed the Solvay process to make soda ash; large amounts were also imported.[10]

Although free chlorine was not a commercial product in the nineteenth century, fixed chlorine in the form of bleaching powder was an important commodity. In Europe, bleaching powder had been made from slaked lime and chlorine since 1798. The chlorine was obtained chemically from hydrochloric acid, a by-product of the LeBlanc soda ash process. As neither the LeBlanc process nor the then commercial method of obtaining chlorine were ever used in the United States, most bleaching powder was imported.[11]

During the 1880s, a number of Americans experimented with electrolytic cells that they hoped would produce caustic soda and chlorine from common salt dissolved in water, or brine. The caustic would be sold to manufacturers of soap and of various other chemicals. The chlorine would be used to make bleaching powder, which was in increasing demand from the textile and pulp industries.

Electrolytic separation of sodium chloride was straightforward. In an energized cell, chloride ions migrate to the anodes and sodium ions are attracted by the cathodes. The major problem addressed by inventors was how to isolate the reaction products to prevent further chemical conversion. Two approaches eventually yielded commercially successful cells: isolation by a porous, chemically resistant diaphragm and isolation by immediate formation of an alloy of sodium and mercury.

In 1890, Ernest A. LeSueur applied for a U.S. patent on a diaphragm cell and successfully operated a pilot plant in a paper mill. In 1893, LeSueur formed the Electrochemical Company and at Rumford Falls, Maine, began the first commercial electrolytic production of caustic soda and chlorine in the United States. The chlorine was used at the same site to make bleaching powder.[12]

Meanwhile, in 1886, American inventor Hamilton Y. Castner had developed a relatively inexpensive chemical method of producing sodium metal. In 1888 he founded (at Oldbury, England) the Aluminium Company to make aluminum by chemical reduction with sodium. But by this time the Pittsburgh Reduction Company was producing aluminum electrolytically and was able to sell the metal for a price less than half that of Castner's company. Seeking even cheaper sodium, Castner developed in 1891 a method of producing the highly active metal by electrolysis of molten caustic soda. To do this he needed very pure caustic soda, so he turned his attention to developing a commercial process to provide it.

By 1892 Castner had invented the first practical mercury cell for electrolysis of brine and was issued a patent two years later. This type of cell went into operation in 1895 at Saltville, West Virginia, under license to the Mathieson Alkali Company. The mercury cell process for the production of caustic soda and chlorine proved to be the most significant result of Castner's efforts to produce aluminum.[13]

The Mathieson plant at Saltville was a technical success, but it was highly electricity intensive and needed cheaper power to be a commercial success. In 1897 Mathieson opened a much larger facility at Niagara Falls and thus became the first representative of the electrolytic alkali industry to locate there. Mathieson's engineers made several key improvements to Castner's mercury cells over the ensuing ten years, and the process remains much the same today.

Firms that tried to use diaphragm cells, on the other hand, were relatively short lived. LeSueur's Electrochemical Company succumbed in 1898 when importers cut the price of bleaching powder by 60 percent. Variants of the LeSueur cell were operated for short periods of time, frequently at paper mills. None, however, became commercially successful.

In 1902, C. P. Townsend, a Washington, D.C., attorney, applied for a patent on an improved diaphragm cell that he had developed with the help of Elmer A. Sperry. They sold their design two years later to a company founded by E. H. Hooker that commercialized promising inventions. Hooker realized that an inexpensive supply of electricity was vital to the commercial viability of any electrolytic process, and by 1906 Townsend cells were being operated by the newly formed Hooker Electrochemical Company at Niagara Falls. The Hooker diaphragm cell used today reflects incremental improvements to the original cell of Townsend and Sperry.[14]

Although most of the chlorine produced in the 1890s was used to make bleaching powder, water and sewage treatment soon proved to be another large market. Before World War I, chlorine also began to be used in the manufacture of various organic chemicals. New markets for caustic soda, however, did not develop as rapidly as those for chlorine. Moreover, electrolysis yields 12.8 percent (by weight) more caustic soda than chlorine per unit of brine. Thus an oversupply of caustic soda accompanied rising chlorine production.

The lime-soda chemical process used a manufactured product (soda ash) and yielded only a single product (caustic soda) which was contaminated with various impurities, principally soda ash itself. The electrolytic process, on the other hand, used a cheap raw material (brine) and yielded not only very pure caustic soda, but also a valuable coproduct (chlorine). These advantages of electrolysis as well as the chronic surplus of caustic soda led to the phasing out of the lime-soda process, beginning around World War I. The total quantity of caustic soda produced annually by electrolysis did not exceed that produced chemically until 1940; by 1968, however, caustic soda was produced in the United States only by electrolysis.[15]

Electrolytic production of caustic soda and chlorine has always been a continuous process. Diaphragm cells and mercury cells both employ steel tanks with graphite anodes immersed in flowing baths of warm, purified brine. Both cells produce chlorine at their anodes; the difference between the two lies in the reactions that take place at their cathodes. In the diaphragm cell, water is decomposed on a steel cathode, producing hydrogen and dilute caustic soda, a cell effluent. An asbestos diaphragm covering the cathode suppresses back mixing of the products and further chemical reaction. In the mercury cell the liquid mercury itself is the cathode, attracting sodium ions. These immediately form an amalgam with the mercury, which slowly flows across the bottom of the tank to a decomposer. There the sodium-mercury alloy is treated with water to form caustic soda, and the purified mercury is recycled.

Caustic/chlorine cells are designed to operate at specific values of voltage and current for extended periods of time — currently on the order of 100 days or more. Operation at other than design conditions is less productive, and interrupted operation can be hazardous. Both diaphragm and mercury cells are run so as to maintain a certain concentration of salt in the brine. Lower-than-normal brine flow rates reduce this concentration and thus decrease brine conductivity, with an attendant loss of electrical efficiency. Further, because conductivity also depends on temperature, changes in operation that upset thermal equilibrium affect operating efficiency.

In the mercury cell, mercury must flow across the steel base in an unbroken sheet. In the event of a break in the mercury surface (caused, for example, by a pump malfunction or electric power interruption), caustic soda will form on the bare steel cathode with a simultaneous release of hydrogen. This condition is hazardous because hydrogen and chlorine can form a highly explosive mixture.[16]

Manufactured Abrasives

Abrasive processes have played an important role in the evolution of industrial civilization. But until late in the nineteenth century, all abrasives were naturally occurring materials such as diamond, corundum, emery, and garnet. The discovery and production of man-made abrasives in the 1890s went hand in hand with the development of large electric furnaces that were capable of making not only abrasive materials, but a variety of other commercial products as well.

Electric furnaces are simply refractory crucibles heated by an electric arc or by electric current. Furnaces are generally classified as arc, resistance, or induction. In the first, heating occurs via an arc established between electrodes and the material being processed, or the material may be heated indirectly by radiation from arcs established between electrodes that are not in direct contact with the material. (Recently, to achieve more concentrated heating, arcs have been confined by electromagnetic fields.) In a direct-resistance or submerged-arc furnace, heat is produced by current flowing between electrodes inserted into the material being processed; an indirect-resistance furnace is one in which the material is heated via a separate high-resistance conductor that may or may not be situated within the material. Finally, in an induction furnace the conducting material being processed forms the secondary circuit of a transformer and is heated by the current induced in it from the primary circuit.

The first commercially successful electric resistance furnaces in the United States were built and operated by Eugene D. and Alfred H. Cowles in the 1880s. These furnaces were designed for continuous production and attained temperatures sufficient to decompose a number of metallic oxides that had resisted decomposition in fuel-fired furnaces. Although the Cowles brothers demonstrated the use of these furnaces in producing a wide range of materials, their aluminum-copper alloys were of greatest commercial interest. In fact, it was probably the Cowles's reputation with these alloys that attracted Charles Hall to the Cowles Electric Smelting Company in 1887. Had Hall not encountered mechanical difficulties with his process during his employment at Cowles, the company might have exercised its option to purchase the Hall patent rights, and the early history of the aluminum industry could well have unfolded somewhat differently.[17]

In 1891 Edward G. Acheson, a former employee of Thomas Edison, attempted to harden a mixture of clay and coke in an electric resistance furnace of his own design. Hard purple crystals were formed, which Acheson correctly identified as silicon carbide. He soon recognized the abrasive character of this product and named it Carborundum. Later that same year, Acheson constructed larger electric furnaces in Monongahela City, Pennsylvania, and organized the Carborundum Company. Although gem polishers were the company's first customers, heavy industry soon recognized the merits of this first man-made abrasive. In 1892 Westinghouse ordered 60,000 grinding wheels made of Carborundum; ten years later annual output of silicon carbide was approximately four million pounds. Meanwhile, in 1895, Acheson's company moved to Niagara Falls and became the cornerstone of the manufactured abrasives industry there.[18]

In 1887 the Austrian chemist Bayer developed a process for preparing relatively pure aluminum oxide (alumina) from bauxite; this plentiful ore immediately became the primary raw material for the production of aluminum. The naturally occurring aluminum oxide called corundum had long been used as an abrasive, and with increased availability of bauxite and of alumina due to Bayer's process came attempts to prepare these materials for abrasive use as well. A submerged-arc electric resistance furnace for fusing calcined, high-grade bauxite or calcined Bayer alumina was developed by Charles B. Jacobs in 1897. The products — "brown" fused alumina and "white" fused alumina, respectively — are excellent abrasives: not quite as hard as silicon carbide, but less brittle. The Norton Company obtained the rights and patents to Jacobs's process in 1901 and constructed at Niagara Falls the first plant to produce fused alumina on a commercial basis.[19]

The growing use of silicon carbide and fused alumina after the turn of the century led to a decline in the use of natural abrasives. These two manufactured abrasives are exceeded in hardness only by diamond and are quite uniform in character; machining, grinding, and finishing tools that employ them are more efficient than tools that rely on natural abrasive materials. At the same time, increasingly harder metal alloys were being developed by metallurgists in order to reduce wear, and many of these could be shaped and finished economically only with synthetic abrasives. Indeed, man-made abrasives were integral to the pre-World War I development of improved machine tools that made possible much more precise fabrication of metal parts.[20]

Furnaces used in the production of silicon carbide are of the indirect-resistance type. High grade silica sand and carbon (petroleum coke or anthracite coal) are mixed in an approximate molar ratio of 1:3 and formed around a cylindrical core of graphite that connects two electrodes. After heating for 36 hours at a core temperature exceeding 2,000°C, the charge is allowed to cool. Newly formed pieces of silicon carbide crystals are crushed and chemically cleaned. Fused alumina is produced in submerged-arc electric resistance furnaces in the form of cylindrical steel shells cooled by water; the water maintains a peripheral layer of unmelted alumina that serves as a refractory lining. Graphite electrodes are inserted into calcined bauxite or Bayer alumina and arcs struck between them. Charge is added as melting occurs, and the electrodes are withdrawn as the level of fused alumina rises. When full after twenty to thirty hours, the furnace is cooled and the ingot is crushed and chemically cleaned.[21]

Electric furnaces have been designed for continuous operation since the resistance furnaces of Heroult and the Cowles brothers in the 1880s. Furnaces are operated at design conditions for extended periods, sometimes called "campaigns." Reducing the power level of an electric furnace and later returning the furnace to full power upsets the thermal equilibrium and decreases production efficiency for an appreciable length of time. For example, if power to a submerged-arc resistance furnace is suddenly curtailed by 50 percent, it could take time equal to as much as 150 percent of the curtailment time to restore the furnace to full production.[22]

Temperatures exceeding 3,000°C are commonly achieved in electric furnaces, almost twice the maximum temperature practically attainable in commercial fuel-fired furnaces. The effects of high temperatures are twofold: the speed of reactions is increased and new states of chemical equilibrium are attained. Indeed, this ability to achieve new equilibrium states is what has made possible the commercial production of such materials as silicon carbide and fused alumina, which were unknown before the advent of the electric furnace.

From the standpoint of overall fuel economy (including the fuel used to generate electricity), electric heating is superior to fuel heating when processes require high temperatures. For example, in indirect-arc furnaces the fraction of the energy that actually serves process needs is three or more times that fraction in fuel-heated furnaces for processing temperatures above about 1,500°C.[23] This advantage of the electric furnace is not a recent development; indeed, as the following statement attests, it has been recognized since before electricity was available commercially.

> The earliest form of arc furnace in which any practical work was done is due to Werner Siemens, who constructed furnaces in 1878 and 1879. Siemens showed (in a 1880 paper before the Society of Telegraph Engineers) that on the basis of fuel economy, the (electric arc) furnace was more economical than the ordinary air furnace and that it would be nearly equal to a regenerative gas furnace.[24]

But even though fuel economy may have been of some importance, it was probably not the major reason for the increasing penetration of electric furnaces after World War I. In his 1923 text *Industrial Furnaces*, W. Trinks of the Carnegie Institute of Technology pointed out that "for many purposes the (electric furnace) is rapidly rising in favor because it offers advantages which cannot be measured in terms of fuel cost."[25] Trinks then described thirteen advantages (and six disadvantages) of electric furnaces with respect to fuel-fired furnaces and went on to define a broader measure — "heating cost" — that better reflected all the costs pertaining to furnace use:

> The word "economy" when used in its true sense in connection with furnaces, has reference to the lowest heating cost per unit weight of finished and perfect product. "Heating cost" involves not only the cost of the fuel, but also the cost of firing and of superintending the furnace, the costs of maintenance

and repairs, and the cost of burned, spoiled, or otherwise rejected pieces. Furthermore, it includes the cost of machining those parts which subsequently fail because of uneven annealing or uneven heat treatment. Finally, it involves the cost of handling the material as it enters and leaves the furnace.

With so many different items entering into the cost of heating, it is quite possible that, in some cases, the most expensive fuel or source of heat energy may be the cheapest in the end, as far as the total heating cost is concerned. As a matter of fact, this very condition exists in many electrically heated furnaces.[26]

It is clear that this expert attributed the penetration of electric furnaces not just to energy savings, but also to the reduction of other direct costs associated with the heating of materials. Further, electric furnaces brought indirect benefits as well as direct cost reductions. Some other advantages of heating electrically mentioned by Trinks in 1923 included (a) greater flexibility in locating furnaces within a plant to increase the flow of production; (b) improved working environment due to the absence of products of combustion; (c) ease of operation and opportunity for more exact, automatic control, leading to higher product quality; and (d) simpler installation, making possible more rapid plant expansion.[27] Thus, manufacturers that installed electric furnaces in place of combustion furnaces for such uses as melting, refining, or heat treating had reason to expect both reductions in direct costs and indirect benefits — in other words, greater output per unit of input. In 1948, engineer Lee P. Hynes[28] confirmed these expectations:

> Very often electric heat actually reduces the unit cost of a manufactured product by improving quality, reducing waste, saving skilled labor, and avoiding hazards. In many processes no other fuel can produce the desired results. The tremendous growth of electric heating in industry is proof that it has its logical place along with fuels. . . . The future of electric heating will not depend on its cost relation to fuels but rather upon how well the responsible engineer analyzes and solves the intricate design problems incident to the full utilization of its inherent possibilities.

Phosphorus

Phosphorus, the twelfth most abundant element, is fairly widely distributed in both igneous and sedimentary rock, mostly as salts of phosphoric acid. A slightly soluble phosphorus compound — "bone phosphate of lime" — is present in bones and in bird droppings, especially the guano produced by sea birds. Indeed, phosphate fertilizers in the form of ground bones or bone meal mixed with guano were used long before the discovery of phosphorus by Brandt in 1669 and of phosphoric acid by Boyle in 1694. Despite limited supplies, bones and guano continued to be the chief sources of phosphorus and phosphoric acid until the middle of the nineteenth century.[29]

In 1842 a British patent was issued to John B. Lawes for treatment of bone ash with sulfuric acid. Lawes had found that the product obtained from this process was a better fertilizer than bone meal, as it was assimilated more quickly by plants. Soon afterward, various grades of phosphate ore were discovered in England. At first these were finely ground and applied directly to the soil. It was soon recognized, however, that Lawes's sulfuric acid treatment worked with mineral phosphate as well, increasing the availability and efficiency of the phosphate for agricultural purposes. The product of the sulfuric acid process was fittingly called "superphosphate." It was so eagerly adopted that by 1862 annual production of superphosphate in England had reached 200,000 tons. In the United States, production was initially limited by raw material availability, but after the Civil War enormous deposits of phosphate rock were discovered in Florida and Tennessee, and the industry grew rapidly.[30]

The first production of elemental phosphorus on a commercial scale — primarily for use in matches — was by the firm of Albright and Wilson at Oldbury, near Birmingham, England, in 1851. The precursor was phosphoric acid made from calcined bone (and later from phosphate rock) by treatment with sulfuric acid. The phosphoric acid was concentrated by evaporation, mixed with charcoal or coke, and dried. This mixture was placed in a bottle-shaped fireclay retort and brought to white heat, liberating phosphorus vapor, which was collected underwater and purified by distillation.[31]

In 1888, J. B. Readman of Wolverhampton, England, was granted a patent for production of white phosphorus in a crude submerged-arc electric resistance furnace. According to George Albright, of Albright and Wilson, "The almost brutal simplicity of the electricide method revolutionized immediately the making of white phosphorus."[32] His firm purchased the Readman patents and a prototype electric furnace in 1890. Why was the electric process so revolutionary?

> It was not merely that one source of heat was exchanged for another, as it might be in a domestic kitchen; the implications were far wider than that. The delicate fire-clay retort, by whatever means it was raised to the temperature necessary, was an intermittent, not a continuous producer of phosphorus. Its charge called for special preparation and, when used up, had to have its residues cleaned out and replaced. The electric furnace opened the way to a continuous process. Into it went not the black powder, itself the result of prior dissolution of calcined bones by sulfuric acid, but the mineral phosphate, untreated save by kibbling into pieces about the size of almond nuts. With the rock went also the two other materials, silica and carbon, which ensured the necessary chemical reaction within the furnace. And out of it came, continuously, a gas from which phosphorus was condensed and a slag which was allowed to escape, at white heat, by the opening of a small hole close to the bottom of the furnace. . . . This slag consisted of a small proportion of ferrophosphorus and over it, of calcium silicate which makes excellent roadmetal.[33]

The prototype plant built by Readman was operated for two years while construction of the first commercial electric phosphorus furnace was proceeding at Oldbury. Production commenced in 1893, and by 1895 the last fuel-fired retort was emptied and its furnace put out. The Oldbury electric furnace, with numerous improvements in detail, produced about 200 tons of white phosphorus annually over the next nine years before being replaced by a new furnace in 1902.[34]

Phosphorus production was originally located at Oldbury because of the inexpensive coal available there. With the coming of the electric furnace, cheap coal was no longer the key to cheap phosphorus. Cheap electricity became the key.

In the United States, white phosphorus had been manufactured by the old chemical method on a very small scale; the principal supply had been imported from England by Albright and Wilson. In 1896 their firm decided to manufacture white phosphorus at Niagara Falls, New York. While the major factor in this decision was probably inexpensive electricity, other factors undoubtedly included new American protectionist policies that made importing more difficult. The Oldbury Electro-Chemical Company leased land along the Niagara River from the Niagara Falls Power Company, constructed a six-furnace plant, and began phosphorus production in September 1897.

Phosphate rock was brought from Tennessee, while coke came from northern Pennsylvania. Initially, the plant's customers were chemical firms that prepared compounds for matches and for dyes. In the ensuing years the plant was expanded and the manufacture of other electrothermal products was added. These facilities, and a 1913 plant for converting white phosphorus into red phosphorus, supplied most of the United States military's need for phosphorus during World War I.[35]

The electrothermal production of phosphorus today differs only in detail and in scale from the Oldbury furnaces. A typical furnace consists of a steel outer shell, cooled by water spray. The floor is of monolithic carbon and the sides of firebrick or cast refractory cement. Baked carbon electrodes penetrate the charge from water-cooled sleeves in the furnace ceiling; their vertical position is adjusted so as to maintain a constant current. As with other electroprocesses, the production of phosphorus is most efficient when the furnace is run continuously at its design conditions (typically 50 megawatts, 1,500°C, and atmospheric pressure). Changes in power and curtailments of operation are avoided: thermal equilibrium is lost and production efficiency is appreciably reduced while the heat balance is being restored.[36]

Phosphate rock nodules, silicon, and coke are distributed evenly throughout the furnace via a series of chutes, and furnace gases containing phosphorus and carbon monoxide are withdrawn. In some plants the phosphorus vapor is burned in excess air to produce phosphorus pentoxide and thence hydrated to form very pure phosphoric acid. In others, the phosphorus is condensed in cooling towers and pumped into tanks.

Elemental white phosphorus is used for making numerous chemicals, in metal treatment, in various military weapons, and in conversion to red phosphorus. The latter form is used for matches and in the manufacture of fireworks. White phosphorus may also be converted to pure phosphoric acid on the site of its use, eliminating the need to transport the acid itself from producer to consumer.

Phosphoric acid made from elemental phosphorus is used in applications that require high purity. These include metal treatment, bonding refractory products, textile dyeing, lithographic engraving, and as a food and beverage additive. Although approximately 85 percent of the white phosphorus is ultimately converted to phosphoric acid, this pure acid represents only 10 percent to 15 percent of all phosphoric acid produced today. The bulk of it is made from phosphate rock using sulfuric acid in what is commonly called the "wet process." Most of the relatively impure acid so produced is used to make phosphatic fertilizers such as ammonium phosphate and triple superphosphate.[37]

NIAGARA: ELECTRIC-INDUSTRIAL COMPLEX

Between 1890 and 1895 promoters of an electric system based on direct current (led by Thomas Edison) and of one based on alternating current (championed by George Westinghouse) engaged in an intense competition often called the "Battle of the Currents." The hydroelectric power system at Niagara Falls — a synthesis of recent advances in electrical engineering in both the United States and Europe — represented a major victory for alternating current. The Niagara system was viewed as "universal": it could supply users of both alternating and direct current, and at any voltage. This universal system proved eminently successful, and utilities across the country increasingly supplied alternating current. Thus 1895 marked the end of the Battle of the Currents, the beginning of the standardization of the nation's electric supply system, and the birth of the first electric-industrial complex.

The initial hydroelectric installation at Niagara Falls consisted of three two-phase 25-hertz 2,200-volt generators, each rated at 5,000 horsepower. Electricity was first generated at the falls on August 26, 1895, and provided by the Niagara Falls Power Company to the Pittsburgh Reduction Company. For the first fifteen months, all electric power sold was stepped down to 100 to 250 volts and delivered to local electroprocess industries. Rotary converters changed alternating current to direct current for use in electrolytic cells. In November 1896, 1,000 horsepower was converted to three phases at 11,000 volts and transmitted to Buffalo, twenty-two miles away.[38]

Niagara's electric power was available continuously, and since electricity is not easily stored, industries that would use power continuously were especially attractive potential customers. So the Niagara Falls Power Company campaigned extensively to bring electroprocess firms and workers to the falls. Even though very favorable long-term electric power contracts were available, additional incentives were offered for relocation to this area. For example, the company owned land near the falls and offered it on attractive terms to prospective customers; a company town, Eschota, was built for incoming workers.[39]

Soon, growing demand led to the completion of the generating facilities in the original powerhouse, bringing installed capacity to 50,000 horsepower. Between 1900 and 1906, a second powerhouse and eleven 5,000 horsepower generators were installed. By late 1902, the company was developing 60,000 horsepower, three-quarters of which went to sixteen local electroprocess firms. Over the next decade the number of these firms grew by about one per year until, in 1910, Niagara Falls was called "the world's greatest center of electrochemical activity."[40]

The Niagara project was economically feasible because a number of electroprocesses had been developed to the point where they were ready to utilize large amounts of power. Other users of electricity — the city of Buffalo and the fabricating industries already at the falls — simply could not use the large amounts of electric power becoming available. So there was a symbiotic relationship between the power company and the electroprocess industries. The company was quite successful in attracting new customers, allowing it to achieve its generating potential; once there, the industries grew and matured.

A number of reasons for the success of this symbiotic relationship can be cited. First and foremost, the price of electricity was relatively low. Between 1895 and 1908 large electroprocess industries at Niagara paid $14 to $20 per kilowatt of demand for a year's service. Hydroelectric power prices in the United States as a whole over this same period ranged up to $25 per kilowatt per year. Prices of electricity generated from coal were over three times higher.[41]

In addition, many of Niagara's power contracts consisted only of a demand charge (i.e., an industry would pay the same whether it used electricity or not). Thus manufacturers had tremendous incentive to operate individual electrolytic cells and electric furnaces with the highest load factor possible — in other words, to run them continuously. Indeed, as described earlier, designs of electroprocesses were optimized for this mode of operation. Moreover, large groups of electrolytic cells and electric furnaces were often arranged and operated so as to maintain an overall constant load.[42]

Second, there was an abundance of power available, enabling firms to build much larger plants at Niagara than elsewhere in order to take advantage of economies of scale inherent in electroprocess equipment. Third, good water and rail transportation was available, making eastern markets and resources readily accessible. The Niagara area had ample deposits of salt and sand, materials needed for making chlorine, caustic soda, silicon carbide, and other electrolytic and electrothermal products. Finally, there were intrinsic benefits associated with the collocation of the industries at Niagara.

Many electroprocess industries were (and are) interdependent. This is because chemicals and primary metals are more often intermediate products than consumer products; that is, one firm's output (or by-product) may be another firm's input. Interdependence among firms in electric-industrial complexes minimizes the transport and transaction costs associated with certain products and input factors. For example, ammonia plants were sited at Niagara Falls primarily because their process used hydrogen — a previously wasted by-product of chlorine manufacture.[43]

Again, calcium carbide was manufactured at Niagara by the Union Carbide Company, beginning in 1898. This new compound was bought by other companies there to make acetylene. Acetylene was used for illumination and for welding; later, it became an important building block for synthetic organic chemicals.[44]

Another benefit of collocation at Niagara was the professional cadre that evolved there. The close collaboration between scientists, engineers, and businessmen was extraordinarily fruitful. It facilitated innovation, the establishment of new firms, and the manufacture of chemicals, metals, and equipment (i.e., carbon electrodes) that had hitherto been available elsewhere only at far greater cost.[45]

DISCUSSION OF PROCESS INNOVATION

The histories of four industries have been recounted separately, as if they were unrelated. However, at least four themes underlie and unite these and perhaps most of the electroprocess industries.

First, one or more important innovations were required before electricity could be used to prepare on a commercial scale the six substances discussed in this chapter. The electrolytic and electrothermal methods that were developed thus embodied new concepts and were inherently less costly than the old techniques. This inherent cost advantage was due to the fact that the new electric processes reduced or entirely eliminated one or more factors of production (capital, labor, energy, materials, or time) that were required by the old methods.[46] The principal innovations and some input factors that were eliminated are summarized in Table 3.1.

Second, the quest for an inexpensive method of producing aluminum occupied many individuals in both the United States and Europe for dec-

Table 3.1   Common Themes of Historical Accounts

	Principal innovation	Factor reduced or eliminated	Principal link to aluminum	Year production began at Niagara
Aluminum	Dissolution of alumina in molten cryolite and subsequent electrolysis	Sodium (in chemical reduction of aluminum chloride)	—	1985
Chlorine and caustic soda	Isolation of electrolytic products by formation of an alloy (mercury cell); isolation of electrolytic products by porous diaphragm (diaphragm cell)	Soda ash (in manufacture of caustic soda); hydrochloric acid (in manufacture of chlorine)	Pure caustic soda needed to produce cheap sodium for chemical reduction of alumina (link to mercury cell)	1897 (mercury cell) 1906 (diaphragm cell)
Silicon carbide and fused alumina	High temperature of electric resistance furnace allows new equilibrium states to be attained	Labor and energy (in procuring—and using—natural abrasives)	Man-made abrasives produced in furnaces originally developed by Cowles to obtain aluminum	1895 (silicon carbide) 1901 (fused alumina)
Phosphorus and phosphoric acid	Electrothermal method is a continuous process	Labor; time	Process developed at same time and place as Castner's efforts to produce aluminum in furnaces	1897

ades. This pursuit also facilitated the development of electrolytic and electrothermal processes for making other chemicals and metals and led to the rise of a number of important industries. Table 3.1 summarizes the possible links between aluminum and the other products that have been discussed.

Third, the paths of development of many of the electroprocess industries came together at Niagara Falls around the turn of the century. There, the availability of large amounts of electricity on favorable terms permitted production on a scale not possible elsewhere, and at lower prices. The industries at Niagara expanded and new ones arose, due in part to their economic and professional interdependence. Table 3.1 notes the years in which various commodities were first produced at Niagara.

Fourth, some chemicals and metals now made electrically had been produced commercially before the advent of electric techniques, but they could be made more cheaply or with far greater purity and uniformity by means of electroprocesses than by means of fuel-based methods. Caustic soda, phosphorus, phosphoric acid, and fused alumina are representative of this class of commodities. Other substances, however, could be produced in commercial quantities only with the advent of large-scale electroprocessing, and most of these — aluminum, chlorine, and silicon carbide, for example — have been of great benefit to society. These new products and their relatively low cost are the most obvious (but not the only) consequences of the use of electricity as the agent of chemical and physical change in bulk processing.[47]

Impact of Electric Unit Drive

Electric unit drive affected both fabrication and bulk processing. The spatial organization of production in the bulk processing industries has always tended to center around large process equipment such as furnaces, electrolytic cells, or chemical reaction vessels. Today bulk solids and fluids are transported to and from this process equipment by means of conveyors, pumps, or compressors. These bulk transport machines are commonly driven by individual electric motors and are located so as to maximize throughput and minimize costs. This usually is achieved when materials flow continuously through the plant from one piece of process equipment to the next; at each piece of equipment, a different physical or chemical change takes place continuously.

During the nineteenth and early twentieth centuries batch processing, not continuous processing, prevailed. In batch processing, materials generally stayed in a single vessel while undergoing the required series of physical or chemical changes. Between each cycle, the vessel had to be emptied, cleaned, and refilled. Sometimes considerable time was needed for reaction conditions to become reestablished.[48] Although some continuous chemical and metallurgical processes were not developed until well after the turn of the century, an important reason for the persistence of batch processing was the lack of flexibility in locating bulk transport machinery.

Conveyors, pumps, and compressors were driven via shafts and belts turned by a central prime mover or via steam piped to the machine and used there for mechanical drive. These methods of providing power to bulk transport machinery involved significant capital and operating costs.

Although it was technically feasible to provide power almost anywhere in a plant, it was not always convenient or cheap to do so.

This constraint disappeared with the advent of electric unit drive in the 1910s and 1920s. Electric motors could be accurately matched to the power requirements of conveyors, pumps, and compressors, and these complete mechanical units could be placed virtually anywhere with relatively little difference in cost. With complete freedom of plant layout, manufacturers could focus on increasing overall throughput. Continuous chemical and metallurgical processes were developed and fed continuously by motor-driven conveyors, pumps, and compressors.[49]

Parallels: Electrified Machine Drive and Electroprocessing

There are a number of parallels between the ways electric motors affected production in the fabricating industries and the ways electroprocesses affected production in the bulk processing industries.

First, in fabrication the substitution of electric unit drive for shaft-and-belt drive led to modest reductions in the energy required to drive machinery and sometimes in the total direct cost of running machinery. The most significant benefits of electric drive, however, came indirectly. A parallel argument was put forth on behalf of electric industrial heating during the 1920s. Some electric furnaces use less energy overall than their fuel-fired counterparts; for many purposes the total heating cost is lower when an electric furnace is used; but often the indirect benefits from the use of electric heat are of prime importance.

Second, indirect benefits associated with switching to electric unit drive included greater flexibility in work space arrangement, an improved work environment, better machine control, and easier plant expansion. There were strikingly parallel indirect benefits associated with using electric furnaces for operations formerly carried out in fuel-fired furnaces, as noted in the section on manufactured abrasives. However, other indirect benefits of electroprocesses have less obvious parallels in fabrication. These include the use of cheap raw materials as inputs rather than more expensive manufactured substances, the production of valuable coproducts, and the greater purity of the desired product itself. Of the industries discussed in this chapter, these indirect benefits pertain especially to the production of caustic soda and phosphorus.

Third, electric motors (as opposed to steam engines with mechanical power distribution) can convert electrical energy into mechanical energy exactly where the conversion is needed — the drive shaft of a machine. This conversion and transfer of energy can be precisely controlled with respect to time. It can be started, stopped, or varied in rate as needed. Electric motors can also be accurately matched to the power requirements of machines. In electric furnaces (as opposed to fuel-fired furnaces), heat is deposited precisely where it is needed — in the workpiece itself. Heat transfer by electric resistance, arc, or induction is extremely rapid compared with radiation and conduction in fueled furnaces, and it can be more accurately controlled with respect to time. There is also no need to heat — and recover heat from — large volumes of gases. Thus electric motors and electric furnaces (and in a slightly different sense, electrolytic cells) exploit the same three aspects of electricity that give it special value in production: the precision in time, in space, and in scale with which energy in this particular form can be transferred.

Finally, electric drive evolved over several decades from electric line shaft drive to electric unit drive. Only in the latter form was the full potential of the electric motor in production realized. This is the normal pattern followed by a totally new technology. In contrast, the electroprocesses discussed here changed relatively little between their invention and their commercial maturity. Yet, in their final forms, neither electric drive nor the various electroprocesses were mere incremental improvements to time-honored techniques. Rather, these electric technologies embodied new concepts in production and even in the commodities produced.

Notes

1. These classifications overlap. For example, in the electrolysis of alumina, migration of aluminum and oxygen ions is possible only in a molten bath of cryolite at a temperature of 950°C. The heat is generated by resistance to the current of the material in the cell, while the cathode and anode attract the aluminum and oxygen ions.

2. Executive Office of the President, Office of Management and Budget, *Standard Industrial Classification Manual*, 1972 (Washington, D.C.: GPO, 1972).

3. Warren D. Devine, Jr., *Electroprocesses in American Manufacturing, 1880–1930*, ORAU/IEA-86-1(M) (Oak Ridge, TN: Oak Ridge Associated Universities, Institute for Energy Analysis, 1986), Tables A.2 and A.3.

4. R. Norris Shreve, *The Chemical Process Industries* (New York: McGraw-Hill, 1945), 302; Robert P. Multhauf, "Industrial Chemistry in the Nineteenth Century," in *Technology in Western Civilization* I, eds. Melvin Krantzberg and Carroll W. Pursell, (New York: Oxford University Press, 1967), 477; Ernest V. Heyn, "Charles M. Hall — Alchemist of Aluminum," *Fire of Genius: Inventors of the Past Century* (Garden City, NY: Anchor Press/Doubleday, 1976), 222–223.

5. Shreve, *Chemical Process Industries*, 302; Heyn, "Charles M. Hall," 225; Martha M. Trescott, *The Rise of the American Electrochemicals Industry, 1880–1910: Studies in the American Technological Environment* (Westport, CT: Greenwood Press, 1981), 127.

6. Heyn, "Charles M. Hall," 234–235; Trescott, *Electrochemicals Industry*," 127, 322–323; Donald H. Wallace, *Market Control in the Aluminum Industry* (Cambridge: Harvard University Press, 1937), 518.

7. Trescott, *Electrochemicals Industry*, 65; Wallace, *Market Control*, 12–13.

8. Wallace, *Market Control*, 9–23. The substitution of aluminum for other materials has proceeded and accelerated for nearly 100 years. Today, applications in building and construction constitute the largest single market of the aluminum industry,

with transportation equipment being the second. The food and beverage packaging industry is the aluminum industry's fastest growing market.

9. Wallace, *Market Control*, 7–8; Sara W. Boercker, *Energy Use in the Production of Primary Aluminum*, ORAU/IEA-78-14(M) (Oak Ridge, TN: Oak Ridge Associated Universities, Institute for Energy Analysis, 1978), 20–23; Warren D. Devine, Jr. and David A. Boyd, *Costs of Electric Power Outages to Manufacturers*, ORAU/IEA-81-9(M) (Oak Ridge, TN: Oak Ridge Associated Universities, Institute for Energy Analysis, 1981), A1–A5.

10. Shreve, *Chemical Process Industries*, 271–276, 282.

11. *Kirk-Othmer Encyclopedia of Chemical Technology*, 3rd. ed. (New York: John Wiley and Sons), I:806, 826 and III:938–939.

12. L. D. Vorce, "Historic Development of Caustic-Chlorine Cells in America," in *Transactions of the Electrochemical Society* 86 (1944): 69–81.

13.Trescott, *Electrochemical Industry*," 66–68; L. F. Haber, *The Chemical Industry 1900–1930: International Growth and Technological Change* (Oxford, England: Clarendon Press, 1971), 77, 80–81.

14. Vorce, "Caustic-Chlorine Cells"; Thomas P. Hughes, *Elmer Sperry: Inventor and Engineer* (Baltimore, MD: Johns Hopkins Press, 1971), 89–93.

15. Shreve, *Chemical Process Industries*, 274; *Kirk-Othmer Encyclopedia*, I:856.

16. Devine and Boyd, *Power Outages*, A13–A17; *Kirk-Othmer Encyclopedia*, I:807–817, 820–821.

17. Heyn, "Charles M. Hall," 234; Trescott, *Electrochemicals Industry*, 3–25; Wallace, *Market Control*, 509–512.

18. Shreve, *Chemical Process Industries*, 321; Trescott, *Electrochemicals Industry*, 24; *Kirk-Othmer Encyclopedia*, IV:520, 528.

19. Shreve, *Chemical Process Industries*, 323–324; *Kirk-Othmer Encyclopedia*, I:33, II:140–144, 218–219, 237–239, and IV:530.

20. Shreve, *Chemical Process Industries*, 321; *Kirk-Othmer Encyclopedia*, I:26–27, 31.

21. Shreve, *Chemical Process Industries*, 321–324; *Kirk-Othmer Encyclopedia*, II:238, and IV:525–529.

22. Devine and Boyd, *Power Outages*, A9; William H. Patchell, *Application of Electric Power to Mines and Heavy Industries* (New York: Van Nostrand, 1913), 321–322; W. Trinks, *Industrial Furnaces* II (New York: John Wiley and Sons, 1923): 309.

23. C. C. Burwell, *Industrial Electrification: Current Trends*, ORAU/IEA-83-4(M) (Oak Ridge, TN: Oak Ridge Associated Universities, Institute for Energy Analysis, 1983), 20.

24. Patchell, *Electric Power*, 314.

25. Trinks, *Industrial Furnaces*, I:3.

26. Ibid., I:47.

27. Ibid., I:309.

28. Lee P. Hynes, "Industrial Electric Resistance Heating," in *AIEE Transactions* 67 (1948): 1361.

29. Shreve, *Chemical Process Industries*, 328–333; *Kirk-Othmer Encyclopedia*, XVII:426.

30. Haber, *Chemical Industries*, 104–106; Shreve, *Chemical Process Industries*, 328–329.

31. Shreve, *Chemical Process Industries*, 337; *Kirk-Othmer Encyclopedia*, XVII:477.

32. Richard E. Threlfall, *The Story of 100 Years of Phosphorus Making* (Oldbury, England: Albright and Wilson, 1951), 94.

33. Ibid., 93–94, 97.

34. Ibid., 97, 99–101.

35. *Kirk-Othmer Encyclopedia*, XVII:478; Threlfall, *Phosphorus*, 260–264.

36. Shreve, *Chemical Process Industries*, 337, 342–343; Devine and Boyd, *Power Outages*, A17; *Kirk-Othmer Encyclopedia*, XVII:478–481.

37. *Kirk-Othmer Encyclopedia*, X:58–76, and XVII:481–487.

38. Robert B. Belfield, "The Niagara System: The Evolution of an Electric Power Complex at Niagara Falls, 1883–1896," in *Proceedings of the IEEE* 64, no. 9 (New York: IEEE, 1976): 1344–1350.

39. Trescott, *Electrochemicals Industry*, 38.

40. Trescott, *Electrochemicals Industry*, 40; Belfield, "Niagara System," (1976); Robert B. Belfield, "The Niagara Frontier: The Evolution of Electric Power Systems in New York and Ontario, 1880–1935," Ph.D. diss., University of Pennsylvania (1981), 73, 80, 91. By the end of the period under consideration in this chapter, the generating capacity of the Niagara Falls Power Company had reached 452,500 horsepower; today, generating capacity at Niagara is over 3 million horsepower.

41. Trescott, *Electrochemicals Industry*, 43.

42. An excellent example is the production during the 1920s and 1930s of silicon carbide in single-phase indirect-resistance furnaces. The electrical resistance of the cylindrical graphite core of the furnaces and of the surrounding charge decreases sharply during operation. This is because carbon has a negative temperature coefficient of resistance, and temperature increases appreciably during production. As a constant power demand was desired, provision was made for initial voltage four to six times final voltage and for initial current one-fourth to one-sixth the final current. It was customary to employ one transformer to serve in rotation a set of four to eight furnaces; operating schedules of the furnaces were arranged so that the single-phase electric load was nearly uniform. A balanced load on the three-phase distribution system was approached by operating the single-phase sets in three groups and connecting the transformers to separate phases. (Federal Power Commission, *Power Requirements in Electrochemical, Electrometallurgical, and Allied Industries* [Washington, D.C.: GPO, 1938], 95).

43. Charles L. Mantell, "Electrochemical Products Emphasize Interdependence of Processes and Industries," *Chemical and Metallurgical Engineering* 35 (August 1928): 486–488.

44. Trescott, *Electrochemicals Industry*, 73; Mantell, "Electrochemical Products,"

45. Trescott, *Electrochemicals Industry*, 92–108; L. F. Haber, *The Chemical Industry During the Nineteenth Century* (Oxford, England: Clarendon Press, 1958), 144.

46. Charles Berg suggests that the inherent cost advantages of most innovative industrial processes, not just electric ones, accrue because the new processes greatly reduce or eliminate certain inherent costs that cannot be reduced any further using current techniques. See Charles A. Berg, "A Suggestion Regarding the Nature of Innovation," in *Energy, Productivity, and Economic Growth*, eds. S. H. Schurr, S. Sonenblum, D. O. Wood (Cambridge, MA: Oelgeschlager, Gunn, and Hain, 1983.)

47. For a discussion of several other possible consequences, see Trescott, *Electrochemicals Industry*, 135–159; and David F. Noble, *American By Design: Science, Technology, and the Rise of Corporate Capitalism* (New York: Alfred A. Knopf, 1977), 194. Trescott argues that the impact of the new electric techniques went beyond the manufacturing of particular chemicals and metals. She claims that the electroprocess industries played a role in the transition — experienced by all bulk processing industries — from a product-specific focus to a new orientation around a relatively small number of procedures called "unit operations" and "unit processes." Noble says this unit concept "laid the groundwork for streamlined mass production" in the bulk processing industries.

48. Lawrence B. Evans, "Impact of the Electronics Revolution on Industrial Process Control," *Science* 195 (1977): 146.

49. Roger Burlingame, *Backgrounds of Power* (New York: Charles Scribner, 1949), 228; Harry Jerome, *Mechanization in Industry* (New York: National Bureau of Economic Research, 1934), 62–63, 102–103, 112–113, 116–117. This subject is treated in a somewhat more general way in Alfred D. Chandler, Jr., *The Visible Hand: The Managerial Revolution in American Business* (Cambridge: Harvard University Press, Belknap Press, 1977), 240–244 and 253–258.

4 | Recent Trends Affecting Process Electrification: Costs, Industry Types, and Processing Options

Calvin C. Burwell

The preceding chapter dealt with early electroprocessing during an era in which electricity, as a newly emerging commercial energy form, was instrumental in the development of wholly new products and industries. The present chapter and those immediately following turn to recent developments in which electricity continues to play an innovative role, but not as dramatic as in the earlier history of electroprocessing. Neither the products nor the industries are new, but old products are being made by new, improved electrically driven techniques, and old industries (as in the case of steel) are being rejuvenated by such techniques.

Chapters 5 and 6 will examine five industries: steelmaking, glassmaking, pulp- and papermaking, petroleum refining, and chemicals, particularly the growing field of organic chemicals production. To explain these choices, and to place these industries in the broader context of manufacturing as a whole, we begin in the present chapter with a broad examination of what has been happening to the use of electricity in manufacturing during recent years.

ENERGY-RELATED COST TRENDS IN MANUFACTURING

Direct Energy Costs

First, what are the comparative costs of the various energy sources used in manufacturing, how have they been changing in recent years, and with what effects on shifts in the use of electricity and fuels?

The most consistent trend evident in Table 4.1 is the rising share of purchased energy supplied by electricity. Natural gas was the dominant manufacturing energy source through 1974 but has lost this position to electricity (measured in energy input terms). The loss by natural gas appears to have been the result of sluggish or even negative growth in some

Table 4.1 Major Sources of Purchased Energy for Manufacturing as a Percentage of Major Sources Total

	1971	1974	1981	1985
Coal	12.7	9.9	11.6	11.0
Gas	42.5	40.3	36.1	33.8
Oil	9.6	10.5	6.2	4.3
Electricity[a]	35.2	39.3	46.1	50.9
Total major sources	100.0	100.0	100.0	100.0
Total major sources as percentage of total purchased energy	92.8	92.4	93.8	n.a.

SOURCES: Computed from U.S. Bureau of the Census, *Annual Survey of Manufactures: Fuels and Electric Energy Consumed*, MC82-S-4 (Washington, D.C.: GPO, 1983), Part 1; and U.S. Department of Energy, *Manufacturing Energy Consumption Survey: Consumption of Energy, 1985*, DOE/EIA-0512(85) (Washington, D.C.: GPO, November 1988), Table 7.

[a] Electricity counted at 10,600 Btu per kWh.

industries that were large natural gas users and of rising natural gas prices both in actual terms and relative to electricity prices.

The dramatically declining price ratio of electricity to gas per million Btu of delivered energy is shown in Figure 4.1. The price ratio declined from more than 40 to 1 in 1935 to a low of about 3.5 to 1 in 1983. Since then depressed natural gas prices, due to a surplus of gas producing capacity, have resulted in a small (and perhaps temporary) increase in the ratio. The rapid increase in oil prices after 1973 has also produced a sharp decline in the price ratio of electricity to oil.

Some brief explanation of the divergent historical trend in electricity and fuel, particularly natural gas, prices is in order. The cost to the customer of natural gas and of electricity each have two main components: a basic energy cost and a capital cost. (In addition there are operation and maintenance costs.) Before World War II natural gas was used primarily in areas near natural gas sources. Long distance transmission of gas was economically prohibitive. However, after World War II, oil pipelines (and pipeline technology) were adapted to natural gas transmission, and it became economic to move the large surplus natural gas resources in the producing areas to eastern markets. Gas prices at the wellhead were low, and improving pipeline technology and increasing volumes reduced transmission costs. Average natural gas prices to the user declined, and gas markets grew.

Electricity can be produced from any primary energy source. Historically, electric utilities have taken advantage of the lowest cost energy source to produce power. Table 4.2 shows shares of the generating market by fuels.

Prior to 1970 and with the exception of regions rich in hydropower, the cost of electricity was not highly sensitive to the type of primary energy source used. The use of gas and coal as generating fuels depended upon

Figure 4.1 Ratio of Electricity to Gas Price for Industrial Energy, 1930–1987

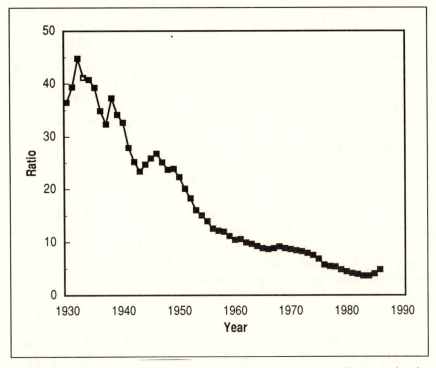

SOURCES: U.S. Bureau of the Census, *Historical Statistical of the United States, Colonial Times to 1970* (Washington, D.C.: GPO, 1975), Part 2, Series S 118 and S 187; U.S. Energy Information Administration, *Annual Energy Review 1987* (Washington, D.C.: GPO, 1988), Tables 73 and 91.

different regional access to fuel resources: gas on the Gulf Coast and coal for the east central states. Beginning in the late 1960s a two-tiered price structure developed among fuels, first with oil and then with gas becoming on average two or more times as costly as coal. These price shifts contributed to the substantial reduction in the use of oil and gas for power generation as well as a decline in their direct use for manufacturing (see Table 4.1).

Increasing efficiencies in generation, transmission, and distribution technology, coupled with gains due to economies of scale and low fuel prices, have led to decreasing unit prices for electricity. The net result of the declining costs of electricity, and the generally rising costs for fuels used in manufacturing, particularly natural gas, is reflected in the sharp fall in the electricity to gas price ratio already shown in Figure 4.1.

Table 4.2   **Sources of Energy for Electric Utility Thermal Power Generation, Percentage of Kilowatthours Generated**

	1962	1971	1974	1981	1985	1987
Coal	65.8	53.0	53.0	59.1	64.1	63.1
Gas and oil	33.9	44.1	39.5	27.2	17.9	16.8
Uranium	0.3	2.8	7.3	13.4	17.5	19.6
Other	0.0	0.1	0.1	0.3	0.5	0.5
Total	100.0	100.0	100.0	100.0	100.0	100.0

SOURCES: Edison Electric Institute, *Statistical Yearbook for the Electric Utility Industry* (Washington, D.C.: EEI, various years).
NOTE: Totals may not add to 100.0 due to rounding.

Other Energy-Related Costs

Direct energy costs are not the only factors influencing decisions to use a particular form of energy. Other energy-related costs also play an important part. These often operate in combination with energy efficiency considerations to affect the prospects for electrical processing.

Among the most important of these other factors are capital costs and raw material costs. The capital costs of an electrified process can be above or below those for a directly fueled technology. Capital costs tend to be lower for electric resistance heating applications than for fuel heating both because of the simplicity of converting electricity to heat and because fuel burning often incurs costs for cleaning up pollutants from both fuel and process materials carried in the flue gases. Conversely, capital costs for special heating techniques such as induction or dielectric heating, and for electromechanical processing are often higher than those for conventional thermal processing. In this case, the decision to electrify would depend upon other advantages favoring the electric technology, such as gains in production efficiency or the lowering of raw material costs.

Raw material costs are a particularly important consideration because they typically exceed the sum of all other operating costs. Thus, if electric technology offers an advantage in raw material costs relative to conventional processing, the electrification step can be more easily taken — even in the absence of an expanding market. There are several ways raw material costs can be different for different processes: a higher fraction of the raw material can be converted into final product; lower grade ores and materials can be utilized economically, as has been the case with electrolytic separation in the processing of low-grade copper ore; recycled materials can displace primary production, as in the increased use of obsolete steel scrap made possible by electric melting; and a more plentiful resource can displace one in diminishing supply, as in the growing use of hardwoods processed thermomechanically to replace the use of softwood pulps for certain paper products.

TYPES OF INDUSTRIES

Manufacturing industries generally have experienced recent increases in the electricity fraction of purchased energy. This trend is detailed by industry in Table 4.3. The industries in that table can be divided into two categories according to the ratio of purchased energy costs to total production wages. The industries in the two groups have the following general characteristics:

Characteristics	Group I	Group II
Labor costs versus energy costs	Labor costs much more important than energy costs	Energy costs generally comparable to labor costs
Product transformations	Intermediate into final	Raw materials into intermediate
Main production operations	Mechanical fabrication and assembly	Thermal processing
Total energy use	Comparatively low	Comparatively high
Degree of electrification	Operations are mechanical and are largely electrified	Operations use heat and are less highly electrified

The separation of manufacturing into these two broad categories is of course arbitrary. There are thermal processing activities in Group I industries and some finished products come from industries in Group II. It could also be argued that several of the Group I industries that convert naturally grown materials into consumer goods (i.e., food, lumber, tobacco) are processing rather than fabricating and assembling these raw materials. On more careful inspection, however, the operations involved (i.e., sawing lumber, packaging fruits, meats, vegetables, and tobacco) are seen to be primarily mechanical and so more akin to fabrication and assembly than to thermal processing.

The purpose here of dividing manufacturing into two main categories is to permit a narrowing down of all manufacturing activities to the few industries selected for detailed study. The Group II industries that rely on energy-intensive thermal processes are the logical candidates for further electrification because they have the characteristics of high energy use, energy costs comparable to labor costs, and comparatively low percentages of electricity in their total energy purchases.

Table 4.3 Purchased Energy (PE) and Electricity in Manufacturing

	Percentage of all manufacturing, 1985		Electricity percentage of PE costs[c]			PE cost expressed as percentage of production wages, 1985	PE cost, percentage of value added, 1985
	PE costs	Total Btu[a]	1974	1981	1985		
Group I							
Food	8.8	7.0	43	42	48	30	5.0
Tobacco	0.2	0.2	48	51	55	14	1.1
Textiles	3.2	2.3	59	56	63	25	9.3
Apparel	0.9	0.4	72	68	75	6	2.0
Lumber	2.2	2.5	43	50	64	16	6.1
Furniture	0.7	0.4	60	62	65	8	2.6
Printing	1.8	1.1	73	71	73	8	1.5
Rubber/plastic	3.4	2.2	62	67	73	21	5.7
Leather	0.2	0.1	56	56	56	9	2.9
Fabricated metals	4.6	2.6	57	56	61	12	4.0
Machines	4.4	2.8	62	66	72	10	2.4
Electrical equipment	4.2	2.4	70	72	76	11	2.3
Transportation equipment	4.7	3.0	61	64	68	8	2.3
Instruments	1.1	0.8	58	63	69	10	1.7
Miscellaneous	0.6	0.3	63	61	66	10	2.4
Group I total	41.0[b]	28.0[b]	57	58	64	13	3.1
Group II[d]							
Paper	9.7	14.8	28	33	41	54	14.3
Chemicals	18.0	18.3	41	44	49	92	11.2
Petroleum and coal	8.8	14.4	26	28	33	208	30.9
Stone/clay/glass	7.0	6.3	29	31	37	51	14.6
Primary metals	15.5	17.9	41	46	54	65	24.3
Group II total	59.0[b]	72.0[b]	36	39	45	74	16.0
Grand total	100.0	100.0	44	46	53	25	6.0

Table 4.3 (Continued)

SOURCES: From following publications of the U.S. Bureau of the Census, *Census of Manufactures* (Washington, D.C.: GPO, various years); *Fuels and Electric Energy Consumed,* M74(AS)-4 (1976), and MC82-S-4, Part 1 (1983); *Statistics for Industry Groups and Industries,* M80(AS)-1 (February 1982); *Value of Product Shipments,* M85(AS)-1 (January 1987); *Hydrocarbon, Coal and Coke Materials Consumed,* M80(AS)-4.3 (February 1982). Also, U.S. Energy Information Administration, *Monthly Energy Review* (January 1987).

NOTE: Group I and Group II are defined in the text.

[a] Includes nonpurchased energy (e.g., wood waste in the paper industry).

[b] Group I industries accounted for 41 percent of manufacturing expenditures for energy purchases in 1985, but they accounted for only 28 percent of all energy including nonpurchased energy (Btus) used by manufacturing. Group II industries accounted for 59 percent of total manufacturing expenditure on purchased energy, but used 72 percent of all energy (Btus) used by manufacturing.

The divergence between the energy cost and the energy quantity percentages is primarily a result of two factors: (a) the higher average price of Btus in the form of electricity, and (b) the fact that Group I industries use a substantially higher percentage of their energy in the form of electricity.

In the energy quantity computations referred to above, 1 kWh is assumed to equal 10,600 Btu. This treatment is standard in this chapter and is approximately the amount of primary energy consumed to produce 1 kWh of electricity.

[c] These columns show the cost of purchased electricity as a percentage of the cost of purchased energy. For some analytic purposes, the percentage on a quantity of energy (Btu) basis is equally or more important. The following tabulation provides some insight into comparisons on both quantity and cost measures:

Purchased Electricity as a Percentage of Purchased Energy: Btu Basis and (Cost Basis)

	1974	1981	1985
Group I total	47	53 (58)	59 (64)
Group II total	32	39 (39)	43 (45)
Paper	27	34 (33)	33 (41)
Chemicals	34	39 (44)	44 (49)
Petroleum and coal	16	25 (28)	32 (33)
Stone/clay/glass	20	25 (31)	30 (37)
Primary metals	45	51 (46)	59 (54)
All industries	37	43 (46)	48 (53)

[d] Four of the five Group II industries obtain substantial amounts of (nonpurchased) energy from waste products and by-products. Unfortunately, data on quantities of waste and by-products are not available for years after 1980. Moreover, the dollar value of these materials as energy sources must be estimated. The following tabulation shows figures for comparison with production wages for 1980 using actual quantities of waste and by-product materials and estimated values:

Purchased and (Total Energy) as a Percentage of Production Wages in 1980

Paper	57	(93)
Chemicals	95	(112)
Petroleum and coal	191	(447)
Primary metals	55	(78)

PROCESSING OPTIONS FOR USING ELECTRICITY

Mechanical Energy

The decreasing cost of electricity relative to fuels carries different implications for manufacturers depending upon whether their major need is for mechanical energy (Group I) or for thermal energy (Group II).

Electricity is nearly fully transformable into mechanical energy in motors. Thus, a requirement for mechanical energy in manufacturing comes down to a choice between purchasing electricity from a utility or purchasing fuel in order to generate steam that would then be used directly (in steam pumps or turbines) or indirectly (through self-generation of electricity) to power motors. In either case, whoever (utility or manufacturer) transforms the thermal energy in fuel into mechanical energy must accept the thermal energy losses (about two-thirds of the thermal energy in the fuel) and pay the capital and operating costs inherent in that energy transformation.

As electric utilities have grown in scale and power generation efficiency, purchasing electricity from these utilities has become the choice of most manufacturers requiring mechanical energy. The exceptions have been manufacturers whose operations generate substantial waste heat (e.g., petroleum refining and petrochemicals) or whose operations create wastes and by-products having fuel value (e.g., coal coking and blast furnace operations in steelmaking, or the bark and black liquor wastes from papermaking). These manufacturers often use the waste heat or combustible wastes from their operations to generate their own electricity (see Table 4.4). Whether purchased or self-generated, however, electricity is the usual energy source for manufacturers engaged in mechanical processing. In Group I industries, electrification is well established.

Table 4.4 Electricity Generation by Manufacturers Expressed as a Percentage of Electricity Purchases

Manufacturing activity	1971	1974	1981	1985
Paper	73.1	67.0	50.8	60.9
Chemicals	19.7	15.1	7.7	13.9
Petroleum	23.7	16.4	16.7	20.9
Primary metals	20.3	13.7	5.5	5.6
All other	3.2	2.5	1.3	1.6

SOURCES: U.S. Bureau of the Census, *Annual Survey of Manufactures: Fuels and Electric Energy Consumed* and *Statistics for Industry Groups and Industries* (Washington, D.C.: GPO, various years).

NOTE: Electricity generation equals electricity generated by manufacturers minus sales.

Thermal Energy

In evaluating the prospects for electrification of the Group II manufacturing activities — those that require heat — the price ratio of thermal energy from electricity to thermal energy from fuel becomes very important. In this case, the cost of the thermal energy as electricity competes directly with the cost of thermal energy as fuel rather than with the cost of a much larger amount of fuel needed to produce mechanical energy. Thus, after allowing for chimney losses of 25 percent of the thermal energy from fuel burning, if electric heat costs four times as much per Btu as fuel, it must in general be three or more times as efficient.

Since the production of heat from electricity has no temperature limit, the use of electricity for process heat is likely to become economic first for high-temperature processing. Electricity's ability to sustain high temperatures can greatly minimize the total energy requirement by minimizing heat loss. For example, scrap steel melts at about 2,500°F; so the hot gases from fuel combustion are not useful for melting unless the gas temperature exceeds at least 2,600°F. When the temperature falls below that level, the hot flue gas must be removed from the process even though it still contains most of the thermal energy created during combustion. By comparison, most of the heat from electricity can be delivered to the process materials above the required temperature, and little heat need be carried away as waste.

Another reason for driving the process at very high temperatures is that the processing takes place quickly, which also minimizes waste. The faster the process conversion occurs, the less energy is lost by heat conduction, convection, and radiation. The fuel used for the process purpose (e.g., to reduce iron ore to iron or to melt sand into glass) is often a small fraction of the thermal energy in the fuel. With fuel-based processing, most of the heat is lost unproductively during the time it takes to heat up the material and to cause the process change.

Table 4.5 Temperature Levels for Process Heat in Energy-Intensive Manufacturing

Manufacturing activity	Percentage of process heat used at temperatures greater than 1,000°C, 1974
Paper	7.8
Chemicals	2.5
Petroleum	0.0
Stone/clay/glass	67.0
Metals	92.0

SOURCE: S. W. Boercker, *Characterization of Industrial Process Energy Services,* ORAU/IEA-79-9(R) (Oak Ridge, TN: Oak Ridge Associated Universities, Institute for Energy Analysis, May 1979).

Applications for high-temperature processing and low-temperature processing conveniently divide the five energy-intensive Group II industries considered in the next two chapters (see Table 4.5). Chapter 5 profiles steel- and glassmaking, which are inherently high-temperature processes. Here the direct substitution of heat from electricity for heat from fuel is the dominant mechanism for electrification. In contrast, Chapter 6 profiles manufacturing groups whose low-temperature requirements for thermal energy are easily met with steam or heat from the combustion of fuels. That chapter will show how the dominant mechanism for electrification in these low-temperature process industries is the substitution of mechanical for thermal energy.

5 | *High-Temperature Electroprocessing: Steel and Glass*

Calvin C. Burwell

Electrification of process manufacturing has advanced most in those industries whose operations require high temperatures. The inherent qualities of electricity make it quite competitive with fuels for applications in which intense heat is needed. This chapter explores why and how electroprocessing is replacing conventional fuel-based methods for high-temperature processing in two major industries: steelmaking and glassmaking.

STEELMAKING

Conventional Fuel-Based Processes

Conventional, fuel-based steelmaking takes place in a large centralized integrated mill complete with coke ovens, blast furnaces, and steelworks. Such complexes are traditionally located so as to be within convenient reach of sources of iron ore and coal.

The conventional technology involves many steps, beginning with the mining of iron ore that is then concentrated, and, for the most part, mechanically pressed into pellets prior to shipment to a steel mill. The composition of the concentrated ore product is about 90 percent iron oxide with a small percentage of impurities. The ore is then smelted with coke — the carbon product from pyrolysis of a particular grade of coal. In the coking step, about 10,000 ft^3 of coke oven gas (containing a heating value of about 500 Btu per ft^3) along with small amounts of other hydrocarbons are released per ton of coal pyrolyzed. These by-product fuel gases are used as energy sources for various purposes throughout steel-making operations.

Coke, ore, and other minerals (principally limestone) react at high temperatures in the blast furnace — a large, brick-lined steel vessel with a hearth up to forty-five feet in diameter. In this step, coke provides both the process energy required and the chemical reducing agent (carbon primarily as carbon monoxide) needed to convert the iron oxides to elemental iron. The other materials fed to the blast furnace, including most of the impurities in the ore and some of the iron, end up as slag that floats on top of the molten iron. Periodically, the molten iron is drawn into a large ladle for transfer to the steel-making step, and the slag is processed for ultimate disposal or reuse.

The molten iron from the blast furnace contains a small amount of carbon. If allowed to solidify, the product, cast iron, while useful for many purposes, would be too brittle to be pressed and drawn into shapes such as strips, sheets, rods or wires. Steel is made by the reaction between carbon in the molten iron and oxygen in order to burn off the excess carbon left in the iron as a result of the blast furnace process.

Open-Hearth Steel

Prior to 1960 almost all steel was made in an open hearth. In the hearth, the iron was kept hot by overfiring with gas burners as long as needed to make the carbon in the metal react with oxygen.[1] Oxygen is introduced along with the gas fuel above the molten material and as iron oxides added to the melt. The source of iron oxides can be iron ore or rusty steel scrap.

The use of fuel in the open-hearth conversion means that energy is available as required to melt any amount of cold scrap added to the molten iron. The amount of scrap used depends on its cost and composition and the specification of the steel to be made. In 1960, about 40 percent of the melt was scrap; today, consistent with changing conditions, scrap makes up over 45 percent of the average open-hearth melt. In 1960, 87 percent of U.S. steel was produced in open-hearth furnaces. By 1984, only 9 percent of U.S. steel was being produced by this method because of the rapid adoption of the basic oxygen and electric furnace processes.

Basic Oxygen Steel

The main difference between the basic oxygen furnace (BOF) and open-hearth processes is that in the basic oxygen process almost pure oxygen is introduced at high velocity into the molten iron. About 2,000 ft^3 of oxygen is used per ton of raw steel produced. In this way, the conversion to steel occurs ten times faster than it does in open-hearth conversion.[2] Because the reaction is exothermic and takes place rapidly (in less than an hour), fuel is not needed to keep the melt from solidifying during its conversion to steel. The basic oxygen process has greatly improved overall productivity in steelmaking. In 1960, 3 percent of U.S. steel was made by the basic oxygen process. By 1976, the percentage had risen to 62 percent, but by 1986 its market share was down to 59 percent as a result of competition from electric melting.

Steel Finishing

Conventional practice for steel casting and finishing operations in integrated steel mills consists of pouring the steel into large ingots. These large ingots are subsequently reheated and, following a series of forming and reheating steps, are made into basic steel products for shipment.

These steel finishing operations represent a substantial part of the overall cost of steelmaking. The costs arise because several separate ingot rolling and reheating operations are required and each carries a cost for labor, energy, and capital equipment and contributes to product losses in the form of cutoff scrap and mill scale. A ton of raw steel can be produced from ore with the use of about 18 million British thermal units (MBtu) of primary energy, but by the time a ton of steel product is shipped, about 40 MBtu of energy has been used.

Electric Processes

In sharp contrast to the conventional fuel-based processes for making steel, electric process technology is characterized by small decentralized facilities drawing on local resources of scrap and producing for local markets. These electric operations are unencumbered by the capital[3] and operating costs associated with coal coking and blast furnace processes and are substantially more productive through the incorporation of continuous casting techniques.

The Electric Arc Furnace

Although electric induction furnaces are in use, particularly for specialty steels, most electric steel is produced in electric arc furnaces. In this process steel scrap is loaded into a brick-lined, water-cooled steel vessel, the cover containing three large-diameter (e.g., 24-inch) graphite carbon electrodes is put in place, and an arc is stuck between the electrodes and the scrap metal. When the charge is molten, oxygen (about 300 ft^3 per ton of raw steel) is injected beneath the surface of the bath. When the melt is ready, the vessel is tapped and the molten steel flows to a ladle for transfer to casting operations.

Unlike the conventional steel processes that begin with iron ore, the electric furnace depends upon scrap as its basic raw material. Some of the scrap is made available by the inability of the basic oxygen process to use more than about 30 percent scrap (along with 70 percent molten iron from the blast furnace) in its charge. This quantity of scrap is barely adequate to absorb the "home" scrap produced in the conversion of large ingots into finished steel shapes for shipment to customers. The excess availability of this and other sources of scrap created the original opportunity for the small electric furnaces. Other sources include "prompt" scrap from industrial operations and scrap reclaimed from discarded or obsolete equipment.

In a sense, therefore, the switchover from open-hearth to basic oxygen processes for steel production opened the way for new producers of electric steel to enter the competition because the oxygen furnace could not include as much scrap in its charge as the open-hearth furnace. Small mills using electric furnaces (which are not tied to the logistics of coal, ore, and limestone supply or the economies of scale associated with blast furnace and coal coking operations) could be built with a relatively modest investment. Such mills were well suited to take advantage of this opportunity.

> The small mill concept became economically viable in the mid-1950s primarily because of two factors: (1) the greater availability of scrap at reasonable prices due largely to increased usage of the basic oxygen furnace with its lower scrap consumption, and (2) escalating freight rates which gave the small regional mill an advantage in obtaining local scrap and also provided a cost advantage over larger mills shipping products into the area from substantial distances.[4]

Thus began the era of the "minimill" and the growth of electric steel production.

Continuous Casting

In continuous casting, molten steel is fed through a funnel into a water-cooled oscillating copper mold where the steel solidifies and is continuously withdrawn from the bottom of the mold at a rate of about ten feet per minute. This ribbon of steel is then cut into appropriate lengths (billets) which, in some plants, are moved immediately through the rolling mill and emerge shortly thereafter as finished shapes ready for shipment.

In principle, all steel could be delivered to finishing operations via continuous casting regardless of the process used to produce the steel. In practice, however, continuous casting operations are most easily and economically introduced in conjunction with new steel-making facilities where the capacity and design of the steel-making equipment can be matched with the capacity and layout of the finishing section and with the line of products planned for the operation. This is precisely the situation of the regional minimill specializing in the production of a small variety of high-volume simple shapes. In Japan, where more than 70 percent of the steel-making capacity came on-line after 1963, more than 80 percent of steel production was continuously cast in 1982. In that year the United States continuously cast only about 25 percent of its steel production.[5]

The Growing Importance of Electric Minimills

Information on the relative importance of the three steel-making processes for the period 1959 through 1986 is given in Figure 5.1 and Table 5.1. At the start of the period, 88 percent of steel was made in open-hearth furnaces and most of the remainder, 9 percent, was made electrically. Throughout the 1960s and the early 1970s, steelmaking with the basic oxygen process rapidly displaced open-hearth production until hitting a peak in the mid-1970s.

The sustained rise of electric steel production began in the mid-1960s. The output from electric furnaces has continued to grow, although some-

Figure 5.1   Trends in Raw Steel Production and Process Shares,
1960–1987

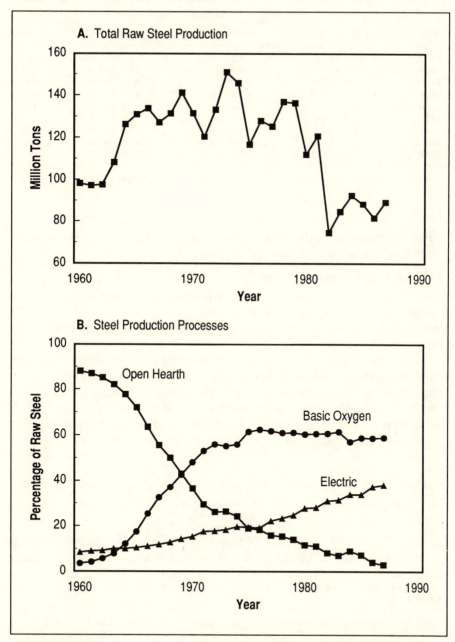

SOURCES: American Iron and Steel Institute, *Annual Statistical Report* (Washington, D.C.: AISI, various years).

Table 5.1 Raw Steel Production for Different Steel-Making Processes for Various Years

	Raw steel production, million tons (percentage)			
Year	Open hearth	Basic oxygen	Electric	Total
1959	81.7 (88.7)	1.9 (2.1)	8.5 (9.2)	92.1
1964	98.1 (77.7)	15.4 (12.2)	12.7 (10.1)	126.2
1969	60.9 (43.1)	60.2 (42.6)	20.1 (14.3)	141.2
1974	35.5 (24.3)	81.6 (56.0)	28.7 (19.7)	145.8
1979	19.2 (14.1)	83.3 (61.1)	33.9 (24.8)	136.4
1984	8.3 (9.0)	52.8 (57.1)	31.4 (33.9)	92.5
1985	6.4 (7.3)	51.9 (58.8)	29.9 (33.9)	88.2
1986	3.3 (4.0)	47.9 (58.7)	30.4 (37.3)	81.6
1987	2.7 (3.0)	52.5 (58.9)	34.0 (38.1)	89.2

SOURCES: American Iron and Steel Institute, *Annual Statistical Report* (Washington, D.C.: AISI, various years).

what irregularly, in both actual tonnage and percentage terms even during periods of declining total steel production.

Efficiency of Materials Use

One useful indicator of production efficiency is the percentage of raw material that ends up as finished product. As shown in Table 5.2, electric steelmaking is about 10 percent more efficient than its competitors in its use of raw material.

Even more impressive is the gain in materials-use efficiency of steel-finishing operations, that is, the fraction of raw steel production that is shipped as finished product. Modern electric mills have production efficiencies in the steel-finishing operations in excess of 90 percent,[6] compared to an average efficiency of steel-finishing operations of less than 70 percent throughout the 1960s. This is primarily the consequence of using continu-

Table 5.2 Production Efficiency[a] of Raw Steel-Making Processes in 1977

	Million tons of materials			
Process	Pig iron consumed	Scrap consumed	Raw steel produced	Production efficiency, percentage
Open hearth	12.5	11.0	20.0	85.1
Basic oxygen	63.9	25.0	77.4	87.1
Electric	0.7	28.2	27.9	96.6

SOURCE: American Iron and Steel Institute, *Annual Statistical Report* (Washington, D.C.: AISI, 1977).

[a] Fraction of iron charged to steelmaking that is converted to raw steel.

ous casting and electric reheating techniques. In 1980, 11 percent of production in integrated mills was continuously cast compared to 43 percent for electric mills.[7] With the growth of electric steelmaking and continuous casting, the average production efficiency in all steel finishing operations is now on the rise (see Table 5.3).

Efficiency of Energy Use

When the energy requirements for each steel-making step are included (i.e., from ore to molten pig iron to steel), the total amount of primary energy per ton of raw steel is about 15 MBtu for the open-hearth and basic oxygen processes compared to about 6 MBtu per ton for electric melting. (The value given for electric steel is the thermal energy in fuel used to generate the needed electricity.) The large difference in energy requirements reflects in part the fact that electric steel is made from scrap, and therefore scarcely any energy is required to reduce iron oxide to elemental iron.[8] In this sense, U.S. resources of iron and steel scrap represent stored energy as well as a raw ore substitute.

For steel-finishing operations, the contrast in energy requirements is even more pronounced. For the industry as a whole, an additional 12 MBtu per ton of raw steel might be used in finishing operations[9] and only 70 percent of the raw steel converted into steel for shipment, while a minimill with continuous casting uses only 2 to 3 MBtu in steel finishing and ships in excess of 90 percent of its raw steel production. Thus, the production energy comparison is about 39 MBtu/ton (15 + 12/0.7) of steel shipped for open-hearth and basic oxygen furnace production and about 10 MBtu/ton (6 + 3/0.9) of steel shipped for electric furnace production with continuous casting. This advantage of up to four to one in primary energy use has contributed to the decline in energy intensity of the industry as a whole shown in Figure 5.2. It also confers a decided cost advantage to all-electric

Table 5.3 Production Efficiency of Steel Finishing

Period	Electric furnace production, percentage of total raw steel production[a]	Production efficiency of steel-finishing operations, percentage[b]
1960–64	9.4	69.4
1970–74	17.8	71.8
1980–84	30.5	75.9

SOURCES: American Iron and Steel Institute, *Annual Statistical Report* (Washington, D.C.: AISI, various years).

[a] Average for the years covered.

[b] Fraction of raw steel produced that is shipped as finished product; average for the years covered.

Figure 5.2   Energy Intensity and Electricity Share, Total Steel Industry

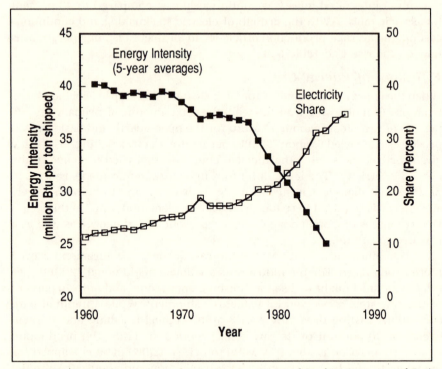

SOURCES: Derived from primary data in American Iron and Steel Institute, *Annual Statistical Report* (Washington D.C.: AISI, various years).

steelmaking, and it is of distinct benefit to the U.S. economy from the standpoints of environmental quality and conservation of energy resources.

Natural Resource Extension

The declining use of pig iron from iron ore in steelmaking reflects the increasing use of scrap iron and steel. "Home" scrap from steel-making operations and "prompt" industrial scrap are recycled immediately. And as a result of the fact that continuous casting reduces the amount of home scrap relative to steel shipments, the use of obsolete scrap is on the rise. Obsolete scrap on average made up 15 percent of shipped products for the 1967 to 1971 period compared with 32 percent for the period 1980 to 1984.

The inventory of recoverable scrap iron and steel in the United States as of the end of 1981 was estimated to be 683 million tons.[10] Recoverable scrap excludes scrap disposed of in a dump or landfill, that is excessively expensive to recover and process, and scrap used sacrificially (e.g., to recover metallic copper from solution). From 1975 through 1981 this recoverable scrap inventory has increased each year by 18 million tons on the average. The amount of obsolete scrap generated each year will continue to increase for years to come because the amount generated is tied to past steel production. Automobiles and major appliances last on the average about ten years, whereas agricultural and industrial equipment lasts typically for twenty-five years. This means that steel discards from the high steel production period that began in the mid-1960s are now entering the scrap inventory. Therefore, it does not appear that the availability of scrap will place limits on minimill production during this century.

The change in resource base that occurred in the 1970s toward a higher fraction of scrap recycled represents a substantial environmental gain. The use of coal has been reduced both because less pig iron is required and because, with electric melting, less energy is required per pound of steel product. Environmental gains stem both from the decrease in pollution from coal combustion and from the increased utilization of scrap.

Steel-Making Outlook

Electric steel production from scrap metal has grown — and continues to grow — both in total quantity and in market share. By comparison, production in open-hearth furnaces has all but ceased, and the market share for steel produced in basic oxygen furnaces has fallen from its high point of 62 percent in 1976 to 59 percent in 1987.

Electric steelmaking is now viable in its own right and is no longer just an adjunct to the historical transition from the open-hearth to the basic oxygen process. The economic advantages of electric steelmaking, in general, and of electric minimills in particular, are clearly established. They stem primarily from gains in energy and production efficiency. Decentralized electric steelmaking is not the mere substitution of electricity for fuel in the production process; it is rather a completely different way, from raw material resource to finished product delivery, of providing steel goods to the consumer.

The future of electricity use in steel production depends on several factors. Perhaps most important is further development of electric steelmaking technology. At present, minimills do not produce a full line of steel products. In particular, they do not produce flat rolled sheet and heavy structural steel of the sort used in automobile production. This limits the fraction of the steel market that they can supply. However, one electric steelmaker plans to install technology developed in Germany that should allow a minimill to produce very thin sheet steel. If this installation proves economically and technically successful, it may point the way for minimills to capture a larger percentage of the total steel market, perhaps including some of the high-quality automotive market.[11]

GLASSMAKING

Making glass consists of mixing a batch of dry ingredients, heating the batch until molten and uniform in composition, forming the molten glass into the desired shape, and annealing, that is, reducing the temperature of the product in a controlled way in order to minimize residual stresses that might otherwise cause breakage. Of the 320 trillion Btu of energy (primary equivalent) purchased by the glass-making industry in 1981, 65 percent was gas, 31 percent was electricity (8.5 billion kWh), and 4 percent was other fuels. The fraction of primary energy purchased by the industry in the form of electricity grew from 15 percent in 1962 to 31 percent in 1981.

Glassmaking is divided into categories according to the different products made. Thus, flat-glass manufacturing (producing sheet and plate of different sizes and thicknesses for car and building windows and partitions) is distinguished from pressed- and blown-glass manufacturing, in which containers, fibers, tubing, and tableware are produced. Eighty percent of energy use is for pressed and blown glass.

Melting Glass with Gas

The dominant process for melting glass is based on the use of gas. Typical equipment is illustrated in Figure 5.3. Shown in the figure are the melting tank or furnace, the regenerators (checkers) for capturing waste heat in the flue gas for use in preheating the air for combustion, and the forehearths for adjusting the melted glass to the exact temperature needed for forming the glass product. Not shown is the forming equipment, which is of several different kinds depending on the type of product being made. The annealing furnace (lehr) is also not shown. It consists of a large conveyor belt enclosed in an insulated structure with provision for adding heat as needed.

The equipment used for melting may be as simple as a ceramic-lined pot that is filled with a batch of ingredients in the evening, which are melted overnight in a gas-fired furnace, and drawn upon during the day as the source of glass gobs to be blown and shaped into desired specialty products. At the other extreme, the equipment may consist of large, multistory structures that include glass-melting tanks with melting areas of 1,000 square feet. These tanks (furnaces) are continuously fed with material to be melted at one end; a stream of up to several tons per hour of flowing glass is continuously delivered at the other end. These large operations are replete with automated machinery for mixing and delivering the raw materials to the melting tank and for drawing, forming, annealing, and handling the glass product taken from it.

Natural gas became the basic source of fuel used in glass furnaces after World War II when interstate pipelines began to carry natural gas from Texas and Louisiana to most industrial areas of the United States. Its flame

Figure 5.3 Conventional Large Glass-Melting Equipment

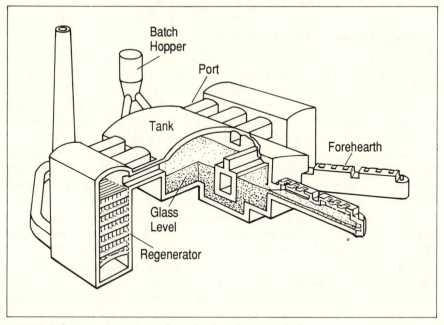

NOTE: Only one regenerator and one forehearth are marked.

temperature is much higher (3,600°F) than the flame temperature of the
producer gas (3,100°F) used prior to that time. As the flame temperature
goes up, the higher thermal driving force (i.e., the temperature difference
between flame and melting glass) increases the rate of production from a
given size glass tank and thereby reduces energy lost per ton of output.
Oxygen enrichment of the air for combustion can increase flame temperatures even further. Since thermal losses dominate energy use in glass melting (only about 20 percent of the fuel energy actually goes to heating and
melting the batch),[12] the payoff in increasing the rate of production in this
way is substantial.

Developments aimed at increasing the output of gas-fired furnaces (i.e.,
through flame impingement on the batch being melted or by raising the
furnace temperature) exacerbate emission problems and material losses in
that these techniques increase the amount of material that is vaporized
from the molten glass and blown out with the flue gas stream. Once
released, these effluents clog airflow passages in the equipment and reduce
its efficiency, or escape from the stack. In either mode, the materials are lost
from the process, thus increasing the cost of the product. To comply with
environmental regulations, glass furnaces are now being equipped with

electrostatic precipitators specially designed to capture the highly resistive submicron-sized particulates that characterize glass-melting emissions.[13]

Melting Glass with Electricity

Electric melting of glass, done by passing an electric current through molten glass, was first practiced in Sweden in the mid-1920s. Initial adoption of electric melting by U.S. industry did not occur until the early 1950s, when molybdenum electrodes became available in a size large enough to do the job and techniques were developed for installing, adjusting, and maintaining them while the furnace was in operation.[14] Glass can be melted completely by electricity, or glass production from fuel-fired melting can be enhanced by electric boosting. In both these techniques, electrodes placed through the melting material add all of the electric heat directly to the molten glass below the surface of the melting materials.

Electric boosting can substantially increase the rate of output above that possible with fuel combustion alone. For example, using this technique, some factories have increased their output by 60 percent.[15] Moreover, electric boosting does not increase emissions or materials losses. Manufacturers can use electric boosting either to increase output and still remain in compliance with environmental regulations, or to maintain output while reducing the effluents per ton of production and thereby come into compliance with regulations. Since the electric heat is added at nearly 100 percent efficiency, it is usual by this technique to reduce the amount of energy used per ton of glass melted even after calculating all the fuel used to generate and deliver electricity to the melt. About half of the container glass manufacturers in the United States now use electric boosters.[16] (Containers account for more than 50 percent of the energy consumption by glass manufacturing as a whole.)

Comparative Melting Costs: Gas and Electricity

Energy Costs

Data comparing energy usage between gas and all-electric melting suggest that for comparable rates of production 1 Btu of thermal energy delivered as electricity is equivalent in melting capability to 2.5 to 4.0 Btu of energy delivered as gas. Despite this greater efficiency of energy use, electricity was not able to compete in cost with fuels until the late 1970s, when fuel costs rose much more sharply than electricity costs did (see Figure 5.4).

Even before this energy cost realignment, glass manufacturers were shifting gradually toward a higher fraction of electricity in their energy mix (Table 5.4). This trend is evident in Figure 5.5, which shows that as the electricity share increased, the overall intensity of energy use for container manufacturing declined.

Figure 5.4 Historical Trends in Electricity-to-Fuel Price Ratios per Contained Btus for Glass Container Manufacturing

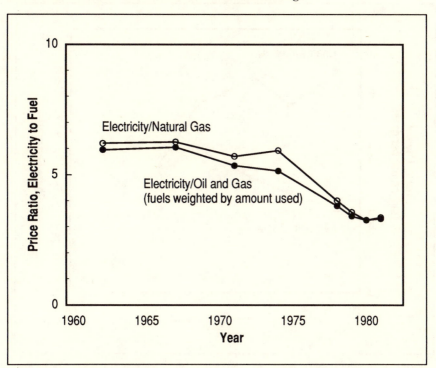

SOURCES: U.S. Bureau of the Census, *Annual Survey of Manufactures: Fuels and Electric Energy Consumed* (Washington D.C.: GPO, various years).

Table 5.4 Energy Purchases for Glass Container Manufacture, 1962–1985

Year	Fuel, 10^{12} Btu	Electricity 10^9 kWh	Fraction electric Btu basis[a]	Cost basis
1962	94.3	1.68	0.16	
1967	106.8	2.33	0.19	
1971	130.3	3.40	0.22	
1974	127.3	3.90	0.25	
1977	128.8	4.59	0.27	
1981	108.8	4.72	0.32	0.32
1985				0.36

SOURCES: U.S. Bureau of the Census, *Annual Survey of Manufactures: Fuels and Electric Energy Consumed* (Washington, D.C.: GPO, various years).
[a] Electricity at 10,600 Btu/kWh.

Figure 5.5   Energy Intensity and Electricity Share, Glass Container
Manufacturing

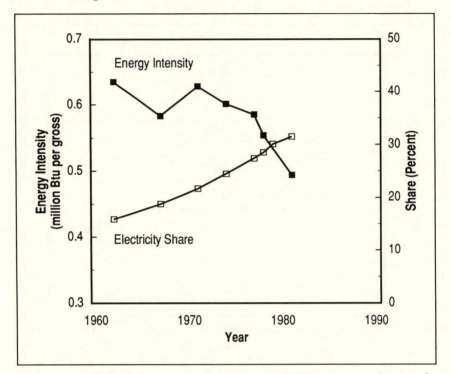

SOURCES: Derived from primary data in U.S. Bureau of the Census, *Annual Survey of
Manufactures: Fuels and Electric Energy Consumed* (Washington, D.C.: GPO, various years);
U.S. Bureau of Economic Analysis, *Business Statistics* (Washington, D.C.: GPO, various years).

Capital Costs

Investment-related costs in glassmaking using gas make up about 12 percent
of total costs. In addition, there is the required investment for emission
control. With emission control, the total capital costs for the melting system
become equal to the energy costs, each contributing about 20 percent
compared with about 60 percent for the raw materials (Table 5.5).

The all-electric melting tank for glass is much simpler than is the fuel-
based system. It does not have flue-gas regenerators, gas ports to and from
the tank, or a flue-gas chimney. The roof to the all-electric tank, if provided
at all, is of much simpler construction, since it is not exposed to high
temperatures. Also not needed for electric melting are the fuel burners,
combustion blowers and fans, and reversing valves for using the regenerators in an alternating fashion. In addition, for a given production rate, the
electric melter is much smaller in physical size because the production or

Table 5.5   Sources of Production Costs in Natural Gas-Fired Glass Melting (With and Without Pollution Control)

	Percentage of total product cost	
	Without pollution control	*With pollution control*
Raw materials and shipping	65	59
Energy	20	18
Operating labor and overheads	3	3
Capital investment-related expense	12	20

SOURCE: U.S. Environmental Protection Agency, *Environmental Considerations of Selected Energy-Conserving Manufacturing Process Options, Glass Industry Report* 11, EPA-600/7/76-034R (Washington, D.C.: GPO, December 1976): 23–25, 31.

pull rate (which is normally expressed in square feet of melting area per daily ton of production) for electric melting is about one-half that for fuel melting. With electric boosting of fuel melters, this difference in pull rate between all-fuel and all-electric melting is less important.

While the all-electric melter has fewer components, the cost of electric equipment is more or less proportional to the size of the required capacity, whereas the cost per unit of capacity for fuel-fired systems decreases with scale-up in size. Thus, the investment cost for the two types of melting, without the cost of emission control equipment, does not clearly favor either system. It is lower for electric melting at low capacity and lower for fuel melting at high capacity.

If, however, the cost of emission control (needed only with fuel firing) is included, the all-electric system is less costly over a much larger range of melting capacity. Fuel melters with abatement equipment were estimated in 1974 to cost $2,000 to $3,000 more per daily ton of capacity than all-electric melters for capacities up to 200 ton per day.[17] Taking inflation into account, the capital cost advantage for all-electric melting might now have increased to as much as $5,000 per daily ton of capacity. The saving in investment-related costs (i.e., including operation, maintenance, taxes, insurance costs in addition to allowances for depreciation and return on investment) corresponds to a 0.6¢ to 0.8¢ per kWh credit that could be allowed for the cost of electricity in computing the break-even ratio in the price of electricity and fuel.

Further, the options available to the glass manufacturers are much more varied than is a choice between new facilities based on either all-fuel or all-electric melting. A range of choices exists from all-fuel firing to fuel firing with electric boosting to "hot top" electric melting with fuel boosting (mixed melting) to all-electric melting. These incremental approaches to substituting electricity for fuel, to managing emissions, and to expanding production are expected to be less costly and more practical to implement than the wholesale replacement of production facilities.[18]

Auxiliary Aspects

Forehearth

Gas heat is traditionally used for the "forehearth," or conditioning trough, which conveys the molten glass from the melter to the glass-forming machines. The gas requirement for this operation is typically 5 percent to 10 percent as large as the requirement for melting.[19] The function of a forehearth is to deliver glass to the forming process at an absolutely uniform temperature. It is more difficult to control glass temperature with heat from fuel because the heat must be transferred to the molten glass from the combustion gases. Thus, there is a temperature gradient from the glass surface to its bottom and there is no exact way of knowing how much of the energy of these gases will be delivered to the glass at depth and at any point in the forehearth. In addition, because of the need for flue-gas ports, heat losses from the forehearth surfaces are difficult to predict.

With electric heating, there are no flue-gas openings, and heat losses from forehearth surfaces are predictable. More importantly, glass temperature is accurately correlated with electric resistance and electric heat is transferred directly to the glass; thus, the amount and location of heat added is known precisely and can be increased or decreased in any desired amount by adjusting electrode voltage. The ability to control the glass temperature from the forehearth within narrow limits improves production efficiency (the weight of glass packed divided by the weight of glass melted).

The authors of two articles[20] covering experience with electric forehearths reach similar conclusions: the electric forehearth uses 10 percent as much energy as does its gas-fired counterpart, and with electric operation, production efficiency is increased 2 percent to 5 percent. Since defective products must be remelted, this small improvement in production efficiency resulting from improved forehearth control can save as much energy at the melting stage as does electrification of the forehearth itself. And the improvement in production efficiency translates into reduced product cost in all other operations as well. The authors cite other benefits of electrification of the forehearth, including a faster response to varying production demands and temporary stoppages, and an improved working environment because the electric forehearth structure is cooler, produces no waste gas fumes, and has a lower noise level.

Lehr

The purpose of the lehr, or annealing furnace, is to control the temperature of the shaped glass products as they cool so that thermal stresses do not lead to excessive breakage. Annealing furnaces for container production comprise a conveyer belt typically several feet wide and perhaps 100 feet in length enclosed in a thermally insulated structure with provisions for burning gas and oil, and for release of the combustion gases. Older models are refractory lined, and the conveyer belt returns external to the furnace. The lehr is an expensive piece of equipment with a relatively long life (e.g., twenty years) for the basic structure and drive components.[21] Thus, its premature replacement can only be justified by substantial savings in energy and/or gains in production efficiency.

In the process of annealing, the temperature of the product is reduced from about 1,000°F in a carefully controlled fashion, either to a few hundred degrees as may be needed for further operations (e.g., decorating) or to room temperature. In many cases, the energy requirements, which with conventional lehrs are typically 5 percent to 10 percent of those for melting, can be largely or completely provided from the heat contained in the glass product at the time it enters the annealing furnace. Thus, it is not so much a question of how to provide needed energy for the operation but rather one of how to control the loss of energy from the product in such a way that it is self-annealing. Energy is added only when needed for efficient production and accurate control of the cooling process.

These requirements are made to order for electric heating. In this technology, the furnace can be completely sealed against convection loss, and heat can be added at any point and in any amount throughout the annealing chamber. According to S. J. Shepard of the Ball Corporation: "Coupled with high-velocity recirculating air and maximum insulation, the new electric lehrs are self-annealing for many production items, requiring no additional heat input after start up."[22]

Experience with electric-heated lehrs parallels that of electrification of the forehearth. Energy savings of 90 percent are reported,[23] and in winter, the waste heat (since it is free of flue gases) can be ducted for additional savings in energy to areas of the factory where heat is required for space heating. Other claimed benefits of the electric lehr include a clean working environment, less maintenance, accurate temperature control, high product quality, and a low reject rate.[24]

Control of Operations

Several benefits are associated with the ability to deposit electric energy directly into the melting glass. This characteristic allows fast heating and cooling of the entire molten glass volume. Thus, if quality problems arise at the forming operation because of error in batch composition or in melting and refining, the problem is short in duration; a 2 percent to 4 percent increase in the fraction of product packed for soda-lime glass, up to 20 percent for lead glass, and a very low loss due to unmelted materials in the glass (0.3%) has been estimated to result from this better process control.[25] (There is also less time lost in restarting operations after holiday cooldown.) The problem of radiant heat transfer into colored glasses (amber, green) from surface fuel firing is also circumvented with electric heating; in addition, colored glass can be produced in deep tanks not designed for its production, thus, giving the manufacturer further flexibility in operations.[26]

All-electric melting confers the additional benefits of reduced noise level, dust-free operation,[27] and easy access to the melting tank for delivering raw materials and for temperature measurement of the melting operation.[28] Since there are no regenerators that gradually decline in performance due to clogging, the all-electric melter can operate at full capacity throughout the campaign.[29]

Recycling

Scrap generated and recycled in the process of manufacture is termed "primary;" scrap recovered after product use by society is termed "secondary." Glass manufacturers purchase secondary scrap for recycling when it is available at a competitive price. This use is advantageous because it reduces the raw material requirement and also reduces energy requirements by lowering the furnace temperature needed for a given rate of production. The recycling of secondary scrap is minor, however, because the value of glass is too low to justify its shipment to centralized locations for melting.

Even so, there are incentives coming from several directions that taken together may ultimately justify expanding the recycling of secondary glass. One is that the cost of waste disposal is a growing burden for metropolitan areas. Thus, at some future time it may become economic to process solid wastes to recover their scrap and energy values.[30] Another reason is that anti-litter campaigns may stimulate new ideas and legislation for litter management. Finally, rising energy costs will enhance glass recycling economics, both because of energy savings at the point of manufacture, and because of energy savings in the production and transportation of raw materials and glass to markets. For example, the energy required to transport raw materials to glass production and to transport throwaway glass products to markets is about 10 percent of the amount used for their production.

The amount of energy to transport glass containers is five to fifteen times larger than it is for competing containers made from paper, steel, or aluminum.[31] Thus, the cost of transportation from centralized glass manufacturing is a factor in the growing market share for containers made from other materials, but it is not by itself sufficient to bring about decentralization.

SHIFTS IN ENERGY SOURCES

This review of the major changes taking place in the steel-making and glass-making industries illustrates why and how electricity is replacing fuels in high-temperature applications.

Energy price trends are one reason. Electricity prices, as discussed in this and the preceding chapter, have been falling relative to those for natural gas. Pollution control is another reason. The growing need for pollution abatement favors the use of electricity over the direct use of fuels that often require costly clean-up measures.

Third and most important from a technological viewpoint, using electricity allows manufacturers to perform operations in ways that would not be practical or even possible using fuels. Electric-based technology that draws on its unique advantages in high-temperature applications has transformed the U.S. steel industry. Similarly, though less dramatically, it has also brought new efficiency and flexibility to the manufacturing of glass.

Such applications are not confined to the steel and glass industries. Wherever a manufacturer using thermal processes requires very high temperatures combined with cleanliness and precise control, some form of electrotechnology is a likely candidate for the job.

Notes

1. In 1960 about 500 ft^3 of oxygen was used per ton of steel in the open-hearth process. Current open-hearth practice uses perhaps three times as much.

2. C. S. Russell and W. J. Vaughan, *Steel Production: Processes, Products, and Residuals* (Baltimore, MD: Johns Hopkins University Press, 1976), 118.

3. F. K. Iverson letter to C. C. Burwell, August 4, 1983. Construction costs for a minimill in the early 1980s were in the range of $200 to $500 per ton of annual capacity depending on the product(s) to be made. Construction costs for an integrated mill exceed $2,000 per ton of annual capacity.

4. F. K. Iverson, "Mini Mills — Capital Cost and Productivity," *Electric Furnace Conference Proceedings* 33 (1975): 9.

5. D. Cosma and F. J. McMulkin, "Energy Efficiency: The Name of the Game in Steel," in *AISI Technical Sessions*, 88th General Meeting (May 21–22, 1980), Waldorf-Astoria Hotel, New York, 9; F. K. Iverson, letter to C. C. Burwell, August 4, 1983.

6. Iverson, "Mini Mills," 13.

7. Shearson Loeb Rhoades, Inc., "Steel Mini-Mills — An Investment Opportunity" (New York: 1980), 22. Continuous casting can be adapted to traditional integrated mills with open-hearth and basic oxygen steel-making processes, but the transition will be slower and not as complete as for electric mills for several reasons: (1) the large scale of production for efficient blast furnace operation and the large steel pours associated with it are not well matched to the small billets delivered by the continuous caster; (2) the existing rolling mills are not easily rearranged to accommodate a new ingot delivery system; (3) even if integrated mills improve efficiency by upgrading their operations to include continuous casting, there is no certainty that the markets lost to electric mills can be recaptured because other attributes of electric mills remain undiminished; (4) slab continuous casters needed for the large-

scale production of flat rolled products have not developed as rapidly as continuous billet casters and when they are ready for general deployment, electric mills will be able to expand their line of products.

8. The extra carbon needed for reducing ore to iron is about 3.2 MBtu per ton. This is about one-third of the difference in energy requirements for making raw steel from ore with fuel compared to making raw steel from scrap with electricity; the other two-thirds of the energy savings is purely the result of the more efficient electric process.

9. This 12 MBtu/ton is calculated by subtracting 15 MBtu for raw steel processing from the total amount of energy used, 27 MBtu/ton of raw steel production for the industry as a whole prior to the mid-1970s.

10. Robert R. Nathan Associates, Inc., "Iron and Steel Scrap," Prepared for Metal Scrap Research and Education Foundation (Washington, D.C.: December 23, 1982), 44.

11. "Is Steel's Revival for Real?" *Fortune Magazine* (October 26, 1987), 96ff.

12. Fay V. Tooley, ed. *The Handbook of Glass Manufacture* (New York: Books for Industry, 1974), 244, 356–358.

13. W. W. Custer, "Electrostatic Cleaning of Emissions from Lead, Borosilicate and Soda Lime Glass Furnaces," in Annual Conference on Glass Problems, *Collected Papers* (Columbus: Ohio State University, 1974), 38.

14. Larry Penberthy, "Recent History of Electric Melting of Glass," *The Glass Industry* (May 1973), 12–13.

15. Larry Penberthy, "Recent History of Electric Melting of Glass," *The Glass Industry* (June 1973), 18.

16. Larry Penberthy, in *The Handbook of Glass Manufacture*, ed. Fay V. Tooley (New York: Books for Industry, 1974), 393.

17. See comparison of estimates in C. C. Burwell, *Glassmaking: A Case Study of the Form Value of Electricity Used in Manufacturing*, ORAU/IEA 82-9(M) (Oak Ridge, TN: Oak Ridge Associated Universities, Institute for Energy Analysis, July 1982).

18. For the period from 1976 to 1986, for all segments of the industry, the use of electric energy remained generally stable, while the consumption of fossil fuels, mainly natural gas, declined somewhat. (Gas Research Institute, "Glass Industry: Opportunities for Natural Gas Technologies," GRI-88/0266 [September 1988], see 13–23).

19. G. Perry and P. J. Doyle, *Energy Audit Series, No. 5, Glass Industry* (London: Departments of Energy and Industry of Great Britain, June 1979), 13.

20. D. A. Hunt, "Electric Conditioning of Glass," *American Glass Review* (March 1977), 6; and Don Hayes, "Forehearth Heating System Switch Pays Dividends," *The Glass Industry* (February 1981), 19.

21. P. W. Roos, "Lehr Priority: Design Concepts to Save Energy," *The Glass Industry* (April 1975), 18–22.

22. S. J. Shepard, "Ball Corporation Modernizes the Asheville Plant," *The Glass Industry* (December 1977), 25.

23. P. W. Roos, "Energy Saving Lehrs," *The Glass Industry* (December 1974), 12–24.

24. Shepard, "Ball Corporation Modernizes," 24–26; see also R. H. Charles, "Electricity in the Glass Industry," *Glass* 52, no. 6 (June 1975), 187.

25. Penberthy in Tooley, *Glass Manufacture*, 393.

26. C. P. Ross, Jr. and C. N. Jewart, Jr., "How Kerr Applies Electric Boost," *The Glass Industry* (June 1978), 12; see also W. H. Fouse, J. W. Jelinek, and B. L. Schmidt, "Application of Electric Melting Energy in Anchor Hocking Glass Furnaces," in Annual Conference on Glass Problems, *Collected Papers* (Columbus: Ohio State University, 1974), 72.

27. Helmut Pieper, "New Developments and Experiences with All Electric Glass Melting Furnaces," in Annual Conference on Glass Problems, *Collected Papers* (Columbus: Ohio State University, 1976), 132–133.

28. R. E. Loesel, "Practical Data for Electric Melting," *The Glass Industry* (February 1975), 8.

29. Penberthy in Tooley, *Glass Manufacture*, 394.

30. H. W. Gershman, "Status Report of Glass Recovery," *The Glass Industry* (October 1976), 24–28; see also *The Glass Industry* Staff Reports, "Recycling Waste" (December 1971), 436–438 and "Glass Industry Learns Recycling" (January 1972), 14–17.

31. Bruce Hannon, "System Energy and Recycling: A Study of the Beverage Industry," CAC Document 23 (Urbana-Champaign: University of Illinois, Center for Advanced Computation, March 1973).

6 | Low-Temperature Electroprocessing: Pulp and Paper, Petroleum Refining, and Organic Chemicals

Calvin C. Burwell

Some manufacturing that uses heat for materials processing requires only low temperatures. Electricity does not have the same advantage here as it does in high-temperature processing, since low-temperature process heat can be provided most economically by fuel combustion. Nevertheless, the electricity fraction of purchased energy is growing even in low-temperature process industries.

One important key to electrification in such industries is a gradual shift away from thermal and toward mechanical processing. In low-temperature thermal processing, heat loss is minimal and most of the energy in fuels can serve process requirements. For electricity to compete with fuels in such applications, its effectiveness must be multiplied in some way to offset the cost advantage of direct fuel use.

Electrotechnology can function as that multiplier. For example, a small amount of electricity powering a motor provides mechanical energy that can accomplish the same work as a large amount of thermal energy from fuel. The substitution of electricity for fuel is indirect as one type of process replaces another.

Mechanical processes driven by electricity are supplementing and in some cases displacing low-temperature thermal processes that rely on fuel combustion. This chapter draws on examples from the pulp and paper, petroleum refining, and chemicals industries to illustrate the shift.

PULP AND PAPER

In principle, paper products can be made from a large variety of plant materials. In practice, they are made primarily from wood because the cost of converting a given weight of biomass into pulp and paper is lowest for wood. Paper products are also reconstituted from paper and cardboard wastes and from certain cloth waste materials. The extent to which wastes are used depends upon the cost to transform them into finished products relative to the cost of using new fiber.

Wood, like cotton, is composed of cellulose fibers. Wood fibers a millimeter or so in length are stuck together by a natural glue called lignin to form the bulk wood structure. To make paper, the bulk wood structure is broken down, physically or chemically, to separate the fibers so they may be reformed into a thin paper sheet.

Pulpmaking

The Basic Processes

The breaking-down process is called pulping, and the result is a thin soup typically containing more than 99 percent water and less than 1 percent fiber, with or without the lignin still attached. Whether or not lignin has to be separated from the wood depends primarily on the type of paper product being made. Lignin left in the paper oxidizes and discolors it; therefore, high-quality paper, such as stationery, is made from delignified pulp.

If lignin can be tolerated as an ingredient of the paper (e.g., as with newsprint for which product lifetime and appearance are not critical), the bulk wood is usually broken down (fiberized) by mechanically grinding it into pulp. In this case, nearly all the wood is incorporated into the paper product, and nearly all the energy requirement for pulping is supplied as electricity to power the grinders.

If the lignin is undesirable in the paper product, the bulk wood is usually broken down with heat and chemicals in processes that modify the lignin to allow its removal from the wood fibers. In this case, only the fiber, usually less than half the wood weight, is incorporated into the paper product. Most of the energy requirements are for steam and high-temperature heat (e.g., for calcining lime) to optimize the chemical process and recover chemicals from the pulping liquor.

In actual practice, different wood pulps are blended to provide the characteristics required by the paper products. Moreover, pulpwood can be partially delignified, and various degrees of mechanical grinding can be combined with various degrees of chemical treatment to produce the desired product from a given wood feedstock. For all practical purposes, a continuum of process options is becoming available as the industry adapts to changing circumstances and increases its efforts to improve the utilization of materials as well as to reduce the environmental impacts of wood pulping. Such modified processes tend to use more electricity because they depend increasingly on mechanical fiberization and on electrically produced chemicals: oxygen, chlorine dioxide, ozone, and sodium hydroxide.

Recent Trends in Mechanical Pulping

In the traditional method of mechanical pulping, the roundwood, after debarking, is pressed against a rotating grindstone in a bath of water, the fibers are torn apart, and an aqueous slurry is produced. This brute force shredding of the wood results in short fibers, which in turn yield paper with limited physical properties. Such pulp is blended with different amounts of higher-quality chemical pulp to make newsprint, container board, and tissue. About 10 percent of U.S. pulp is made this way.

Despite the low quality of conventional groundwood pulp, it is used whenever suitable because it is inexpensive. Cost is low because the pulp yield is very high (about 95 percent of the roundwood weight fed to the grinder ends up as pulp) and because it avoids the extra costs that arise with chemical pulping for purchasing chemicals, reprocessing spent liquor from pulping, and treating noxious wastes prior to disposal. These advantages of mechanical pulping — high yield and low environmental impact — provide strong incentive to improve the quality of mechanical pulps so that they can be used for an increasing fraction of paper and paperboard products.

The general direction taken toward improving mechanical pulping is to treat the wood thermally, chemically, or in both ways before mechanically separating the fibers. In these techniques (called thermomechanical and chemimechanical pulping), the roundwood is first chipped and then presteamed before being ground. Since the feed material has been reduced to small chips, the grinding itself is performed by feeding the chips between two grinding discs that rotate in opposite directions. As a result of the wood having been presoftened, the fibers separate more easily during grinding and sustain less physical damage. The fibers thus produced are perhaps twice as long as are those made by conventional grinding.

These pretreated pulps have better physical properties with broader applications in papermaking. For example, newsprint can be made entirely from thermomechanical pulp instead of requiring a blend of perhaps 20 percent chemical pulp with 80 percent groundwood. Because mechanical pulp is used as part of the pulp blend for many paper products, the higher-quality thermomechanical pulp has extended the range of application for mechanical pulps in these paper products as well.

The longer fiber length that results from thermomechanical pulping compared to conventional grinding means that, with some chemical pretreatment, usable pulps can be produced with mechanical pulping processes from low-density hardwoods, such as aspen and poplar, instead of softwoods. Usable hardwood is less expensive, contains a lower lignin fraction, and pulp made from it has superior softness and moisture absorption properties. Therefore, with thermomechanical pulping, hardwood pulp can be lower in cost and higher in quality than conventional groundwood pulp

made from softwoods. These improvements have expanded traditional markets and created new markets for thermomechanical pulps.[1]

Recent Trends in Chemical Pulping

Conventional chemical pulping is accomplished by steeping wood chips in a
chemical bath containing sulfur compounds under controlled conditions of
temperature, pressure, and time until the lignin that glues the wood fibers
together is sufficiently modified that the chips disintegrate when blown
from the chemical digester. The resulting wood fibers are longer than are
those produced from comparable roundwood by mechanical grinding, and
because some of the lignin is removed along with the spent digestion liquor,
the chemical pulp has better physical properties and color stability than
does mechanical pulp. Color and color stability are relatively unimportant
for many products (e.g., corrugated cardboard boxes); so the chemical pulp
can be used at this stage in its unbleached form. The disadvantages of
chemical pulping are that to the extent the lignin is removed, the yield from
roundwood is reduced, the capital and operating costs incurred in reprocessing the spent liquor raise the cost of the pulp, and the effluents from
reprocessing negatively affect the environment.

If bright, color-stable paper is required, the chemical pulp from the
digester is bleached with chlorine. The effluent from the chlorination contains a high level of corrosive chlorides that cannot be recycled within the
pulping system. The effluent also contains chlorinated organic compounds
unsuitable for release into natural bodies of water, which makes the disposal of bleach plant effluent the most difficult environmental problem
facing the industry.

The new approach to chemical pulping substitutes oxygen followed by
lignin extraction with sodium hydroxide[2] for the troublesome environmental chemical, sulfur, in the pulping process. Oxygen pulping is done at low
temperatures compared to pulping with sulfur and is less damaging to the
lignin, which raises the possibility of future uses for the lignin as an organic
chemical feedstock.

Thermomechanical pulping *followed* by oxygen-alkali treatment to
remove lignin lends itself better to continuous processing and is less costly
in capital than is chemical pulping.[3] These delignified thermomechanical
pulps, while not as yet on a par in quality with the highest grade chemical
pulps, can increasingly take the place of chemical pulps in many applications for intermediate grade material. Thus, it seems likely that as needs for
new pulping capacity arise, a growing share will be committed to thermomechanical technology.

Oxygen is also being used to displace chlorine in pulp bleaching. The goal
is to produce a bleach plant effluent that can be recycled for use in pulping
or be sent to the chemical recovery system. The approach takes a variety of
forms, including an oxygen delignification step prior to the initial chlorination stage, use of chlorine dioxide to replace part of the chlorine in the

chlorination stage, use of oxygen in the first extraction stage, and use of oxygen and ozone to treat the effluent. These various techniques are successful both in reducing the quantity of chlorine required and in improving the color of the treated water, and reducing their residual biological demand for oxygen.[4]

Papermaking

To make paper and paperboard products from pulp, a continuous sheet of a dilute (e.g., 1 percent) mixture of wood pulp in water is formed at the head end of the paper-making machine, the moisture is removed in a series of steps, and the dry sheet is coiled into a large roll of finished product. The series of steps includes atmospheric draining, vacuum draining, pressing in vacuum, and evaporative drying on steam-heated rolls. Most of the water in the pulp (98 percent) is removed electromechanically; even so, the last 2 percent (about 3 pounds of water per pound of paper product) takes most of the energy because of the high energy requirement for water evaporation.

About a third of the thermal energy used to evaporate water in the dryer actually serves that purpose. The other two-thirds is rejected in the steam condensate, in the sensible heat of the water-saturated air leaving the dryer, and through radiation and conduction from the drying equipment to the surroundings.[5] This relative inefficiency in the use of thermal energy for evaporative drying, along with the relatively large amount of energy required, gives incentive to improve on the energy efficiency of the evaporative-drying section of the paper-making process.

The low thermal efficiency of evaporative drying can be improved if heat can be added to the evaporative section without increasing the heat losses. One technique is to use jets of superheated steam that, after saturation following contact with the wet paper, is recompressed and reused in the steam-dryer rolls.[6] Another technique is to heat the air used for drying with electric or fuel-fired heaters. The heat added is used at nearly 100 percent efficiency, and even if the cost of the heat added is high compared to the cost of heat for the unboosted condition, the operation can be cost effective. The throughput can be increased a few percentage points, while heat losses remain nearly the same. In addition, the higher productivity of the process resulting from the increase in machine throughput also means that unit costs of labor and capital decline.

Boosting with electric heat in the form of electromagnetic waves (radio frequency or dielectric heating) offers further advantages. With radio frequency (RF) heating, the energy increment is added to the wettest part of the paper web rather than to the paper surfaces. As a result, the moisture profile becomes uniform across the cross section of the paper, and the paper (thus improved in strength) can be removed from the dryer at a higher average moisture content than is possible otherwise. In addition, it is reported that more moisture is removed with RF drying than can be accounted for by the total added electric energy, the so-called "bonus effect." The explanation is that RF heating moves moisture from the paper center to the paper surface where the conventional steam rolls can evaporate it more effectively. In some instances, the contained moisture is vaporized so rapidly that the vapor forcibly ejects water particles from the paper surface without evaporation. Because of these effects RF drying permits a large (e.g., 20 percent) increase in machine throughput.[7]

Energy Use

The industry is a major energy user, consuming approximately the same amount of energy as such industries as primary metals, petroleum, or chemicals. Moreover, it is surpassed only by petroleum refining in the proportion of its energy requirements produced internally (see Table 6.1).

Purchased Fuels and Electricity

In 1986, the pulp, paper, and paperboard industry purchased some 43 billion kWh of electricity and over 800 trillion Btu of various fuels in other forms.[8] The cost of purchased energy nearly equaled the cost of production labor for pulp, paper, and paperboard mills (see Table 6.2).

Table 6.1 Sources of Energy for the U.S. Pulp, Paper, and Paperboard Industry

	Percentage		
	1975	*1980*	*1985*
Source of energy			
Purchased electricity	13.5	17.0	18.0
Purchased fuel	46.2	39.5	32.1
Residue fuel and hydro	40.3	43.5	49.9
Total	100.0	100.0	100.0
Use of energy			
Purchased electricity	13.5	17.0	18.0
Self-produced electricity	12.3	11.4	13.4
Total electricity	25.8	28.4	31.4
Nonelectric energy	74.2	71.6	68.6
Total	100.0	100.0	100.0

SOURCES: American Paper Institute, *U.S. Pulp, Paper, and Paperboard Industry Estimated Fuel and Energy Use* (New York: April 20, 1987), and U.S. Bureau of the Census, *Annual Survey of Manufactures: Fuels and Electric Energy Consumed* (Washington, D.C.: GPO, 1975 and 1980), and *Annual Survey of Manufactures: Statistics for Industry Groups and Industries* (Washington, D.C.: GPO, 1985).

NOTE: All electricity is counted at 10,600 Btu per kilowatthour. Assumes that energy sold was not electricity. The part of the paper industry that manufactures secondary products such as bags from the basic products is not included.

Industrial production of paper and paperboard exceeded 72 million tons in 1986.[9] If electricity is counted at its average primary input energy content (10,600 Btu per kWh), 17.6 million Btus of primary energy were purchased by the pulp and paper industry for each ton of production. Electricity purchases increased at a compound annual rate of over 3.3 percent per year between 1972 and 1986, with the most rapid growth occurring in the last half of the 1970s (Figure 6.1). During this same period output grew at a compound annual rate of 1.7 percent per year. Thus, on

Table 6.2 Production Costs in Pulp, Paper, and Paperboard Mills in 1985 (Millions of Dollars)

	Materials	Purchased electricity and fuels	Production wages
Pulp mills	2,095	384	413
Paper mills	13,930	2,859	3,182
Paperboard mills	6,172	1,508	1,241
Total	22,197	4,751	4,836

SOURCE: U.S. Bureau of the Census, *Annual Survey of Manufactures: Statistics for Industry Groups and Industries*, M85(AS)-1 (Washington, D.C.: GPO, 1985).

Figure 6.1 Electricity Purchases by Pulp, Paper, and Paperboard Mills

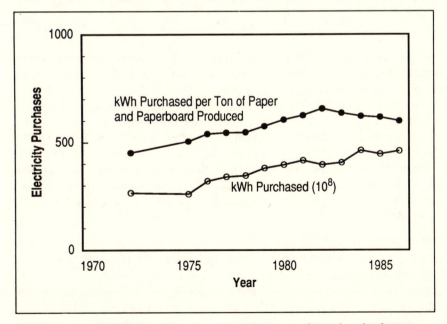

SOURCE: American Paper Institute, Inc., *U.S. Pulp, Paper and Paperboard Industry's Energy Use* (New York: 1987).

Figure 6.2   Fuel Purchases by Pulp, Paper, and Paperboard Mills

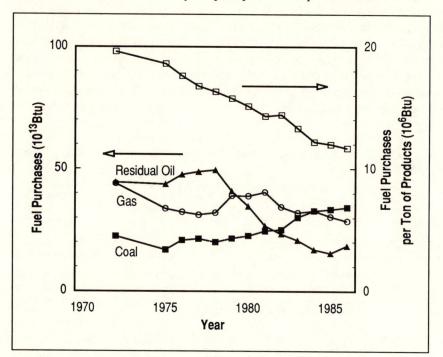

SOURCE: American Paper Institute, Inc., *U.S. Pulp, Paper and Paperboard Industry's Energy Use* (New York: 1987).

average, about one-half of the growth rate in purchased electricity can be attributed to growth in industry output and the other half to the increasing intensity of purchased electricity use.

The use of purchased fuels followed a pattern opposite that of electricity as shown in Figure 6.2. Purchased fuels per unit of output declined at a compound rate of 4 percent per year during the period from 1972 to 1986. The quantity of coal used declined during the 1960s and first half of the 1970s and began to increase in the mid-1970s to replace first oil and then gas. The shift away from the use of oil and gas and toward the use of coal and electricity is consistent with overall trends in the cost of energy (see Figure 6.3).

Self-Produced Energy

Because wood is a fuel, the energy requirements of the pulp and paper industry could, in principle, be furnished by the use of wood-derived fuels. In practice, however, about half of the current energy requirement of the pulp and paper industry is met through the purchase of fuels and electricity (see Table 6.1).

Figure 6.3 Purchased Energy Price Ratios, Electricity to Fuel, for the Pulp and Paper Industry

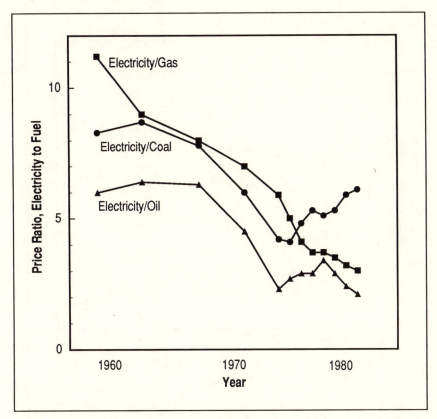

SOURCES: U.S. Bureau of the Census, *Annual Survey of Manufactures: Fuels and Electric Energy Consumed* (Washington, D.C.: GPO, various years).
NOTE: Electricity is measured at 3,412 Btu/kWh.

As for wood-derived fuels, the growth in total use of energy from process wastes per ton of pulp output in recent years has not been large. The increase that has occurred is due largely to growth in the use of hog fuel (see Figure 6.4). Hog fuel is ground wood that comes primarily from chipping wood residues (i.e., tree boles and branches) in the forest and transporting them to the paper mill for use as fuel. The wood wastes are used because it is economical to harvest some of the wood residues and to recover the chemicals in the spent liquor, and because disposal of the bark in other ways is usually more costly than using it as fuel. There is however no strong incentive to use fiber sources directly for fuel because they have been more costly than the purchase of fossil fuels (see Table 6.3).

Figure 6.4   Wood-Derived Energy per Ton of Wood Pulp Production

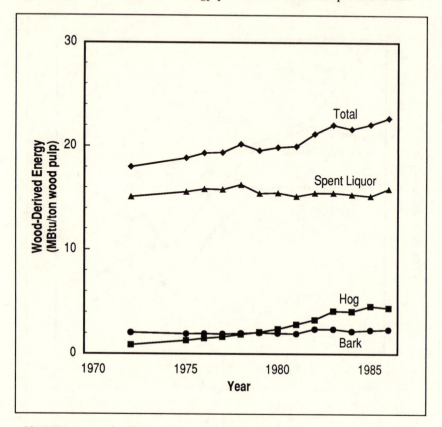

SOURCES: Energy data from American Paper Institute, Inc., *U.S. Pulp, Paper and Paper-board Industry's Energy Use* (New York: 1987). Wood pulp production from U.S. Depart-ment of Commerce, *Survey of Current Business*, various issues.

Table 6.3   Relative Costs of Materials as Fuels in Pulp, Paper, and Paper-board Mills (Dollars per Million Btu)

	1967	*1971*	*1977*	*1981*
Wood at 11 MBtu/cord	1.95	2.26	3.71	n.a.
Wastepaper at 17 MBtu/ton	1.77	2.22	3.95	n.a.
Coal at 26 MBtu/ton	0.30	0.44	1.15	1.66
Residual oil at 6.3 MBtu/bbl	0.36	0.55	2.06	4.70
Natural gas at 1.03 MBtu/10^3 ft^3	0.28	0.36	1.62	3.46

SOURCES: U.S. Bureau of the Census, *Census of Manufactures* (Washington, D.C.: GPO, various years).

Table 6.4 Electricity for Pulp, Paper, and Paperboard Mills, 1974–1986

	Electricity use, 10^9 kWh				Electricity use, 10^9 kWh		
Year	Purchased	Self-generated	Total	Year	Purchased	Self-generated	Total
1974	30.9	27.0	57.9	1981	42.2	26.4	68.6
1975	29.7	24.2	53.9	1982	41.1	27.1	68.2
1976	33.9	25.3	59.2	1983	41.5	32.0	73.5
1977	35.1	27.3	62.4	1984	42.8	32.0	74.8
1978	35.7	26.8	62.5	1985	41.7	32.0	73.7
1979	36.3	27.0	63.3	1986	43.1	n.a.	n.a.
1980	39.8	26.3	66.1				

SOURCES: For the years through 1981—U.S. Bureau of the Census, *Annual Survey of Manufactures: Fuels and Electric Energy Consumed* (Washington, D.C.: GPO, various years).
  1982–1985—U.S. Bureau of the Census, *Annual Survey of Manufactures: Statistics for Industry Groups and Industries* (Washington, D.C.: GPO, various years).
  1986—American Paper Institute, Inc., *U.S. Pulp, Paper, and Paperboard Industry's Energy Use* (New York: American Paper Institute, 1987).

Generation of electricity by the industry grew until the early 1970s but not apace with output. From the mid-1970s through the early 1980s, almost all new demand for electricity was met by purchases from utilities (Table 6.4). For example, between 1974 and 1981 electricity purchases increased 35 percent, but self-generation of electricity remained essentially unchanged. After 1981 the level of purchases stabilized, and such growth as occurred was supplied by self-generation. All this growth in self-generation occurred between 1981 and 1983, with both purchased and self-generation being stable through 1985.

Total Energy Use

Two key aspects of the aforementioned trends in energy use — energy intensity and electricity share — are summarized in Figure 6.5. Total energy used per unit of output (i.e., purchased and self-produced fuels and electricity per ton of paper and paperboard produced) declined approximately 9 percent between 1975 and 1986, while the electricity share increased 22 percent. The latter measure represents the fraction of all the energy used by the industry that is used in the form of electricity.

PETROLEUM REFINING AND ORGANIC CHEMICALS MANUFACTURING

Petroleum refining and organic chemicals manufacturing are considered together because the types of processing required and the technological trends in processing are similar for both industries.

Figure 6.5   Energy Intensity and Electricity Share; Pulp, Paper, and Paperboard Manufacturing

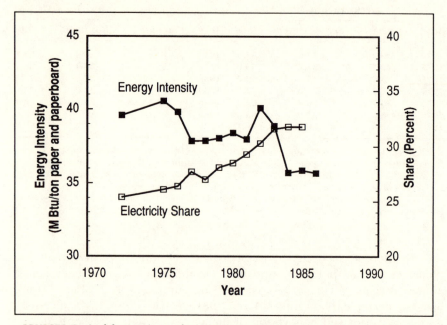

SOURCES: Derived from primary data in American Paper Institute, Inc., *U.S. Pulp, Paper, and Paperboard Industry's Energy Use* (New York: American Paper Institute, 1987). Self-generated electricity from U.S. Bureau of the Census, *Annual Survey of Manufactures: Fuels and Electric Energy Consumed* and *Statistics for Industry Groups and Industries* (Washington, D.C.: GPO, various years).

Petroleum Refining

Trends in Processing

Petroleum refining converts the mixture of organic molecules in crude oil primarily into gasoline for automobiles and distillate oil for jet and diesel engines and heating. Other products and by-products include lubricating oils and greases, asphalt felts and coatings, and petroleum residuum and coke. Various thermal processes are used to maximize the value of a market-determined set of products made from crude oils containing different mixtures of hydrocarbons and different levels of impurities.

Following the initial fractional distillation of crude oil, unwanted light distillates can be reformed into heavier, higher-value mixtures using temperatures and pressures that favor the desired mixtures. Alternately, unwanted heavy mixtures can at elevated temperatures be thermally "cracked" to yield a mixture of lighter and still heavier fractions. If cracked in the presence of hydrogen, the light fraction yield can be increased and certain impurities can be removed as hydrogen compounds (e.g., H_2S); without hydrogen, the heavy residuum can be sold as a low-grade fuel or pyrolyzed to coke. Finally, if conditions warrant, the residuum or coke can be gasified by its reaction with oxygen and steam to yield carbon monoxide and hydrogen gases useful for synthesizing higher value fuels or organic chemicals.

The amount of processing required and the choice of processes used depends on the quality of the crude oil and the mix and quality of products desired. Both have changed in major ways over time.

During the 1950s, motor gasoline quality improved substantially. Octane ratings increased about eight points, volatility increased, and sulfur content was halved.[10] These changes were all responses to the performance requirements of the internal combustion engine. After 1960, performance attributes of motor gasoline changed little until the late 1960s when environmental concerns forced refiners to reduce lead alkyl octane boosting, while maintaining octane ratings near previous levels. The lead content of leaded gasolines was halved, and unleaded gasolines made up half of all production by 1981.

Sulfur recovery by refiners increased dramatically after 1970 (Figure 6.6). Environmental requirements to reduce sulfur emissions in flue gases forced refiners to clean their own emissions and to put cleaner fuels on the market. The rising sulfur content of crude oils added to refiners' sulfur recovery burden. Sulfur recovery at petroleum refineries in 1985 was 0.7 long tons per thousand barrels of crude oil input, up from 0.25 long tons in 1971; recovered sulfur from refineries and natural gas plants accounted for about one half of the total domestic output of sulfur in all forms.

Besides the lead and sulfur restrictions already noted, other environmental restrictions on solid, liquid, and gaseous effluents forced refiners to clean up emissions. Rising contaminant levels in crude oils (e.g., heavy metals) forced additional decontamination.

The most complex of the changes in refinery output is in the mix, or slate, of products produced. The product slate constantly shifts in response to relative changes in demand for the various products and regulatory and other incentives. No major oil product has experienced more output changes over the last three decades than residual oil. Residual oil production decreased steadily until the early 1970s, then rose quickly to peak late in the decade before declining again (Figure 6.7).

Residual oils are the leftovers of petroleum refining; their production can be reduced by using lighter crude oils or by additional processing, which converts residual oils into higher-valued products. Added processing accounted for most of the pre-1970 decline. Then, during the 1970s, many simple refineries were built so the average level of processing declined and residual oil production began to increase. Refiners also were forced to resort to heavier crudes, further increasing residual oil production.

Figure 6.6   **Sulfur Recovery at U.S. Refineries and Natural Gas Plants,**
1966–1985

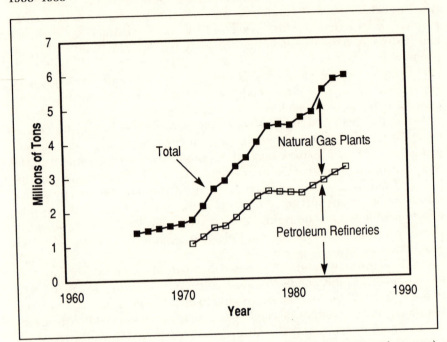

SOURCES: U.S. Bureau of Mines, *Minerals Yearbook* (Washington, D.C.: GPO, various years).

Since the late 1970s, refiners have emphasized cleaning up residual oils and converting them to other products. This upgrading was accomplished by more extensive and intensive application of technologies already widely used. From 1979 to 1984, while crude oil distillation capacity dropped 8 percent, capacity for upgrading residuals increased: vacuum distillation by 7 percent, hydrogen processing (hydrocracking, -refining, and -treating combined) by 12 percent, hydrogen production capacity by 26 percent, and coke production capacity by 47 percent.[11] Production of marketable coke rose from 1.4 percent of input crude in 1979 to 2.1 percent in 1983.[12] These statistics reflect conventional upgrading processes applied to more of the residuals and/or applied more intensively to accommodate higher contaminant levels. Vacuum distillation separates additional light products from the crude (atmospheric) distillation residuum; hydrogen processing improves the yield of lighter hydrocarbons and removes sulfur, metals, and other contaminants; and coking cracks the residuals into lighter oil and coke.

Estimates differ widely but generally indicate that the basic separation-distillation process for petroleum refining is less than 5 percent mechanical (calculated at 10,600 Btu/kWh for electrically produced mechanical energy).

Figure 6.7 Residual Fuel Oil Production as a Percentage of Crude Oil

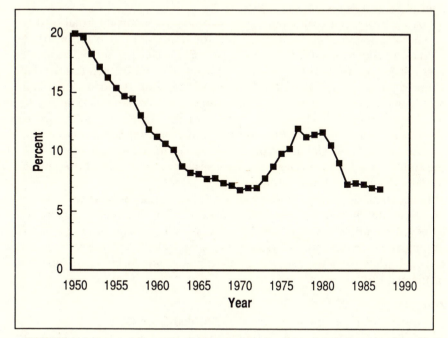

SOURCES: U.S. Energy Information Administration, *Petroleum Supply Annual* and predecessor publications (Washington, D.C.: GPO, various years).

However, energy use for distillation is growing as the need for distillation processing under vacuum of heavy feedstocks increases. Important transformation processes — catalytic cracking, reforming, and hydrogen processing — are more than 10 percent mechanical, and these operations are now used for an increasing fraction of crude input.[13] Hydrogen processing is more than 10 percent mechanical even when the hydrogen used is produced from natural gas. The amount of mechanical energy used is nearly doubled if hydrogen is produced by the steam reforming of residual oil. Catalytic cracking uses a high proportion of its energy in mechanical form, but it also produces waste refinery fuel because some of the feedstock forms coke that inhibits catalytic action and must be burned off of the catalyst. In recent years, some reactors use combustion air enriched in oxygen[14] to burn off the coke.

In this way the regenerative capacity of the converter is increased, catalyst regeneration and reactivity are improved, and a fraction of the coke removed from the catalyst is converted to carbon monoxide, a valuable synthesis and fuel gas, instead of carbon dioxide. This shift to the production of carbon monoxide reduces the amount of waste heat generated, and it has potential for steam production at the expense of extra mechanical energy for enriching air in oxygen.

The electromechanical production of oxygen is important in all of these processing trends and is responsible for a substantial fraction of the increasing requirements for mechanical energy. Oxygen is needed to convert residual oil to synthesis gas and to produce the hydrogen required for sulfur removal, as well as for the hydrogen processing of crude oil distillates and heavy fractions from crude oil reforming. The same is true for the production of organic chemicals. Oxygen for these purposes is, at least for the major companies, produced on site and is not reported separately as oxygen production. The quantities of oxygen required for petroleum refining and the production of organic chemicals are several times the amount of oxygen produced for sale to others.

To summarize, in nearly every case, the direction of change is toward higher quality products from lower quality feedstocks. Higher quality was achieved by additional or more severe processing: more cracking and coking, more reforming and alkylation, more hydrotreating, and more sulfur removal. All of these operations require a higher and increasing proportion of mechanical energy.

Energy Use and Price Trends

Per barrel, electricity use in refining doubled from the mid-1950s to the early 1980s. Purchased electricity supplied the entire increase, with refiners' self-generated electricity remaining nearly constant over the period. Thus, electricity purchases became a larger share of the total, and amounted to 5 of every 6 kWh in 1981. More recently per barrel use of both purchased and self-generated electricity has increased. Figure 6.8 confirms the long-term upward trend for purchased electricity and demonstrates that it has not been a smooth rise. Periods of stability alternate with periods of increase.

Fuel consumption per barrel (bbl) moved in the opposite direction (Figure 6.9). Again, the trend was not a smooth one; much of the decrease came between 1974 and 1977. Use of nonpurchased fuels (included within the total) remained near 350,000 Btu/bbl; as was the case with electricity, most of the change in fuels consumption occurred in the purchased component. In 1962, roughly 1 out of every 10 Btus of energy purchased was in the form of electricity; by 1981, 1 out of every 4 Btus purchased was electricity (counted at 10,600 Btu per kWh — the approximate number of Btus required to produce a kWh of electricity).[15] As in other cases examined in this study, the intensity of overall energy use relative to output declined in conjunction with the rise in electricity's share of total energy (Figure 6.10).

Energy costs, for both fuel and electricity, have greatly increased since 1970 and have constituted an increasing share of refining costs.[16] Figure 6.11 compares cost indices for all refinery operating costs with those for

Figure 6.8 Sources of Electricity in Petroleum Refining

SOURCES: U.S. Energy Information Administration, *Petroleum Supply Annual,* and predecessor publications (Washington, D.C.: GPO, various years), and U.S. Bureau of Census, *Census of Manufacturers* (Washington, D.C.: GPO, various years).

energy for the years 1981 and 1986. The rapid increase in energy costs relative to other refining costs provided strong incentive for refiners to conserve energy between 1970 and 1981. Although refinery fuel and power costs relative to other costs have declined somewhat since 1981, they are still a larger share of total refining costs than in 1970. Actual savings of energy per barrel are probably greater than suggested by Figure 6.9 because refiners are processing lower grade crudes than in earlier years and, as discussed previously, these require more processing.

The fact that electricity prices had been falling relative to fuels is less well known. In 1954, a Btu of electricity delivered to refiners (that is, at 3,412 Btu/kWh — the approximate number of Btus provided by 1 kWh of electricity) cost 14.5 times as much as a Btu of energy as fuel. By 1971, the ratio had fallen to 7.9 and, by 1981, was only 3.6 (Figure 6.12). Compared with refiners' fuel prices, electricity in 1981 was only one-fourth as expensive as in 1954. More recently low oil and gas prices have increased that ratio, but it is

Figure 6.9   Energy Use in Petroleum Refining

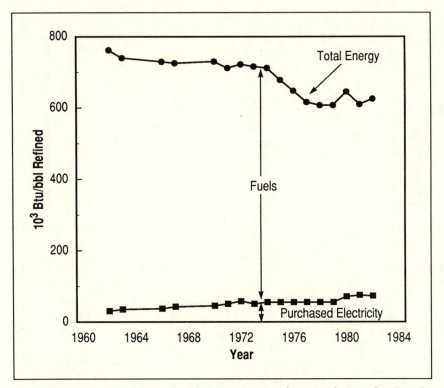

SOURCES: U.S. Energy Information Administration, *Petroleum Supply Annual,* and predecessor publications (Washington, D.C.: GPO, various years). Total includes electricity at 10,600 Btu/kWh.

NOTE: The fuels numbers include purchased fuels as well as still gas, petroleum coke, and other components of refiners' raw materials that are consumed as fuels. Also included are small amounts of purchased steam.

likely to turn downward again within five to ten years. Clearly, over the long term, the declining cost of electricity relative to fuels has supported the growing electricity share in refinery energy input trends.

Price-induced energy conservation may itself increase the fraction of energy used as electricity if the opportunities for conserving electricity are more limited than are the thermal conservation opportunities. In electricity use, only minor motor and pump efficiency improvements are possible; other conservation measures such as the recovery of mechanical energy via hydraulic turbines and turbo expanders were practiced to an extent before 1970, and variable frequency drive systems for motors are not yet widely used.[17] By contrast, thermal energy conservation options include the additional recovery by heat exchange, additional insulation, better control of

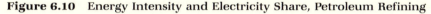

Figure 6.10 Energy Intensity and Electricity Share, Petroleum Refining

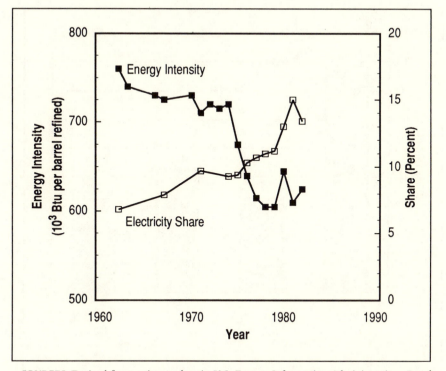

SOURCES: Derived from primary data in U.S. Energy Information Administration, *Petroleum Supply Annual,* and predecessor publications (Washington, D.C.: GPO, various years).
NOTE: Energy intensity includes purchased electricity counted at 10,600 Btu/kWh, purchased fuels, and still gas, petroleum coke, and other components of refiners' raw materials that are consumed as fuels. Also included are small amounts of purchased steam.

fired heaters, and numerous housekeeping options. The American Petroleum Institute reports a greater number of refiners' projects directed toward thermal energy conservation than toward mechanical energy conservation.[18]

Observed trends toward using less total energy while acquiring more energy in the form of purchased electricity are consistent with refining cost trends. Since 1970, energy costs increased substantially more than the other costs of refining; since 1954, fuel prices have risen four times as fast as electricity prices.

Outlook

The expanding use of electricity in petroleum refining and related industries is an evolutionary change and not attributable to any particular

Figure 6.11   Refinery (Current Dollar) Cost Indices (1970 = 100)

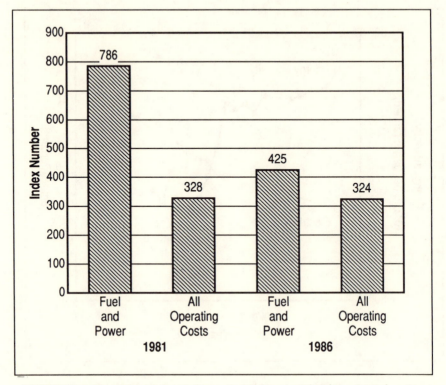

SOURCES: Based on Nelson-Farrar Refinery Cost Indices in *Oil and Gas Journal,* various issues.

process or processes. It appears to be driven by two major long-term trends. First, compared historically with the cost of fuels, electricity has been becoming a cheaper source of mechanical energy. Second, the work of refining is getting more difficult as poorer grade feedstocks are used to produce upgraded products. The extent and complexity of process operations have increased so that these operations now require a higher fraction of mechanical energy compared with thermal energy.

The technology and process improvements being adopted in petroleum refining are similar to those occurring in the production of organic chemicals. Indeed, in many instances and by means of the same general process operations, the production of organic chemicals picks up where petroleum refining leaves off in the further separation and conversion of refinery products into specific organic compounds. Organic chemicals manufacturing is explored in the next section.

Figure 6.12   Electricity-to-Fuel Price Ratios in Petroleum Refining

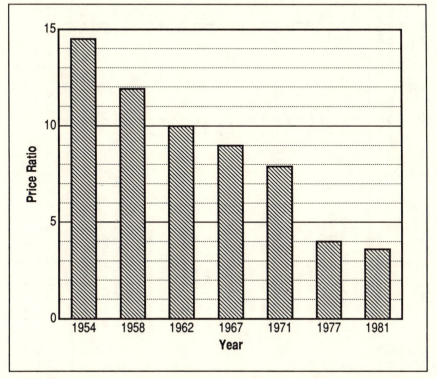

SOURCES: U.S. Bureau of the Census, *Census of Manufactures* (Washington, D.C.: GPO, various years).

Chemicals Manufacture

The chemicals industry is enormously varied and includes, in addition to the manufacture of specific chemical end products, the manufacture of semifinished products. For example, ethylene glycol is both a chemical raw material and used as antifreeze for car radiators, and plastics are semifinished products made primarily from organic chemicals. The industry is the largest of all manufacturing industries both in terms of tonnage of products and energy requirements. The top fifty large tonnage chemicals are listed in Table 6.5 along with production quantities for 1981 and annual production growth rates between 1971 and 1981.

Table 6.5 The Top Fifty Chemicals in 1981

#	Chemical	Production — Billion pounds, 1981	Production — Annual rate of change 1971–1981, percentage	#	Chemical	Production — Billion pounds, 1981	Production — Annual rate of change 1971–1981, percentage
1	Sulfuric acid (E)	81.35	3.4	26	Ethylene oxide (B)	5.11	3.6
2	Ammonia (C)	38.07	2.7	27	Hydrochloric acid (E)	4.89	1.5
3	Nitrogen (D)	37.31	11.2	28	Ammonium sulfate (C)	4.22	1.5
4	Lime (E)	35.99	−0.4	29	Ethylene glycol (B)	4.06	2.8
5	Oxygen (D)	34.93	2.9	30	p-Xylene (A)	3.70	8.3
6	Ethylene (A)	28.87	4.6	31	Cumene (A)	3.31	4.4
7	Sodium hydroxide (D)	21.30	1.0	32	Butadiene (A)	3.05	−0.7
8	Chlorine (D)	21.12	1.2	33	Carbon black (C)	2.73	−1.0
9	Phosphoric acid (E)	19.83	5.2	34	Acetic acid (B)	2.71	3.3
10	Nitric acid (C)	18.08	1.7	35	Phenol (B)	2.55	3.9
11	Ammonium nitrate (C)	17.58	2.8	36	Aluminum sulfate (E)	2.41	0.7
12	Sodium carbonate (E)	16.56	1.5	37	Sodium sulfate (E)	2.33	−1.5
13	Urea (C)	14.97	8.1	38	Acetone (B)	2.17	3.5
14	Propylene (A)	14.02	7.4	39	Acrylonitrile (B)	2.01	7.4
15	Toluene (A)	10.32	5.0	40	Vinyl acetate (B)	1.93	7.6
16	Benzene (A)	9.91	2.3	41	Calcium chloride (E)	1.83	−2.8
17	Ethylene dichloride (B)	9.17	2.0	42	Propylene oxide (B)	1.81	4.3
18	Methanol (C)	8.41	5.5	43	Cyclohexane (A)	1.75	0.1
19	Carbon dioxide (D)	7.96	11.5	44	Isopropyl alcohol (B)	1.64	−0.4
20	Ethylbenzene (A)	7.85	4.7	45	Titanium dioxide (E)	1.50	1.0
21	Vinyl chloride (B)	6.72	4.5	46	Sodium silicate (E)	1.48	1.5
22	Styrene (A)	6.61	3.5	47	Sodium tripolyphosphate (E)	1.37	−4.0
23	Xylene (A)	6.43	3.8	48	Acetic anhydride (B)	1.25	−1.9
24	Terephthalic acid (B)	6.35	10.6	49	Adipic acid (B)	1.21	−0.8
25	Formaldehyde (C)	5.86	2.6	50	Ethanol (B)	1.16	−3.4

SOURCE: *Chemical and Engineering News* (June 14, 1982), 33.

Category A: Eleven organic chemicals: hydrocarbons from petroleum refining and coal coking. *Category B:* Fifteen organic chemicals: oxidation products of hydrocarbons. *Category C:* Eight chemicals derived from methane. *Category D:* Five inorganic chemicals made electrolytically and electromechanically. *Category E:* Eleven inorganic chemicals derived in part from natural minerals and reaction with sulfuric acid.

Recent Trends in Relative Use and Cost of Fuels and Electricity

The use of purchased electricity for chemicals manufacturing increased by about 10 percent for the period 1974 through 1981, while the use of purchased fuels declined about 13 percent (Table 6.6). The use of coal remained essentially constant over the period; thus, oil and gas together suffered the entire decrease in fuel use, and electricity's share of purchased primary energy grew from 30 percent to 35 percent. The self-generation of electricity for production of chemicals declined 45 percent from 1974 to 1981, and by 1981 self-generation supplied only about 9 percent of the industry's electricity requirements (Table 6.7).

The recent changes in fuel and electricity use patterns are consistent with changes in their relative prices. As shown in Figure 6.13, the cost of energy delivered as electricity relative to the cost of energy contained in both oil and gas fell nearly by half from 1971 to 1981. The energy contained in coal deliveries in 1981 was less expensive compared to electricity than it was in 1971, which helps to explain the recent slight increase in the energy market share for coal in chemicals manufacturing.

Table 6.6 Fuel and Electric Energy Purchases for the Manufacture of Chemicals[a] in 1974 and 1981

Energy purchases	1974		1981		1981/1974 percentage change
	10^{12} Btu	Percentage	10^{12} Btu	Percentage	
Fuel	2,500	70	2,179	65	− 12.8
Electricity[b]	1,052	30	1,156	35	+ 9.9

SOURCES: U.S. Bureau of the Census, *Annual Survey of Manufactures: Fuel and Electric Energy Consumed* (Washington, D.C.: GPO, 1976 and 1983).
[a] Excludes energy for gaseous diffusion plant operations.
[b] At 10,600 Btu/kWh.

Table 6.7 Sources of Electricity Consumed for the Manufacture of Chemicals in 1974 and 1981

Source of electricity	1974		1981		1981/1974 percentage change
	10^9 kWh	Percentage	10^9 kWh	Percentage	
Purchased[a]	95.4	83.6	109.1	91.5	14.4
Self-generated	18.7	16.4	10.2	8.5	− 45.5

SOURCES: U.S. Bureau of the Census, *Annual Survey of Manufactures: Fuel and Electric Energy Consumed* (Washington D.C.: GPO, 1976 and 1983).
[a] Excludes electricity for gaseous diffusion plant operations.

Figure 6.13   Electricity-to-Fuel Price Ratios in Chemicals Manufacturing

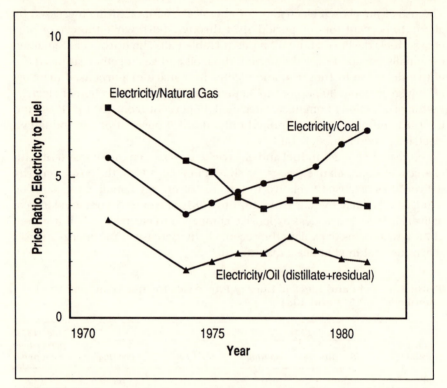

SOURCES: U.S. Bureau of the Census, *Annual Survey of Manufactures: Fuels and Electric Energy Consumed* (Washington, D.C.: GPO, various years).

Central Role of Organic Chemicals

The eight U.S. Bureau of the Census chemical categories are listed in Table 6.8 along with their purchases of fuels and electricity in 1974 and 1981. In addition to purchased energy, an amount of thermal energy equivalent to about 30 percent of fuel purchases was derived from process materials in the production of organic chemicals.

All eight categories are important economically, but the primary focus of the discussion that follows is on the production of the organic chemicals. The reason is that our purpose in this chapter is to understand the role of technological change and its tie to the use of electricity in transforming basic materials, for example, making paper not stationery, glass not table-ware, steel not galvanized roofing. Given this purpose, our treatment can be confined largely to dealing with the production of high-volume basic organic chemicals as opposed to those chemical manufacturing activities

Table 6.8 Energy Purchases for the Eight Census Categories of Chemical Manufacturing for 1974 and 1981, 10^{12} Btu

Description	1974			1981		
	Fuels	Electricity[a]	Electric fraction	Fuels	Electricity[a]	Electric fraction
Inorganic chemicals[b]	472.9	447.3	0.486	356.8	391.1	0.523
Plastics, resins, rubber and fiber	430.9	182.3	0.297	345.7	225.8	0.395
Drugs	8.6	35.2	0.375	61.0	49.5	0.448
Detergents and cosmetics	47.7	21.5	0.311	46.9	24.3	0.341
Paints and allied products	14.0	9.5	0.404	13.3	9.3	0.412
Organic chemicals	1,058.4	244.6	0.188	939.5	319.1	0.254
Agricultural chemicals	308.4	84.0	0.214	333.5	104.9	0.239
Miscellaneous products	108.9	27.2	0.200	81.9	31.8	0.280
Total	2,500.0	1,052.0	0.296	2,179.0	1,156.0	0.347

SOURCES: U.S. Bureau of the Census, *Annual Survey of Manufactures: Fuels and Electric Energy Consumed* (Washington, D.C.: GPO, 1976 and 1983).

[a] Electricity counted at 10,600 Btu/kWh.

[b] Excludes energy for gaseous diffusion plant operations.

NOTE: Totals have been rounded; electric fraction represents electricity percentage of "fuels" plus "electricity."

that are primarily engaged in converting these basic chemicals into finished products.

Thus, manufacturers of plastics and resins use organic chemicals such as ethylene as the starting point; detergent manufacturers begin with organic alcohols; paint manufacturers mix organic resins with pigments; most drugs are complex highly purified organic compounds (i.e., aspirin is acetyl salicylic acid), and so on. The discussion also includes agricultural chemicals. This is because the production of ammonia (the heart of the agricultural chemicals industry) begins with the production of hydrogen from natural gas. This process is identical to that used to produce hydrogen for the petroleum refining industry.

Twenty-six of the fifty largest tonnage chemicals are organic chemicals and another eight are made from methane (see Table 6.5). From 1974 to 1981 production of these twenty-six organic chemicals increased about 15 percent (from 52 million to 60 million tons), use of purchased electricity increased 32 percent (from 23 billion to 30 billion kWh), and consumption of purchased fuel decreased 11 percent. Growth in the use of electricity corresponds to an increase of about 60 kWh per ton (i.e., from 440 to 500 kWh per ton); the electric share of the total primary energy purchased increased from 19 percent to 25 percent. Along with the downstream products made from them, organic chemicals production requires about 80 percent of the energy used for chemicals manufacturing as a whole.

We omit discussion of the production of inorganic chemicals. One reason is that five (see Table 6.5, Category D) of the sixteen large-tonnage inorganic chemicals are already made electrically (i.e., alkalies and chlorine) or electromechanically (i.e., industrial gases[19]); thus, the issue of further electrification is relatively unimportant. A second reason is that in contrast to the field of organic chemicals and industrial gases, the production of several inorganic chemicals is stagnant or declining as is their use of fuels and electricity. Third, the eleven (see Table 6.5, Category E) large-tonnage inorganic chemicals not made electrically occur in nature or are made in part by reacting sulfuric acid with natural minerals. In this sense their production is more a process of mining or leaching of natural minerals rather than it is one of material transformations.

Industrial organic chemicals and the synthetic resins and plastic materials made from them are the fastest growing portion of the large-tonnage chemicals business. The new applications for plastics are apparent everywhere, from plastic pipe instead of steel for household plumbing to plastic for automobile parts, which grew from 33 pounds per car in 1965 to 168 pounds per car in 1976 and is estimated to exceed 300 pounds per car in 1990. The chemical industry itself is using plastics to replace metals in pumps, valves, gears, pipes, tanks, ducts, and even chemical reactors. Continued growth seems assured as new products find application: electrically conductive plastics for electromagnetic shielding and battery electrodes, and plastic composites with inorganic fibers for high-temperature, high-pressure tanks and vessels.[20]

The production of organic chemicals and synthetic materials from them is a multistep process analogous in large degree to petroleum refinery operations. It often begins with crude oil fractions that are first reformed to produce a more desirable mix of chemicals. These are then separated into relatively pure chemicals, which are in turn polymerized or undergo further reactions to yield specific compounds or materials. Thus ethane from petroleum or natural gas refining operations is reformed to yield a mix of chemicals high in ethylene content; the ethylene is separated from the mix and subsequently partially oxidized to yield ethylene glycol for use in automobile radiators or polymerized to yield polyethylene plastic sheeting for use in sandwich bags.

This sequence — reforming, separating, and partial oxidation or polymerization into final products — generally typifies organic chemicals manufacturing for the vast majority of products. More or fewer steps may be required in any particular instance. For example, cyclohexane can come directly from refining or be produced by the hydrogenation of benzene from coal coking operations, and ethylbenzene can come directly from refining or be produced by combining ethylene and benzene. The most important chemicals from refining and reforming are ethylene for aliphatic (straight chain) organic compounds and benzene for cyclic (ring) organic compounds.

A large fraction of organic chemicals production is used to make synthetic materials. For the 1974 to 1981 period, production increased from 35 to 44 million tons compared to 52 to 60 million tons for the total of organic chemicals. This set of products drives the large-tonnage chemicals business and to a lesser extent offsets growth in other basic materials industries, such as steel, glass, and paper. The production of these semifinished products depends on mechanical operations (e.g., spinning fiber, extruding plastic sheet) to a larger extent than does production of the basic chemicals. Thus, it is not surprising that electrical usage is high per ton of product (i.e., nearly 1,000 kWh per ton).

When an allowance is made for the production of the precursor chemicals (including petroleum refining) used to manufacture these synthetic materials, electricity consumption totals in the neighborhood of 1,500 kWh per ton of product. The 1,500 kWh per ton requirement is substantially higher than for most other basic manufactures used in society. For example, the production of glass, steel, and paper requires electricity in amounts ranging from 400 to 600 kWh per ton.

NEW APPLICATIONS IN PETROLEUM REFINING AND ORGANIC CHEMICALS PRODUCTION

We have seen that processing trends are similar in petroleum refining and organic chemicals production. To reiterate, the use of electricity is expanding, due partly to relative price shifts. Even more important from a technological viewpoint, the use of poorer grade feedstocks to make upgraded products requires a greater degree of processing, which has shifted the type of energy needed from thermal to mechanical. The growing share of mechanical energy goes primarily for pressurizing gas streams as necessary for operating the process reactors and for separating the resulting mix into salable products.

Concurrently these industries have been successful in reducing their use of fuel per unit of output, which means that less surplus thermal energy is available for meeting process demands for mechanical energy either by steam turbine drive or by self-generation of electricity. Thus, much of the increased use of electricity goes to power motors required to effect the higher degree of processing that has become economic. There is little evidence of widespread conversion of existing prime movers to electric motors;[21] rather, as new needs for mechanical energy arise, motors are used and the electricity needed for power is purchased. For example, motor drive is now commonplace on new regenerator blowers for catalytic crackers, an application formerly dominated by steam turbines.[22] The growing trend to efficient heat integration in industry reduces the demand for low-pressure steam exhausting from steam turbines, and thus increasingly favors the economics of using electric motors and purchased electricity.

Using more electricity to provide mechanical energy is, however, not the whole story. Electric energy is also being employed in applications formerly performed with fuel-derived thermal energy. The technologies to be discussed are not specific to any particular feedstock, chemical, or process. Rather they are in the nature of unit operations or unit processes that can be applied not only in the petroleum refining and organic chemicals industries, from which the first set of examples are taken, but also where similar operations arise in other industries.

Niche Technologies

These applications are small in scale when compared with the magnitude of the total thermal energy requirements for process manufacturing, but they illustrate special niches where the unique characteristics of electricity offset its high cost as a source of low-temperature heat. A case in point is the use of electrical spark ignition to replace the pilot light on flare stacks.[23] In this application, the electric technique virtually eliminates the use of fuel.

Another niche application is the electric regeneration of activated charcoal as used widely in manufacturing and elsewhere to absorb impurities. When no longer effective, the inactive carbon is sometimes discarded or sometimes regenerated in gas-fired furnaces operated by businesses offering that service. In either case, carbon losses in discard, handling, and combustion during regeneration can exceed 30 percent. Electric furnaces are now available for on-site carbon regeneration at reported operating costs one-tenth that formerly experienced; overall carbon losses are less than 5 percent per cycle.[24] Individual installations have a minuscule impact, but the many applications are measurable in the aggregate. This furnace technology typifies similar development of small, high-temperature electric furnaces with a range of applications such as the on-site destruction of toxic wastes.[25]

A final example having very broad application throughout process manufacturing is the use of electric heating tape to replace steam tracing in the prevention of pipeline freeze-up. Electric heating tapes are available to provide a specific temperature. This contrasts with steam tracing, which delivers heat at a fixed temperature according to the pressure of the tracing system (e.g., 150 psi steam has a temperature of 358°F). The use of a steam temperature higher than needed is wasteful of energy and in some cases can degrade the liquids being handled.

Electric heating tapes are now available that are self-regulating in temperature. If the temperature rises, the electrical resistance of the tape increases to reduce current flow.[26] Since line tracing is often a back-up system needed only in the case of unplanned interruptions in pipeline flow, the electric tape provides heat automatically and uses electricity only if needed.

Electric line tracing eliminates major problems with steam tracing.[27] If a steam tracer leaks, the steam condenses in the pipe insulating material, causing it to lose its effectiveness and incurring expense for its repair. Also, at the end of each steam tracer is a steam trap — a device that collects and intermittently discharges steam condensate into the condensate system. Steam traps are prone to failure, either by leaking steam or by failing to discharge condensate. Steam leakage into the condensate system simply wastes energy, but failure to deliver condensate can result in trap and line freeze-up and interrupt operations until pipeline flow can be restored. To avoid the consequences of trap failure in extreme weather, it is common practice to bleed steam to the atmosphere at each trap. Under these conditions, the heat in the bled steam is wasted, operations must be performed in a cloud of condensing steam released from the bypassed traps, and the condensed steam freezes on handrails and walkways, creating an unsafe working environment.

Electromechanical Technologies

The use of mechanically energized processes instead of thermal techniques to separate various substances from one another is a much more important development in the growing use of electricity in low-temperature processing. There are many examples, such as the use of centrifuges instead of thermally enhanced settling ponds to hasten the separation of sluggish mixtures, and the use of vacuum pumps to replace inefficient steam ejector systems.[28] These are conventional technologies that are finding increasing application.

Separating Mixtures

Two new mechanically driven technologies are finding widespread use in separating gas mixtures in refining and chemical processing: membrane separation and pressure swing adsorption (PSA). Membrane separation depends on differential permeation rates through the membrane material. PSA involves selective adsorption by molecular sieves at high pressure, with the adsorbed gas being released later at low pressure. Applications include air separation, acid gas removal, and hydrogen separation, purification, and recovery. The major significance of these new technologies is that they reduce the cost of many separation tasks and so will open up new markets.

Some of the most important and far-reaching effects of mechanical processing now emerging come from the large-scale application of semipermeable membranes. Packaged in the form of a shell and tube bundle of thousands of hollow polysulfane fibers, these devices substantially improve the efficiency of separating gas mixtures. Current applications include the recovery of hydrogen from refining waste streams and from purge gas in the production of ammonia; the separation of acid gas (carbon monoxide and hydrogen sulfide) and carbon dioxide (for tertiary oil recovery) from natural gas; and the enrichment of air in nitrogen or oxygen.[29]

The possibility of impregnating the semipermeable membranes with specific materials greatly expands the potential applications for the technology. The Amoco Research Center has obtained good separation of ethylene or propylene from methane streams with cellulose acetate fibers impregnated with silver halide.[30] A joint venture between Solvay and Cie (Brussels) and SNIA Fibre (Milan) has been working with membranes containing enzyme-loaded organisms. This concept allows for simultaneous product reaction and separation.[31]

Mechanical energy for vapor recompression is also displacing thermal energy in separating liquids whose components have close boiling points.[32] Butane splitting is an example. In the conventional process to separate normal butane from isobutane, heat is applied to the liquid in the bottom of the distillation tower and removed by condensing the vapors exiting the top. Butane splitters use high reflux ratios, so condenser losses per unit of product are quite high.

The vapor recompression alternative works as follows. Mechanical compression raises the temperature of overhead vapors, which are then cooled in the reboiler where the heat of vaporization, normally lost in a condenser, is transferred to the bottom of the tower for reuse. Most of the overhead product is then used as reflux as it is in the conventional process.

Another mechanically driven separation process is freeze concentration. Whereas evaporation and distillation achieve separation through the application of thermal energy, freeze concentration achieves separation by cooling until some component(s) crystallize. The crystallized material can then be separated from the remaining liquid. Frequently, separation by freeze concentration using electric motors consumes substantially less energy than thermal separation processes. Processing of black liquor in the pulp and paper industry and benzene-toluene-xylene fractionation in the petrochemical industry are potential large volume applications for electricity-based freeze concentration technology. Many smaller volume applications, such as milk and whey separation in the dairy industry, also exist.

New Uses for Oxygen

The electromechanical separation of oxygen from air is creating a chain of new applications. One example is the use of oxygen instead of air to oxidize (burn) or partially oxidize carbonaceous materials and thereby provide high-temperature heat internal to the process reactor. This technique represents a major improvement in process technology with broad present and future implications as the use of this oxygen helps minimize the use of fuel in thermal processing operations. The increasing uses for oxygen were referred to earlier, and that discussion is expanded here with examples of new, highly productive, applications.

If heated to a high temperature, carbonaceous materials of all sorts can be gasified by their reaction with steam (steam reforming). Under such conditions, the oxygen in the water vapor shifts to carbon atoms to produce carbon monoxide (partial oxidation), which is a desirable fuel gas or a useful chemical reactant. The hydrogen from the water vapor can later react with heavy hydrocarbons to yield light hydrocarbons or react with impurities in the hydrocarbons, such as sulfur, so they can be removed from the product stream.

In conventional steam reforming processes, the carbonaceous materials to be reformed are first heated in a variety of ways by burning other fossil fuels or a portion of the reactor products in air. Much of the heat from fuel combustion is carried away by the nitrogen in the combustion air. The process is limited in temperature because of operating problems associated with handling hot materials external to the reactor or because the carbonaceous material heats up and cools down as it is alternately blown with air and steam. Thus, the process is relatively inefficient, capital costs are relatively high and, of course, a large volume of flue gas is released.

In steam reforming with oxygen, oxygen reacts along with steam and carbonaceous material within the reactor. Some of the carbonaceous material is partially oxidized, with the heat of combustion being released inside the reactor and providing the thermal energy needed for steam reforming. No fuel is burned external to the reactor, less process equipment is required, and no flue gas need be created. Because the temperature is less restricted by materials handling problems or intermittent operations, reactor temperatures are as high as 2,000°C and reactions take place rapidly. Moreover, reactor time, temperature, and pressure can be optimized to enhance the yield of desired products. Because of these advantages, the thermal energy losses in processing can be reduced by as much as 90 percent.

Processing with oxygen is applicable to all carbonaceous materials. Current applications are primarily in the production of the synthesis gases, hydrogen and carbon monoxide, from low-grade fuels such as petroleum residuum, asphalt, petroleum coke, and coal instead of high-value materials such as methane and naphtha. Hydrogen and carbon monoxide are the basic starting materials for the production of a large number of high-value organic chemicals and fuels, which suggests the future importance of the technique. Such a process is used in South Africa to produce gasoline from coal.

The advantages of the process are exemplified by "crude-oil cracking" technology to produce ethylene, announced jointly by Union Carbide Corporation and Kureka Chemical Industry, Limited. Deasphalted crude oil is partially oxidized in the presence of steam at 2,000°C. The advantages of the high-temperature conversion are that the yield of ethylene and coproducts is increased from about 40 percent to about 65 percent, and that residence time in the cracker is reduced from 75 milliseconds to 17 milliseconds compared to conventional steam reforming with external combustion fuels in air. Other applications include Exxon's oxo-alcohol plant in Baton Rouge, Louisiana, a joint venture between DuPont and U.S. Industrial Chemicals Company in Deer Park, Texas, and the Tennessee Eastman plant in Kingsport, Tennessee, for the production of organic chemicals including methanol, acetic acid, and vinyl acetate. The incentive is to use heavy residuum from petroleum refining and coal for the production of high-value chemicals rather than using more expensive natural gas.[33] Elsewhere in the world, partially oxidized asphalt and residuum are being used instead of high-cost naphtha in the production of hydrogen for ammonia synthesis.[34]

SHIFTS FROM THERMAL TO MECHANICAL PROCESSING

The petroleum refining and organic chemicals industries, and the pulp and paper industry, illustrate an overall trend toward the use of mechanical processes driven by electricity to supplement or replace many low-temperature processes that rely on the direct combustion of fuels. Electromechanical processes have enabled them to turn low-grade raw materials into high-quality products, and, in pulping operations, to reduce damaging environmental impacts.

Applications cover a broad range. They run the gamut from changing one step in a series of steps (ammonia production) to improving an entire process (wood pulping). In addition, mechanically produced oxygen is now being used for a great many purposes in manufacturing.

Taken together, these multiple shifts from thermal toward mechanical processing add significantly to the growth of industrial electrification.

Notes

1. The softwoods (spruce, pine, and fir) are preferred for both lumber and pulpwood because the natural fiber length of softwoods is about twice as long as that of hardwoods. The long fiber adds strength to lumber used for building construction and to pulp used for paper products. Reflecting this quality, softwood cordage is valued about 20 percent higher than hardwood.

As the softwood resource becomes increasingly constrained, sawmill wastes (sawdust and mill shavings) are being incorporated into construction products, such as chipboard and flakeboard, substitutes for plywood in such applications as floor underlayment, and wall sheeting. Faced with growing competition from the lumber industry for both timber and the mill wastes from converting timber into lumber, the pulp and paper industry has increased the fraction of hardwood used from 7 percent to 25 percent since 1940.

2. R. D. McKelvey, V. J. Van Drunen, and G. A. Nicholls, "Oxygen Pulping of Hardwoods and Softwoods Under Oxygen-Rich Conditions," *TAPPI* 61, no. 12 (December 1978): 40–42; J. S. Fujii and M. A. Hannah, "Oxygen Pulping of Hardwoods," *TAPPI* 61, no. 8 (August 1978): 37–40; K. Hata and M. Sogo, "Oxygen-Alkali Semichemical Pulping of Wood Chips," *TAPPI* 58, no. 2 (February 1975): 72–75; H. Chang, J. S. Gretzl, and W. T. McKean, "Delignification of High-Yield Pulps with Oxygen and Alkali," *TAPPI* 57, no. 5 (May 1974); J. L. Minor and N. Sanyer, "Factors Influencing the Properties of Oxygen Pulps from Softwood Chips," *TAPPI* 57, no. 5 (May 1974): 120–126.

3. R. Marton, A. Brown, and S. Granzoe, "Oxygen Pulping of Thermomechanical Fiber," *TAPPI* 58, no. 2 (February 1975): 64–67.

4. A. R. Jones, "How Bleaching Hardwood Kraft Pulp with Oxygen Affects the Environment," *TAPPI* 66, no. 12 (December 1983): 42–43; G. Forsum, B. Lindquist, and L. E. Persson, "Full Bleaching of Kraft Pulps Delignified to Low Kappa Number by Oxygen Bleaching," *TAPPI* 66, no. 12 (December 1983): 60–62; R. P. Singh, "Ozone Replaces Chlorine in the First Bleaching Stage," *TAPPI* 65, no. 2 (February 1982): 45–48; "Pulp-Bleaching," *TAPPI* 62, no. 12 (December 1979): 9; W. H. Rapson, "Pulp Bleaching as of Today," *TAPPI* 62, no. 6 (June 1979): 14–17; L. Almberg, I. Croon, and A. G. Jamieson, "Oxygen Delignification as Part of Future Mill Systems," *TAPPI* 62, no. 6 (June 1979): 33–35; S. W. Eachus, "Atmospheric-Pressure Oxygen Bleaching," *TAPPI* 58, no. 9 (September 1975): 151–154; S. Rothenberg, D. H. Robinson, and D. K. Johnsonbaugh, "Bleaching of Oxygen Pulps with Ozone," *TAPPI* 58, no. 8 (August 1975): 182–185; C. J. Myburgh, "Operation of the Enstra Oxygen Bleaching Plant," and A. G. Jamieson and L. A. Smedman, "Mill-Scale Application of Oxygen Bleaching in Scandinavia," *TAPPI* 57, no. 5 (May 1974): 131–136; H. Makkonen, M. Pitkanen, and T. Laxen, "Oxygen Bleaching as a Critical Link Between Chemical Fiberization and Fully Bleached Sulfite Pulp," *TAPPI* 57, no. 2 (February 1974): 113–116.

5. S. I. Kaplan, *Energy Use and Distribution in the Pulp, Paper, and Boardmaking Industry,* ORNL/TM-5884 (Oak Ridge, TN: Oak Ridge National Laboratory, 1977), 51.

6. S. T. Han, "The Dryer Section of the Future," *TAPPI* 60, no. 7 (July 1977): 14–15.

7. L. Gottsching, "New Developments in Papermaking in Western Europe," *TAPPI* 63, no. 3 (March 1980): 50–54.

8. U.S. Bureau of the Census, *1982 Census of Manufactures: Fuels and Electric Energy Consumed,* MC82-S-4 (Washington, D.C.: GPO, 1983).

9. U.S. Bureau of Economic Analysis, *Business Statistics 1981* (Washington, D.C.: GPO, annual).

10. Ella Mae Shelton, M. L. Whisman, and Paul W. Woodward, "Trends in Motor Gasolines: 1942–1981," DOE/BETC/RI-82/4 (Bartlesville, OK: Bartlesville Energy Technology Center, June 1982).

11. *Oil and Gas Journal,* Annual Refining Issue (usually no. 12), various years.

12. U.S. Energy Information Administration, *Petroleum Supply Annual* and predecessor publications (Washington, D.C.: GPO, various years).

13. C. E. Jahnig, "Petroleum (Refinery Processes, Survey)," in *Kirk-Othmer Encyclopedia of Chemical Technology,* 3rd ed. (New York: John Wiley and Sons, 1982) 17: 183–256; V. O. Haynes, *Energy Use in Petroleum Refineries,* ORNL/TM-5433 (Oak Ridge, TN: Oak Ridge National Laboratory, September 1976).

14. Frank Macerato and Sidney Anderson, "O_2 Enrichment Can Step Up FCC Output," *Oil and Gas Journal* (March 2, 1981), 101–106; Frank J. Elvin and Ray Milne, "FCC Heat Balance Key to High O_2 Enrichment," *Oil and Gas Journal* (February 28, 1983), 84–86; T. S. Hansen, "Study Shows Oxygen/Resid Relationships in FCC Operations," *Oil and Gas Journal* (August 15, 1983), 47–52.

15. U.S. Bureau of the Census, *Census of Manufactures,* various years.

16. A distinction is made between a price index and a cost index; the latter acknowledges productivity changes. See W. L. Nelson, "Here's How Operating Cost Indexes Are Computed," *Oil and Gas Journal* (January 10, 1977), 86–88.

17. S. S. Braun, "Recovery of Mechanical Energy from Refinery Process Streams," in *Proceedings of the Refinery Division,* American Petroleum Institute (Washington, D.C.: API, 1973), 335–346; also, A. P. Krueding, "Cat Cracker Power Recovery Techniques," *Chemical Engineering Progress* (October 1975), 56–61; F. D. Fishel and C. D. Howe, "Cut Energy Costs With Variable Frequency Motor Drives," *Hydrocarbon Processing* (September 1979), 231–236.

18. American Petroleum Institute, *Energy Efficiency Improvement and Recovered Material Utilization Report* (Washington, D.C.: API, annual). For instance, the 1982 report lists only five firms undertaking pressure recovery projects and seventeen firms upgrading pump and compressor efficiency. By contrast, there were forty-one firms with insulation projects, thirty-five adding heat exchangers, and seventy-seven making efficiency improvements for boilers and process heaters.

19. Oxygen is an important chemical product for our purposes not in the sense of process changes that affect its production, but because it is being used widely to improve other manufacturing processes (e.g., lime and cement calcining, copper smelting). As such its use represents an indirect substitution of electricity for fuels. Oxygen produced for sale other than for making steel grew 47 percent from 1974 to 1981.

20. "Why Polymers Have Become Red Hot," *Business Week* (February 18, 1985), 138a–138c; Jayadev Chowdhury, "Growth of Composites is Bolstered by New Fibers," *Chemical Engineering* (August 20, 1984), 37–41; "Chementator," *Chemical Engineering* (June 25 and May 28, 1984), 19; Gordon M. Graff, "Engineering Plastics: Primed for Key CPI Role," *Chemical Engineering* (August 23, 1982), 42–45; J. Albert Ralston, "Fiberglass Composite Materials and Fabrication Processes," *Chemical Engineering* (January 28, 1980), 96–98; "Plastics Move to Make a Bigger Dent in Cars," *Chemical Engineering* (July 4, 1977), 7–9.

21. One such conversion is cited by G. Alan Petzet, "U.S. Refiners Still Seeking Ways to Use Energy More Efficiently," *Oil and Gas Journal* (March 15, 1982), 27–31.

22. "Refiners Push FCCs with New Catalysts, Techniques," *Oil and Gas Journal,* NPRA, Q&A-1 (February 8, 1982): 100; S. S. Braun, "Recovery of Mechanical Energy From Refinery Process Streams," in *Proceedings of the Refinery Division,* American Petroleum Institute (Washington, D.C.: API, 1973), 335–346; also, A. P. Krueding, "Cat Cracker Power Recovery Techniques," *Chemical Engineering Progress* (October 1975), 56–61.

23. Petzet, "U.S. Refiners;" Charles G. Guffey, "New Systems Save Gas Plant Energy" *Oil and Gas Journal* (August 3, 1981), 117–120.

24. "Furnace Economically Regenerates Spent Carbon," *Chemical Engineering* (February 6, 1984), 35–36.

25. "New Ways to Destroy PCBs," *Chemical Engineering* (August 10, 1981), 37–38.

26. Eugene Fisch, "Winterizing Process Plants," *Chemical Engineering* (August 20, 1984), 128–143.

27. Stafford J. Vallery, "Steam Traps — The Quiet Thief In Our Plants," *Chemical Engineering* (February 9, 1981), 84–86; Joseph T. Lonsdale and Jerry E. Mundy, "Estimating Pipe Heat-Tracing Costs," *Chemical Engineering* (November 29, 1982), 89–93. According to W. F. Lair, a construction manager for Procter and Gamble Manufacturing Company, Procter and Gamble now specifies electric tracing for all process lines and storage tanks. This has led occasionally to the elimination of the need for steam altogether (and the boiler house to provide it) at manufacturing sites (personal conversation with C. C. Burwell).

28. C. Churchman and J. Martinez, "Economic Energy Savings in New Phosphoric Acid Plants," *Phosphorus and Potassium*, no. 117, (January–February 1982): 33.

29. Allan M. Watson, "Use Pressure Swing Adsorption for Lowest Cost Hydrogen," *Hydrogen Processing* (March 1983), 91–95; Valdis Berzins and Niels R. Udengaard, "Revamp Your Hydrogen Plant," *Hydrocarbon Processing* (May 1983), 65–67; W. J. Schell, C. D. Houston, and W. L. Hopper, "Membranes Can Efficiently Separate CO_2 From Mixtures," *Oil and Gas Journal* (August 15, 1983), 52–56; B. C. Price and F. L. Gregg, "CO_2/EOR: From Source to Resource," *Oil and Gas Journal* (August 22, 1983), 116–122; Ronald Schendel, "Process Can Efficiently Treat Gases Associated with CO_2 Miscible Food," *Oil and Gas Journal* (July 18, 1983), 82–86; W. J. Schell and C. D. Houston, "Process Gas With Selective Membranes," *Hydrocarbon Processing* (September 1982), 249–252; Gerald Parkinson, "Membranes Widen Roles in Gas Separations," *Chemical Engineering* (April 16, 1984), 14–19; Mark D. Rosenzweig, "Unique Membrane System Spurs Gas Separations," *Chemical Engineering* (November 30, 1981), 62–66.

30. "Chementator," *Chemical Engineering* (April 20, 1981), 20.

31. "Chementator," *Chemical Engineering* (October 3, 1983), 17.

32. J. Barnwell and C. P. Morris, "Heat Pump Cuts Energy Use," *Hydrocarbon Processing* (July 1982), 117–119; Charles Drake and Jack Jameyson, "Gas-Liquid Fractionation Plant is Fuel-Efficient," *Oil and Gas Journal* (May 31, 1982), 133–137; Charles G. Guffey, "New Systems Save Gas Plant Energy," *Oil and Gas Journal* (August 3, 1981), 117–120.

33. John C. Davis, "Crude-Oil Cracking Gains," *Chemical Engineering* (June 6, 1977), 78–79; "Chementator," *Chemical Engineering* (May 9, 1977), 85–86.

34. Guy E. Weismantel and Larry J. Ricci, "Partial Oxidation in Comeback," *Chemical Engineering* (October 8, 1979), 57–59.

7 | *Emerging Electrotechnologies: A Range of Innovations*

Calvin C. Burwell

Various new technologies based on electricity are now emerging. Like those examined in the preceding chapters, they rely on the unique advantages of electric energy in production operations. This chapter presents a sampling of such innovations. Some operate only on a small scale, filling niches in the production process, while others have the potential for major impacts on the very nature of manufacturing operations.

THERMAL PROCESSING OF MATERIALS

We have already considered applications for electrothermal processing to materials like glass and steel, where the high-temperature potential of electricity could be achieved just by imposing a resistance to its flow. Because of its relative simplicity, the technology for such applications was developed very early and was easily adaptable to conductive materials. For high-temperature processing it was quite efficient.

Electric resistance heating is not suitable, however, for direct heating of nonconducting materials. If used external to the material being processed, it loses much of the energy efficiency and high-temperature advantages it has relative to heating with fuels. So the energy for thermal processing of nonconducting materials has until recently been provided almost exclusively by fuel combustion. Now, however, some new electrical techniques, based on unique properties of electricity, are being introduced.

Heating with Electromagnetic Waves

One emerging electrothermal technology is the use of electromagnetic waves for heating. With this technique (called dielectric, radio frequency, or microwave heating), electromagnetic waves and the energy they contain are absorbed volumetrically in the material to be heated. The wave length generated can be designed to match the absorption characteristics and thickness (volume) of the material being heated so that heating takes place efficiently. The cost of owning and maintaining equipment for wave generation is high relative to other conventional heating techniques. Thus the technique is not used unless its unique heating characteristics substantially improve the overall productivity of the process.

This technique is suitable for a broad range of temperatures but since other techniques are competitive for very high-temperature processing, the major emerging applications are for the rapid processing of nonconducting materials that cannot withstand high temperatures without damage, such as food, paper, and rubber. A few examples (in addition to the use of RF heating for paper drying mentioned in Chapter 6) serve to illustrate the scope of application for heating materials with electromagnetic wave energy. What matters most in these applications are the gains in through-put, product quality, and waste reduction achieved through the use of electromagnetic waves.

Microwave heating in a vacuum is being used to remove a small amount of moisture from soybeans which, in addition to drying the grain uniformly, "causes unwanted bean hulls to be released more quickly and completely and without overheating, than with conventional forced hot air systems." In contrast, conventional hot air heating "yields a hot, nonuniformly dried product that must be held in tempering bins (where it unavoidably cools) for five to ten days to equalize moisture content" prior to being reheated for further processing.[1]

Other food processing applications include proofing and frying dough-nuts, meat thawing and tempering, and pasta drying.[2] (Proofing is the proc-ess of warming the yeast so that the dough will rise.) Conventional air-heated proofers are hard to keep clean, maintain, and control and they require thirty to forty minutes for proofing compared to four minutes in the microwave ovens. In tempering, meat is warmed from a hard frozen state to a temperature just below freezing where it can be further proc-essed more easily and then refrozen without harm. With conventional tem-pering, temperature control is difficult, and the time requirement can exceed eight hours for a sixty pound block of beef and can result in as much as a 15 percent loss of meat juices. Microwave tempering cuts the time to five minutes — thus minimizing drip loss and bacterial contamination. Con-ventional pasta drying requires five hours in moist warm air to avoid skin hardening and later "checking" (cracking) as the trapped moisture breaks through the hard surface. Microwave units dry pasta evenly in thirty min-utes under conditions that keep bacteria counts low.

Microwave energy is also being used to supply the heat for rubber vulcan-ization. Heat transfer in rubber is a slow process with conventional surface heating despite the use of specially designed machinery to aid in heat trans-fer. The volumetric heating available with microwave technology greatly speeds up the process, thereby enhancing productivity and reducing wastes. Firestone Tire and Rubber Company now uses microwave heat for heavy thick sections of rubber, and General Tire and Rubber Company is using microwave heating for making large offroad tires.[3]

The Goodyear Company is using a microwave-based technique for rubber devulcanization as well.[4] The proper use of microwave energy can sever the carbon-sulfur and sulfur-sulfur bonds without depolymerizing the rubber (i.e., breaking the carbon-carbon bonds) as occurs with thermal or chemical devulcanizing processes. With this process, up to 25 percent of devulcanized material can be substituted for new rubber, compared to five percent to ten percent with the other techniques. Other advantages of reusing rubber devulcanized with microwave energy include reduced materials costs and compounding time, since filler and oil are already dispersed in the recycled feed.

Heating with Plasma

This technology takes advantage of the unique property of certain gases to ionize (thus becoming conductive) at the high temperatures achievable in an electric arc. Because of the high temperature, the gases must be confined by magnetic fields, water-cooled containers, or some combination thereof. If compatible with reactants, the plasma can be directed into a reactor and become an internal source of heat for the reaction; alternatively, reactants can be introduced along with the carrier gas into the plasma arc where the desired reaction takes place instantly.

Plasma arc technology has been in commercial service since 1940 by Huls at Marl, in West Germany, in the production of acetylene and ethylene. The high-temperature technique is appropriate for the production of acetylene because above 1,500°C, acetylene is more stable than any other hydrocarbon. Huls, SKF, and AVCO have developed this technology to produce acetylene and synthesis gas directly from coal.[5] In addition to the advantage of operating at very high temperatures, the carbon monoxide-to-hydrogen ratio in the synthesis gas can be shifted by changing the fraction of the energy requirement that is provided electrically. Plasma production of acetylene from coal is estimated to be capable of producing cheaper vinyl chloride than is its production from ethylene.

The high-temperature capability of plasma arc technology is also well suited to the processing of inorganic materials. For example, the process is being developed for the direct reduction of iron ore. In this way the blast furnace can be bypassed altogether and the iron can be produced at the mine site. Plasma arc heating is also being developed to boost the output of cement kilns and iron-melting cupolas. These applications are analogous to boosting the output of glass-melting furnaces by the addition of electric heat.

There are also benefits in addition to increasing the furnace capacity. For example, for the iron-melting cupola used in foundries to make metal castings, most of the heat comes from the plasma, which means that gas flows for fuel combustion can be reduced by two-thirds. The lower gas velocity permits the use of low-cost forms of metal scrap, such as borings and turnings, that otherwise could not be directly recycled. In addition, the plasma heat can easily double the blast temperature and create conditions favoring the reduction of silicon. This in turn reduces the loss of silicon to oxidation and makes it possible to use sand as the source of silicon instead of expensive ferrosilicon. These savings in the cost of materials make up 90 percent of the cost advantage of plasma cupola operation; savings in energy costs make up the remainder.[6]

These two emerging technologies, dielectric and plasma arc heating, when added to electric resistance heating techniques, extend the range of applications for electrothermal processing to all materials — conductive or nonconductive, heat sensitive or insensitive, fluid or solid.

OTHER INNOVATIVE APPLICATIONS

Electromagnetic Wave Technology

There are many other applications for the special effects achievable through the use of electromagnetic radiation beyond the thermal processing of bulk materials. Examples of techniques that have broad application are described below.

Laser beam energy is being used for metal cutting.[7] The precise control and high-energy deposition rate possible with laser beams make them particularly suitable for cutting hard materials such as the titanium alloys used for aircraft components. Laser cutting speeds are typically twenty-five times those of mechanical sawing. In softer materials like steel the gain in cutting speed is five to ten times that of mechanical sawing.

The gain in cutting speed reduces labor costs by more than 60 percent. Moreover, the laser beam can be focused to a diameter of only a few thousandths of an inch, so that the amount of material removed is much less than it is with mechanical sawing. This means additional savings in raw materials and waste recycling.

Today's laser generators are relatively inefficient and their use increases the energy requirement for metal cutting. The extra energy cost is easily recovered, however, through savings in labor costs for justified applications.

Electron beams have high penetrating power and can be magnetically focused to a very small area (.001 to .002 sq. inches). These characteristics are ideally suited for welding thick metal sections. For example, metal plates six inches thick can be welded in a single pass, compared to twenty passes with conventional arc welding techniques. Moreover, the heat affected zone is much smaller and is uniform over the weld cross section, thus producing superior welds without distortion of the material compared to arc welding. As a result, operating and labor costs are typically reduced by 75 percent.

In contrast to laser generators, electron guns are highly efficient. This efficiency in combination with nearly 100 percent absorption of the incident beam in the welded material leads to energy savings of over 90 percent when compared with arc welding.[8]

Radiation curing of coatings and adhesives through the use of ultraviolet (UV), infrared, and electron beam generators is much faster and uses less coating material than does oven drying. In addition, the techniques can be used to cure coatings on heat-sensitive materials that would be adversely affected by oven drying. Curing time can be reduced from minutes to seconds, and the coating materials can be 100 percent solids instead of the 35 percent solids typical for oven curing. The elimination of the need for a solvent in the coating material, which is necessary for oven-dried coating, removes the need for a ventilation system for recovering or disposing of the solvent released during the oven-curing process.[9]

Ultraviolet light is being used to break molecular bonds to induce or catalyze chemical reactions. Ultraviolet light also excites chemical oxidants, such as ozone, thus increasing their power to oxidize. These UV techniques operate at room temperature, reduce processing time, and have low capital and operating costs. The technique is effective in the destruction of dioxin, polychlorinated biphenyls (PCBs), and other chlorinated hydrocarbons, as well as organisms and viruses in waste water.[10] In another application, high-dosage UV radiation successfully destroys the DNA in the bacteria that feed on the iron content of irrigation water. Rapid multiplication of bacteria is a basic cause of the clogging of holes[11] in drip irrigation piping systems.

Particle Charging

A high-voltage field across the flow path of a nonconducting fluid can electrically charge impurity particles, which will then move to the oppositely charged electrode, where they can be collected and removed from the fluid. This technique has been extensively used for several years to remove fine fly ash from the flue gas from coal-fired power stations. More recently the technique is being refined and applied across a broad spectrum of manufacturing processes to clean product and waste streams.

Because very fine particles can be separated with the technique (i.e., 0.01 micron), purity levels can be achieved that are much higher than are possible with conventional mechanical methods, such as filtration and centrifugation. Particles in submicron and micron sizes are the most difficult to remove and are most harmful to human health because they lodge in the lungs. The technique can also be applied to clean a wide range of fluid stream densities, from heavy viscous fluids (e.g., the removal of catalyst fines from the hydrogenation of vegetable oils) to water sprays in wet-scrubbing applications to aerosols. It can even catch the finely divided, highly resistive heavy metal particles carried out with the flue gas from gas-fired glass melting.

Particle charging is normally applied in addition to conventional filtration systems in order to substantially increase the overall efficiency of particle removal. The benefits justifying this increased use of electricity include a higher quality product, the lower release of unwanted wastes to the environment, and the improved recovery of valuable process materials for recycling.[12]

In agriculture, electrostatic charging of pesticide sprays has proved to be three to four times more efficient in use of pesticides.[13] The charged particles adhere to the underside of plant leaves where pests often thrive but where they are difficult to reach with conventional spray techniques.

Ultrasonic Spraying

Sonic-atomized fluid has several advantages over pressure spray systems. The sonic spray is low velocity, which virtually eliminates overspraying and related material losses. It also produces very small droplets that are more efficient in applications such as spray-drying and humidifying. Since the liquid is not forced through a small aperture, the sonic technique can handle a variety of liquids and slurries at widely varying flow rates without clogging.

In a specific application, sonic water sprays are being used to increase lime kiln throughput by 10 percent. Water spray is used to help cool the exhaust gas stream from the kiln and thus reduce the need for dilution air. Because the sprayed water particle size is small, it evaporates before lime in the exhaust gas can react with the water. This reaction, if allowed to occur, would clog the downstream baghouse filter system.[14]

ELECTROCHEMICAL PROCESSING

The uninformed visitor to a coal or petrochemical complex is hard pressed to find or assign importance to the reactor that is the heart of the conversion process — surrounded as it is by a city-block-sized three-dimensional array of piping and connected equipment that is needed to separate, recycle, and recover the mixture of chemicals the reactor delivers. It is this circulatory system that dominates the capital and operating costs of chemicals production and that demands large-scale centralized operations for economic viability.

Except for a few special products that are costly to produce thermally, such as aluminum, and chlorine or high-purity chemicals such as phosphoric acid, the cost of electricity generally restricts the use of electrochemical processing to the production of high-value, low-volume chemicals having high molecular weights. Some chemicals such as adiponitrile and tetraethyl lead have been produced electrically for decades; ten to twenty other electroorganic processes have reached the commercial stage in recent years; and another ten or so are in various stages of development.[15] Much of the development is taking place outside the United States, where economics for electroprocessing are more favorable. For example, a variety of chemicals are produced electrically in India because the scale of operations there is much smaller than for more developed countries. Electroprocessing is often practical at low levels of production, and in this way India can meet its requirements indigenously.

Frequently electroprocessing is developed or adopted because it confers advantages not available from conventional techniques. For example, the conventional halogenation route to anisic alcohol leaves contaminants with unpleasant odors. This is a problem if the alcohol is used to impart fragrance, but it does not arise with the electrochemical method. Again, electrosyntheses of isocyanates and carbamates eliminate the use of highly toxic phosgene needed otherwise. This "process cleanliness" is also a major advantage of the electroorganic route in the production of vitamin C. In this case, the use of a hypochlorite oxidizing agent, which is troublesome to dispose of, is eliminated.

Electron beam technology is also being used to synthesize organometallic compounds, such as boron trichloride, at yields ten times those for conventional manufacturing. In a vacuum, the energy intensity of the beam vaporizes target metal, which then reacts at high yield with organic vapor fed into the reactor. New chemicals are also being produced, such as tris-butadiene complexes of molybdenum and tungsten, which show promise for use as polymerization catalysts.[16]

Electrolysis of acid leach solutions is now used for 10 percent of U.S. primary copper production. This process is preferred to fuel smelting when the copper ore concentration is less than 0.3 percent, and U.S. ores now average only about 0.4 percent copper. This hydrometallurgical route can be followed economically at a small scale because the capital cost is a small fraction of that for thermal smelting. In addition, the process is environmentally benign in that all chemicals are recycled.[17]

The brief sampling of emerging electrotechnologies offered in this chapter gives some sense of the scope and breadth of these developments. Some of these technologies, like those discussed in Chapters 5 and 6, may provide more efficient ways to produce existing products. Others, like the first wave of electrotechnologies that revolutionized industry in the past, may point the way to entirely new products and industries in the future.

Notes

1. "Chementator," *Chemical Engineering*, (September 7, 1981), 17.

2. P. M. Kohn, "Microwave Technology: Penetrating CPI Markets," *Chemical Engineering* (January 3, 1977), 52.

3. Ibid., 51–53.

4. "Chementator," *Chemical Engineering* (May 21, 1979), 83; Robert S. Glaubinger, "Devulcanized Rubber — Not Ready Yet For Tire Use," *Chemical Engineering* (March 10, 1980), 84–86.

5. Richard J. Zanetti, "Plasma: Warming Up to New CPI Applications," *Chemical Engineering* (December 26, 1983), 14–17; "Chementator," *Chemical Engineering* (July 11, 1983), 17, and (September 10, 1979), 87.

6. "The Plasma Torch, Revolutionizing the Foundry Fire," *EPRI Journal* (October 1986), 12–19.

7. P. S. Schmidt, *Electricity and Industrial Productivity: A Technical and Economic Perspective*, EPRI EM-3640 (Palo Alto, CA: Electric Power Research Institute, March 1985), 10–11.

8. Ibid., 11–9.

9. Ibid., 16–9.

10. Robert W. Legan, "Ultraviolet Light Takes on CPI Role," *Chemical Engineering* (January 25, 1982), 95–100; "UV System Lightens Cost," *Chemical Engineering* (February 11, 1980), 71–72; Mark Lipowicz, "Rapid Oxidation Destroys Organics in Waste Water," *Chemical Engineering* (November 2, 1981), 40–41; also see "Chementator," *Chemical Engineering* (September 5, 1983; September 7, 1981; April 20, 1981; July 6, 1979; May 7, 1979; and August 1, 1977).

11. Robert E. Flatow, "Iron Bacteria — The Invisible Threat to Drip Irrigation Systems," *Irrigation Journal* (May/June 1985), 30–32.

12. Larry J. Ricci, "Electric Spark of Ionizers Hikes Scrubber Efficiency," *Chemical Engineering* (September 26, 1977), 52–57; "Electrostatic Precipitator Packs a Higher Charge," *Chemical Engineering* (March 14, 1977), 95–96; "Particle Charging Aids Wet Scrubber's Sub-Micron Efficiency," *Chemical Engineering* (July 1, 1975), 74; "Hybrid Scrubber Cuts Cost," *Chemical Engineering* (April 28, 1975), 66; James M. Ballinger, "New Filtration Concept Uses Electrodeposition," *Chemical Engineering* (May 14, 1984), 38–39; "Chementator," *Chemical Engineering* (March 15, 1976).

13. "Chementator," *Chemical Engineering* (October 11, 1976), 67.

14. "Chementator," *Chemical Engineering* (October 27, 1975), 55.

15. Gordon M. Graff, "Electroorganic Routes Make Bid for a Comeback," *Chemical Engineering* (March 21, 1983), 14–19; John C. Davis, "New Spark Comes to Electroorganic Syntheses," *Chemical Engineering* (July 7, 1975), 45–47.

16. "Chementator," *Chemical Engineering* (January 6, 1975), 56–57.

17. S. D. Atchison, "Copper's Hope for Climbing Out of the Pits," *Business Week* (March 10, 1986), 86H.

Part III
Electricity in
Nonmanufacturing: Its
Various Roles

Introduction

In Part III we turn to four nonmanufacturing sectors. Two of them, coal mining and agriculture (Chapters 8 and 10), round out our coverage of electricity use in the principal industrial sectors of the American economy, while Chapter 9 on transportation and Chapter 11 on residential uses both depart from the general theme of electricity in industrial production. Transportation is covered because early in the twentieth century electricity was very important in providing transportation services, and there is a possibility that it may be again, particularly if technological developments leading to a practical battery-powered automobile are successful. The residential sector is included partly because household uses add up to an important part of total national electricity consumption, but principally because this sector vividly illustrates the strong connections that have existed throughout the twentieth century between electricity use and technological change in everyday life.

Like the manufacturing sectors covered in Parts I and II, the next four chapters span a long period of years. Underground coal mines and city street railways adopted electricity in the 1880s, as soon as its generation on a reasonably large scale became practical. Electricity has remained the main energy source in coal mining to this day, but has, with few exceptions, become almost negligible in transportation.

Electricity was also introduced into homes in the 1880s, and by the beginning of the 1930s practically all urban dwellings had electric service. But less than 10 percent of the nation's farms received central station electricity in 1930, and electric service did not reach the majority of farms until almost mid-century.

In Chapters 8 through 11 we examine the various roles of electricity in these four diverse sectors. We look at the extent to which functions are or have been dependent upon the use of electricity and at the connections between electrification and technological progress.

Chapter 8 considers both underground and surface coal mining. Although 60 percent of the bituminous coal and lignite produced now comes from surface mines, underground mines were the major source of production until 1974. Electricity's importance to mechanization within the coal indus-

try has differed widely depending on whether the operations are conducted underground or above ground.

Until the 1880s, mining was a craft and underground mining tasks were
done essentially by hand. The distribution of mechanical power to mining
machinery from steam engines entailed great difficulties and risks, and
power distribution via compressed air was highly inefficient. Electricity had
such strong advantages over other means of providing power to machinery
underground that the dominance of electric mining technology was virtually assured from the time it first became available. Indeed, electricity was
the essential means of mechanization in underground coal mining and
today accounts for 90 percent of all the energy used in that activity.

Large-scale surface coal mining did not begin until the development of
appropriate excavating machinery in the 1920s. Digging and earth-moving
equipment was initially powered by steam produced from coal, but today
diesel fuel accounts for two-thirds of the energy used in surface mining.
Electricity, however, is used to drive the very largest machines — draglines,
shovels, and giant haulage trucks — despite the need for a trailing cable or
on-board generation. Distributing power aboard a large vehicle by wire is
relatively simple compared with distribution via the massive and complex
mechanical system required for straight-diesel power.

Transportation, which is considered in Chapter 9, is largely the domain of
liquid fuels, not electricity. Only fragments of today's transportation system
depend on central-station electricity: urban rapid transit systems, a handful
of trolley bus networks, and a few railroad lines. Yet electricity was very
important to transportation earlier in the twentieth century.

Electric streetcars, interurbans, and trolley buses were common, and there
were a number of electrified railroad lines. Electric automobiles were also
of some importance in the early history of the motor car, but the high
degree of mobility and the range required in its operation were more
readily achieved with liquid fuels and the internal combustion engine than
with batteries and the electric motor. The growing use of gasoline-powered
private automobiles and motorbuses after World War I also led to the eventual demise of many, but not all, of the nation's electrified public transportation systems. When a new wave of technological progress led to the
transformation of the nation's railroad network away from coal-powered
steam locomotives after World War II, the instrument of change was the
diesel-electric locomotive, whose costs were generally more favorable than
those for completely electrified lines.

Various ways in which energy is used in agricultural operations are covered
in Chapter 10. In those activities requiring mobility — agricultural field
operations, for example — the internal combustion engine using liquid fuels

has provided the essential basis for mechanization. Nevertheless, electricity has become increasingly important in particular task areas such as irrigation and livestock production. In activities such as these, requiring little or no mobility, roughly half the energy used is purchased as electricity.

The largest single use of electricity is in pumping irrigation water, which substantially increases crop yields and overall agricultural productivity. Electricity performs many other tasks in livestock and poultry operations in connection with feeding, providing sanitation, and maintaining suitable temperatures and humidity. In effect, much agricultural production has evolved over the past century from being totally dependent on natural inputs (land, sunlight, rain) to substantial dependence on liquid fuels and electricity for control of the growth environment.

In Chapter 11 we deal with the wide range of activities that are synonymous with residential electricity use. Homes were wired initially for electric lighting, one of the most important of all technological advances in its social and economic effects. Lighting continued to be the major residential electricity service for many decades. Mostly, though, Chapter 11 is concerned with the astonishing proliferation of electrical applications in the home that account for the vast growth in residential electricity use. All of them represent technological advances, large and small.

The use of electricity in radios, irons, and washing machines became important during the 1920s, and each subsequent decade has witnessed the introduction of new and improved electrical appliances to serve human needs and wants. Services that are mechanical in nature, such as refrigeration, air conditioning, vacuuming, and clothes washing, and those that are electronic, such as audio-video entertainment and information processing, are from a practical standpoint almost completely tied to electricity as the essential energy form. On the other hand, there are several important electrical applications that are highly sensitive to comparative energy costs, such as space heating, water heating, and (nonmicrowave) cooking. In all of these, natural gas continues to be a widespread competitive alternative.

Overall, then, the extent of electrification varies considerably among the nonmanufacturing sectors surveyed in Part III. Underground coal mining and residential applications depend heavily on the use of electricity, whereas transportation is now mostly fuel powered and agriculture relies on both liquid fuels and electricity. The diverse profiles presented in these chapters round out our descriptive analysis of electrification and technological progress in twentieth-century America.

8 | Coal Mining: Underground and Surface Mechanization

Warren D. Devine, Jr.

Since the 1880s much of the technological progress in the bituminous coal-mining industry has been made possible by the availability of electricity. Because of the great difficulty and hazards associated with transmitting power mechanically from steam engines, electricity was the only feasible way to mechanize underground mining. Today, too, underground mining remains highly electrified. Most of the primary energy used in these mines and at preparation plants is purchased as electricity.

Large-scale surface coal mining was not practical until the development of appropriate digging and earth-moving machines that were initially powered by steam. Although diesel and gasoline engines eventually replaced these steam engines, electricity has been used to power the largest machines almost since surface mining first became feasible. As machine scale has increased, so also has the application of electric power. Today, electric power drives ever-larger machines, even though surface mining still relies mainly on diesel fuel.

The most important emerging role for electricity in coal mining may be in monitoring, controlling, and integrating production. Both underground and surface mining are moving toward electronic computer-related technologies. Indeed, the century-old association between electrification and technological progress in coal mining continues today.

UNDERGROUND MINING: MECHANIZATION THROUGH ELECTRIFICATION

1890–1950: Mechanization of Miners' Tasks

The Craftsman-Miner

Until the late 1870s, all coal mining was done by hand. Coal was removed from beneath the surface by the "room-and-pillar" method — a standard approach to underground coal mining even today. Room and pillar mining involved opening main entries horizontally into the coal seam from the bottom of the shaft or slope opening. Side entries were then driven perpendicular to the main entry at intervals of several hundred feet. "Rooms" were cut at right angles to the side entries, leaving thick "pillars" of coal between the rooms to support the roof of the mine. The principal mining activity occurred in these rooms, which were typically 12 to 40 feet wide and 250 to 400 feet long at completion, depending on the distance between side entries.

Prior to mechanization, each room was the province of a single miner and his helper. Their first task was to undercut the coal face, done by making a deep horizontal or wedge-shaped slit at the bottom of the seam with a pick. This was the most time-consuming part of their work, taking three to six hours depending on the width of the room and the hardness of the coal. Next, the miners had to blast the coal loose, so that it would fall and expand into the undercut. To do this it was necessary to bore a five-foot deep hole in the coal face and to carefully insert blasting powder and a fuse. A single explosion would typically shatter a ton or more of coal.

The miner and his helper then obtained empty mine cars and loaded the shattered coal into them by shovel. The loaded car was hauled away by animals or pushed from the room by the miners. Sometimes the mining of a seam was carried to completion by extracting the coal in the pillars, allowing the roof to fall. In view of these burdensome tasks, it is surprising that productivity was as high as it was: an average of 2.5 tons of coal per miner per day.[1]

The knowledge, experience, and dexterity required of the pick-and-shovel coal miner were acquired by serving an apprenticeship of up to several years duration. Once trained and assigned to a room, the miner worked with little supervision: one foreman for 100 or more miners and helpers was typical. Indeed, the coal miner was an independent craftsman and mining was a craft occupation with counterparts in other industries of the nineteenth century.[2]

Steam engines were used to pump water from mines and were later applied to mine ventilation and to the transport of men and materials between the surface and working level. But mechanization of the tasks of the individual miner had to await methods of transmitting power through the underground labyrinth. Transmission from a steam engine near the mine entrance via rotating shafts and belts was not practical due to the size and complexity of the mechanical system required to reach the working faces. It was equally impractical to employ small engines that used steam generated in boilers located within the mine. Generally, boilers could not be situated in mines because of the danger of explosion from the mixture of methane or coal dust and oxygen present in most coal mines or because of the noxious combustion products from firing the boiler.

Mechanical Coal Cutting

The first successful mechanical coal undercutter was marketed in 1877. Called the "cutter-bar machine," it was essentially a toothed disc rotated via a chain drive and compressed-air engine. Several hundred were in use before percussive-type machines called "punchers" were introduced beginning in 1880. In 1889 an electric motor was substituted for the compressed-air engine on the cutter-bar machine, and for about five years these electric undercutters competed with the compressed-air punchers.[3]

Large compressor plants — boiler, steam engine, air compressor — were situated on the surface, and under favorable conditions smaller units in the mine were convenient. But compressed air was not an efficient way to distribute power in mines. Numerous measurements showed that rarely was more than 25 percent of the power indicated in the steam cylinder of the compressing engine supplied to the coal cutter. Loss was due to friction and to leakage of air from joints and drain cocks.[4]

Electricity had been used at mines for illumination and for pumping of water since the early 1880s, but its use at the working face was delayed because it was felt that sparking from the early direct current motors on cutters posed too great a threat. By the early 1890s, however, the ease and small power loss with which electricity could be distributed from generators on the surface to scattered work sites in the mine had become apparent. In zones where it was deemed unsafe to operate electric equipment, compressed air engines could still be used: an electric motor-driven air compressor was simply positioned as close to such a zone as was considered practicable.[5]

In 1894 several manufacturers introduced the electric motor-powered "breast chain cutter." These mobile machines were much simpler, faster, and more easily operated than any of their predecessors and soon completely displaced the compressed-air punchers. Improvements in design early in the twentieth century enabled the machine to cut continuously across the working face and to cut at almost any height above the floor.[6] Before World War I more than half of the bituminous coal mined underground was cut by machine; by mid-century less than 10 percent was still undercut by hand (see Figure 8.1).

Mechanical cutters were taken from room to room by a group of workers called "machine runners." As these men cut the coal themselves, undercutting ceased to be part of the craftsman-miner's trade.[7] Eventually, other customary tasks of the miner also passed to specialists: firing the explosives, installing timber roof supports, and laying track.[8]

Despite the almost complete mechanization of coal cutting and the trend toward specialization, average tonnage produced per miner-day increased very slowly before the middle of the twentieth century (see Figure 8.2). One reason productivity did not increase faster is that the growing division of

Figure 8.1   Portion of Total Coal Mined Underground That Was Cut by
Various Methods, 1891–1978

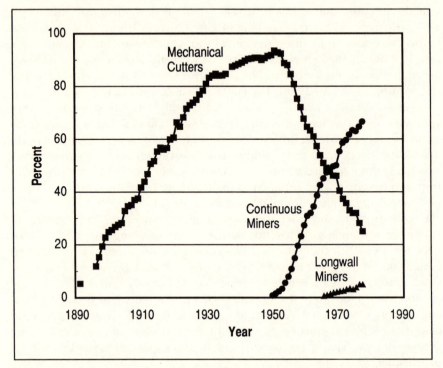

SOURCES: U.S. Bureau of the Census, *Historical Statistics of the United States, Colonial
Times to 1970, Bicentennial Edition* (Washington, D.C.: GPO, 1975) Series M93 and M105; U.S.
Bureau of Labor Statistics, *Technology, Productivity, and Labor in the Bituminous Coal Industry, 1950–1979,* Bulletin 2072 (Washington, D.C.: GPO, 1981); U.S. Energy Information Administration, *Coal Data: A Reference,* DOE/EIA-0064(84) (Washington, D.C.: GPO, 1985), Table
6, and *Coal Production 1987,* DOE/EIA-0118(87) (Washington, D.C.: GPO, 1988), Table B1.
Data are missing for some years.

labor underground was not accompanied by adequate coordination of activities. For example, a miner would often be idle while specialists came to his
room and performed tasks there that were formerly his own responsibility.[9]
Significantly, the main job of the craft miner eventually became that of
loading; and loading was done almost entirely by hand until the mid-1920s
(see Figure 8.3).

Mechanical Loading and Hauling

Commercial use of loading machines began in a few mines in 1918. One of
the first widely used types was the "pit-car loader" — a mobile, motorized
conveyor rising the short distance from the floor to the top of a mine car.
With this machine the miner needed only to shovel the coal onto the bottom

Figure 8.2 Tons of Bituminous Coal Produced per Miner-Day, 1914–1987: Underground and Surface Mines

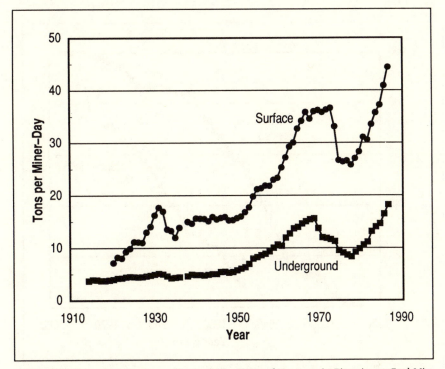

SOURCES: Howard N. Eavenson, "Seventy-Five Years of Progress in Bituminous Coal Mining," in *Seventy-Five Years of Progress in the Mining Industry, 1871–1946,* ed. A. B. Parsons (New York: The American Institute of Mining and Metallurgical Engineers, 1947), 226; Harold Barger and Sam H. Schurr, *The Mining Industries, 1899–1939: A Study of Output, Employment and Productivity* (New York: National Bureau of Economic Research, 1944), Table A.9; U.S. Bureau of Labor Statistics, *Technological Change and Productivity in the Bituminous Coal Industry 1920–1960,* Bulletin No. 1305 (Washington, D.C.: GPO, 1961); U.S. Energy Information Administration, *Coal Data: A Reference,* DOE/EIA-0064(84) (Washington, D.C.: GPO, 1985), Table 8, and *Coal Production,* DOE/EIA-0118 (Washington, D.C.: GPO, various years), Tables 42 and 43. Data are missing for some years.

of the conveyor. In the late 1920s, use of the pit-car loader gave way to the "mobile-loader." This device had a pair of claws rotated by motor that reached into the pile of broken coal and drew it onto its separately motorized conveyor for delivery to the car. Later improvements, especially caterpillar traction, rendered the machine more mobile and led to the elimination of the job of hand shoveling.

The mechanization of hauling was made possible by the availability of electricity in the mine. The first electric locomotives, introduced between 1888 and 1896, derived their power by trolley from 500-volt direct current

Figure 8.3 Portion of Total Coal Mined Underground That Was Mechanically Loaded and Portion That Was Produced in Mines with Conveyors, 1923–1978

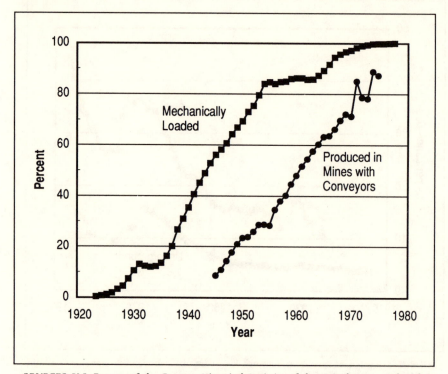

SOURCES: U.S. Bureau of the Census, *Historical Statistics of the United States, Colonial Times to 1970, Bicentennial Edition* (Washington, D.C.: GPO, 1975), Series M93 and M106; U.S. Bureau of Mines, "Coal—Bituminous and Lignite," in *Minerals Yearbook* (Washington, D.C.: GPO, annual); U.S. Energy Information Administration, *Energy Data Report, Coal—Bituminous and Lignite in 1976*, DOE/EIA-0118/1(76) (Washington, D.C.: GPO, 1978), and *Bituminous Coal and Lignite Production and Mine Operations*, 1978, DOE/EIA-0118(78) (Washington, D.C.: GPO, 1980), and *Coal Data: A Reference*, DOE/EIA-0064(84) (Washington, D.C.: GPO, 1985), Table 6, and *Coal Production 1987*, DOE/EIA-0118(87) (Washington, D.C.: GPO, 1988), Table B1; U.S. Bureau of Labor Statistics, *Technological Change and Productivity in the Bituminous Coal Industry, 1920–1960*, Bulletin No. 1305 (Washington, D.C.: GPO, 1961), and *Technology, Productivity, and Labor in the Bituminous Coal Industry, 1950–79*, Bulletin 2072 (Washington, D.C.: GPO, 1981).

lines. But it was neither safe nor practical to extend these fixed, bare wires to the continually advancing working faces. In 1902 a locomotive was introduced with a trailing cable wound on a reel driven from the axle or by a separate motor. This "cable-reel" locomotive could be operated safely in gassy places and gained wide acceptance. After World War I, reliable battery-powered locomotives were developed; because they eliminated wires altogether, they were deemed safer than either trolley or cable-reel locomotives.[10] By 1920 the greater part of the shift from human or animal power to electric power for hauling was complete.

Increased production resulted in larger amounts of fine coal dust and methane in the atmosphere of underground mines, with the greater probability of explosion. This hazard called for improved mine ventilation. By World War I, high-speed electric fans were becoming common in mines, replacing ventilation by natural draft, by furnace suction, and by low-speed steam-driven fans.[11] Here the introduction of one electric technology (mechanized undercutting) increased the need for another service (ventilation) that also turned out to be best provided electrically.

The fact that mechanization in underground coal mines was predominantly based on electric power is reflected in the electricity intensity of the industry as a whole. Electricity used per ton of coal produced increased from zero in the early 1880s to 4.7 kWh in 1929 and to around 9 kWh in 1950 (see Figure 8.4).

The mid-twentieth century marks the end of what might be called the era of mechanization of miners' tasks. Mechanization began with undercutting and soon spread to include haulage. By 1923 when mechanical loading was just becoming significant, over two-thirds of the coal mined underground was cut mechanically, and almost all coal was hauled by machine. But in 1949, one-third of the coal taken from underground was still loaded by hand, precluding in many mines the integration of mechanical cutting and hauling with loading and the adjustment of the tempo of these processes with one another so as to form a continuous cycle. Between 1890 and 1950, the accompanying improvement in labor productivity for underground mining was substantial. But it would be more rapid later when integration of mechanical processes became more complete.

1950 to Date: Toward Integrated Production

The "Continuous" Miner

In 1948 several mines began testing the first models of an innovative machine called the "continuous miner" that would remove and load coal from the solid face in a single operation. These electrically driven machines (supplied with power by a trailing cable) have a drum studded with steel teeth that rotates about once per second. This drum is driven into the top of the coal seam and moved downward, shattering the coal. Gathering arms pull the fallen coal onto a central conveyor that discharges it onto a haulage system.[12] Figure 8.1, a classic illustration of the substitution of one technology for another, shows the rise of the continuous miner between 1949 and 1969. By integrating cutting and loading into a single, continuous operation, the continuous miner eliminates the separate intermediate steps of drilling and blasting and the bottlenecks that occur even in fully mechanized mines when any one step in the mining sequence falls behind.

Figure 8.4   Electricity Intensity: Total Bituminous Coal and Lignite
Mining Industry, 1929–1982

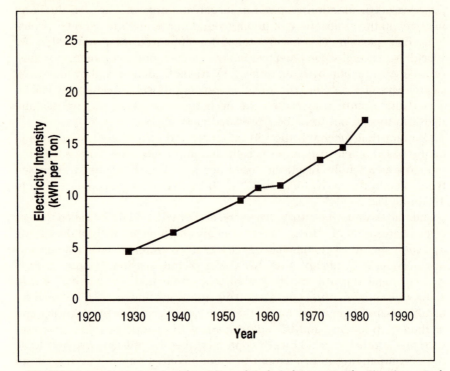

SOURCE: Warren D. Devine, Jr., "Electricity and Technical Progress: The Bituminous Coal
Mining Industry, Mechanization to Automation," ORAU/IEA-87-1(M) (Oak Ridge, TN: Institute
for Energy Analysis, Oak Ridge Associated Universities, 1987), Table A-2.

The increased production rate made possible by continuous miners
required concurrent improvements in haulage to prevent the accumulation
of loosened coal from interrupting operations. After World War II, self-
propelled rubber-tired (trackless) "shuttle cars" became increasingly popular for linking the moving coal face with a central belt conveyor.[13] Shuttle
cars derived power from an electric cable or from batteries, having evolved
from the earlier mine-car-gathering electric locomotives that ran on tracks.
Rubber-tired mine cars or trailers hauled by battery-powered rubber-tired
tractors were also used. These trackless systems were flexible and permitted the more nearly continuous availability of empty cars at the face. The
increasing use of conveyors — principally as the link between the trackless
haulers and mainline haulage — is illustrated in Figure 8.3.

As mechanization and production rates increased, so also did the need for
improved mine ventilation. Continuous mining machines operated at high

speed, generating greater quantities of coal dust and liberating more methane than earlier techniques.[14] Greater use of fans was only part of the answer because the time-honored method of roof support — a framework of heavy timber — restricted the flow of air.

By the later 1940s a new method of support was being introduced: roof bolting. This involved drilling holes four to six feet deep into the roof of mine passages in a grid-like pattern with intervals of four feet. Steel bolts inserted into the holes were held in place by a mechanical expansion shell or, more recently, by polyester resin. These bolts bond together the various rock strata overlying the coal seam, supporting the roof. The probability of collapse was lessened and restrictions to airflow were removed, enhancing ventilation. This is a second example of how new electric machinery (continuous miners) created conditions that led to improvements in a totally different part of the mining operation.[15]

In mines with continuous miners, trackless haulage, and conveyors, mechanically integrated (but not necessarily continuous) production was finally possible. In 1959, output per miner-day in mines using continuous mining machines exclusively was 24 percent greater than in mines employing undercutting machines and mobile loaders, and 144 percent greater than in mines where coal was loaded by hand.[16] As continuous miners became more prevalent and as experience was gained with fully mechanized mining, productivity for the underground segment of the coal industry increased appreciably. Over the twenty-year period ending with 1969, the average annual increase in output per miner-day was over four times that of the preceding sixty-year period (Figure 8.2), and an all-time record production rate of 15.6 tons per miner-day was achieved.

Coal Cleaning

Mechanization affected the purity of the coal produced. Reliance on mechanical loaders and on continuous mining machines eliminated the individual miner's ability to avoid loading refuse along with coal. The fraction of refuse in raw coal increased from 0.09 in 1928 to 0.18 in 1956 and is generally 0.25 to 0.30 today.[17]

In the early 1920s, less than 5 percent of total raw coal was cleaned, almost all of it destined for making coke.[18] This fraction increased steadily until 1960, when nearly two-thirds of all raw coal was subject to sizing and subsequent cleaning (see Figure 8.5). (One reason the fraction falls after 1960 is that by that year surface-mined coal represented nearly one-third of total production, and relatively little coal taken from the surface is cleaned.)

During the mechanization of coal mining, users of coal were mechanizing boiler feeding and improving burners, leading to demands for coal of more uniform size and of greater purity. Thus, the sizing and cleaning of coal by the producer outside of the mine — formerly done manually and restricted

Figure 8.5 Fraction of Total Bituminous Coal and Lignite That Was Sized and Cleaned, 1906–1978

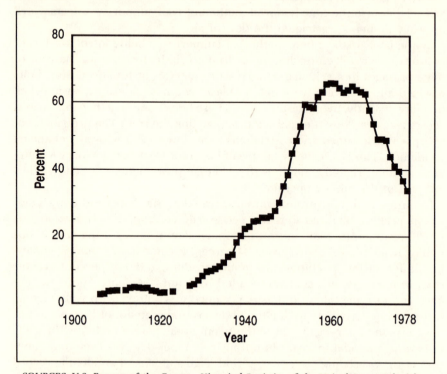

SOURCES: U.S. Bureau of the Census, *Historical Statistics of the United Sates, Colonial Times to 1970, Bicentennial Edition* (Washington, D.C.: GPO, 1975), Series M93 and M104; U.S. Bureau of Mines, "Coal—Bituminous and Lignite," in *Minerals Yearbook* (Washington, D.C.: GPO, annual); U.S. Energy Information Administration, *Energy Data Report, Coal—Bituminous and Lignite in 1976*, DOE/EIA-0118/1(76) (Washington, D.C.: GPO, 1978), Table 1, and *Bituminous Coal and Lignite Production and Mine Operations, 1978*, DOE/EIA-0118(78) (Washington, D.C.: GPO, 1980), and *Coal Data: A Reference*, DOE/EIA-0064(84) (Washington, D.C.: GPO, 1985), Table 6. Data are missing for some years.

to coking coal — was greatly expanded. This trend was especially strong during the fifteen years immediately following World War II.[19]

According to Table 8.1, over 90 percent of the energy used at coal preparation plants is electricity. This should come as no surprise because coal cleaning involves crushing and screening as well as cleaning, and these are all mechanical processes. Mechanical coal cleaning is one more example of how one set of electric technologies (for underground mining) enhanced the need for another operation (cleaning) that turned out to be best provided electrically.

Table 8.1 Energy Fractions, Energy Intensity, and Electricity Intensity: Bituminous Coal and Lignite Mining, 1972

	Establishments exclusively operating		
	Underground mines	Strip mines	Preparation plants
Fraction of total primary energy as			
Purchased electricity	0.94	0.29	0.91
Coal	0.01	—	—
Diesel fuel and oil	0.04	0.67	0.08
Gasoline	0.01	0.04	0.01
Total primary energy, 10^{12} Btu	12.98	8.24	4.48
Tons mined or prepared, 10^6	54.8	44.0	88.3
Energy intensity, 10^3 Btu/ton	237	187	51
Electricity intensity, kWh/ton	19.7	4.8	4.1

SOURCE: Warren D. Devine, Jr., "Electricity and Technical Progress: The Bituminous Coal Mining Industry, Mechanization to Automation," ORAU/IEA-87-1(M) (Oak Ridge, TN: Institute for Energy Analysis, Oak Ridge Associated Universities, 1987), Tables A-3, A-4, and A-5.

Longwall Mining

During the 1970s, new electric cutting and loading technologies continued to displace older methods. Longwall mining machines began to be used where coal beds are relatively flat and thick. This electric-powered machine consists of a plow or shearer that moves back and forth across a block of coal 300 to 1,000 feet wide and as much as 6,000 feet long, an integral belt conveyor, and a mechanical roof support system. The shearer-conveyor is followed by a row of large hydraulic jacks topped with broad steel beams or shields that prevent the mine roof from falling on the machine and its operators. After each pass of the cutter across the face, the support jacks momentarily release their pressure against the ceiling, move forward, and press into a new supportive position. The roof is allowed to collapse thirty to fifty feet behind the working face.[20]

Longwall mining operations — cutting, loading, and hauling — are truly integrated and in the mid-1970s could produce 750 tons per shift compared to 400 tons with a continuous mining machine and 300 tons with conventional machinery.[21] Moreover, longwall mining can recover up to 95 percent of the coal in a bed versus an average of around 50 percent with room-and-pillar techniques.

Again, newer continuous mining machines and longwall systems operating at greater depth, at higher speeds, and producing a finer product resulted in increased liberation of gas and dust, raising the requirements for ventilation.[22] And, again, progress toward automatic production underground increased some coal users' demands on coal preparation plants; these facilities automated, some adopting computer monitoring of materials flows.[23]

The continuing application of electric technologies underground is reflected in statistics for the coal mining industry as a whole. Electricity intensity rose from 9.6 kWh per ton in 1954 to 13.5 kWh per ton in 1972, and the fraction of total energy used for electricity increased from 0.57 to around 0.65 over the same period (Figures 8.4 and 8.6, respectively).

But labor productivity underground — after having climbed steadily since the onset of mechanization — declined markedly over the interval from 1970 to 1978, then returned to its pre-1969 pattern of increase. Sometime prior to 1972, the energy intensity of the industry as a whole began to increase after having decreased at least since 1954.[24]

Figure 8.6 Energy Intensity and Fraction of Total Primary Energy Used to Generate and Distribute Electricity: Total Bituminous Coal and Lignite Mining Industry, 1954–1982

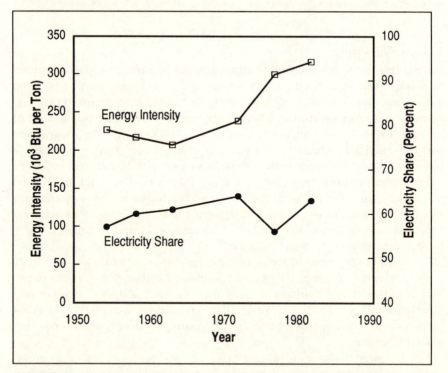

SOURCE: Warren D. Devine, Jr., "Electricity and Technical Progress: The Bituminous Coal Mining Industry, Mechanization to Automation," ORAU/IEA-87-1(M) (Oak Ridge, TN: Institute for Energy Analysis, Oak Ridge Associated Universities, 1987), Table A-2.

What was responsible for this relatively brief but serious sag in labor productivity and for this change in the energy intensity trend?

Federal Standards Impact Underground Mining

The Federal Coal Mine Health and Safety Act of 1969 required the coal industry to comply with standards that covered almost every aspect of the day-to-day operations of underground mines.[25] Immediate changes were necessary in most processes in order to meet standards on methane, respirable dust, ventilation, and roof support. For example, face crews were required to take time from producing coal to check methane concentrations in the air, to retard the presence of coal dust within fifty feet of the face, and to expand the ventilation system. A roof-bolting crew had to proceed with each mining crew according to a previously prepared roof control plan. In fact, the largest single occupation in underground mines in 1976 — by far — was that of roof bolter.[26]

Establishments that operated continuous mining machines were severely affected. The machines had to stop cutting and alter their position much more frequently than before. By 1976, continuous miners were actually cutting an average of only 20 percent of the time. In a number of cases, continuous mining became more expensive and less productive than conventional mining, with its various specialized machines operating in sequence.[27]

During the 1970s there was unusual, widespread agreement among government, business, and union leaders that the requirements of the 1969 Health and Safety Act were largely responsible for the lower output per miner-day.[28] Several retrospective studies show a strong association between the imposition of the standards and the productivity decline.[29] The workers who were employed in order to comply with the new regulations were simply not needed to produce the coal. The machines used to meet the standards and the energy that drove them — much of it in the form of electricity — were also superfluous to actual production. Thus, coal output decreased relative to all inputs; more labor, more capital, and more energy were needed per ton of coal produced after 1969 than had been required prior to that year.

Current Developments in Electricity Applications Underground

By the early 1980s, coal output per underground miner per day was on the upswing again. A number of reasons for this turnaround have been offered. For example, the U.S. Bureau of Labor Statistics noted that the labor force of the coal-mining industry was older and more experienced. Mining companies had learned to work effectively within the constraints of the safety, health, and environmental regulations after a number of years of experience dealing with them. And new electric technology — much of it developed in response to the regulations — was beginning to affect production.[30]

The following paragraphs provide a brief overview of several new electric and electronic computer-related technologies that are currently being used on a limited basis. These technologies, which involve monitoring, integrating, or controlling certain operations, could significantly raise productivity in the underground coal mining industry.

Computer monitoring. Computer-controlled, remote monitoring systems were introduced into underground mines in the United States in 1981. Currently, systems are available that can continuously monitor a variety of mine environmental conditions and machines, and display pertinent information in a surface office. Data typically available to supervisors include methane, carbon monoxide, and oxygen concentrations; air velocities and pressures; levels of humidity, dust, and smoke; conveyor belt speeds and alignments and motor voltages and bearing temperatures; and power being used by ventilation fans. Monitoring systems can have a significant impact on safety. Electronic sensors eliminate the need for people to take environmental readings in hazardous areas. Further, a U.S. Bureau of Mines study found that conveyor monitoring systems — through immediate identification and location of problems — can increase coal production by 10 percent per shift.[31]

Remote operation and computer-aided control. Continuous miners have been built that can be run via a cable or radio remote control unit. The operator can be up to fifty feet away from the machine. An advantage of remote control is that without an operator on board, the machines can be advanced considerably further into the coal face before they must be withdrawn for installation of roof supports. By reducing interruptions in this way, one mine boosted production nearly 20 percent on each shift.[32]

New mechanized operations. Removal of coal pillars in a room-and-pillar mine is a particularly hazardous operation due to the possibility of premature, uncontrolled caving of the mine roof. Roof support during "retreat mining" is provided by miners manually setting posts, hydraulic props, or roof bolts. Many of these devices are not recoverable and thus become part of the cost of extracting the coal. To minimize danger to miners, to lower equipment loss, and to speed the retreat operation, the U.S. Bureau of Mines has developed a machine for providing temporary, retrievable roof support. The Mobile Roof Support (MRS) consists of four electrohydraulic jacks that form capped columns between floor and roof, each providing 50 tons of support. The jack assemblies are mounted on a chassis with motor-driven crawlers. Power is derived from an electric cable and operation is controlled remotely via radio. During testing at the Southern Utah Fuels Company No. 1 Mine, the MRS increased productivity during retreat mining by over 10 percent.[33]

Continuous, automated mining. Between 1983 and 1987, average output in longwall mines increased from 56 tons per miner-day to 113. This is due in part to improvements in the equipment. A system now under testing would completely automate all operations at the face. Microprocessor-based sensors on the shearer can determine its position and guide it across the face. As it cuts the coal, the shearer itself can initiate the advancement of the roof supports and the conveyor. Sensors installed in the supports can monitor pressures from the roof and floor and facilitate adjustments. Eventually, one operator — on the surface — may be able to command an entire underground mining operation.[34]

SURFACE MINING: SUPPORTING ROLES FOR ELECTRICITY

1918–1950: The Emergence of Surface Mining

One of the most important changes in the coal industry is shown in Figure 8.7 — the growth in surface mining relative to total production. Since 1974, surface mines have become the source of the majority of coal production.

Surface mining consists of three main operations: loosening and removing the "overburden" (topsoil and underlying dirt and rock) in order to expose the coal seam; loosening the coal and loading it for shipment; and replacing the overburden and restoring the land close to its former condition. This basic sequence has been followed since the beginning of large-scale surface mining, but the last operation was not emphasized until the 1970s.

The particular technique employed depends on the terrain. "Area mining" takes place on large blocks of flat or gently rolling land. Overburden, averaging about 100 feet in depth in eastern states and about 35 feet in western states, is piled into "spoil banks" at the sides of excavations or "cuts." "Contour mining" requires the removal of the overburden along a hillside or completely around a hill and sometimes results in a long or circular shelf with spoil material cast below. As mining progresses into the hillside the amount of overburden that must be removed increases. When doing so becomes uneconomic, coal can still be recovered by boring horizontally into the side of the exposed seam with large augers. "Auger mining" was introduced in the late 1940s and has become an important supplement to contour mining in some areas.[35]

Surface-mined coal did not exceed 1 percent of total coal produced until the full-revolving caterpillar-mounted steam shovel was developed around World War I.[36] In 1925 "draglines" were introduced to supplement shovels in overburden removal. These machines have a bucket or scoop suspended with a cable from a boom. The scoop can be swung further than shovels can reach and can drag material along the ground.[37] Because of the range of the scoop only limited mobility is required.

Steam was the most important power source for shovels and draglines until 1935. By this time, most new excavating machines were being equipped with diesel or gasoline engines, but some were being built with

Figure 8.7   Bituminous Coal and Lignite Production, 1890–1987

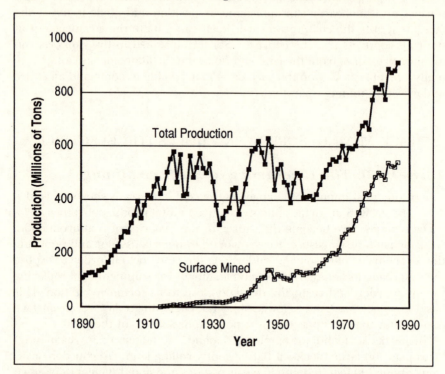

SOURCES: U.S. Bureau of the Census, *Historical Statistics of the United States, Colonial Times to 1970, Bicentennial Edition* (Washington, D.C.: GPO, 1975), Series M93; U.S. Energy Information Administration, *Coal Data: A Reference* DOE/EIA-0064(84) (Washington, D.C.: GPO, 1985), Table 6, and *Coal Production 1987*, DOE/EIA-0118(87) (Washington, D.C.: GPO, 1988), Table B1.

electric motors powered by a trailing cable or by an on-board diesel-powered electric generator. By mid-century only forty-two steam-powered shovels and draglines were still in use (see Figure 8.8).

Surface mining as we know it today did not exist before mechanization. During the period from 1918 to 1950, the basic techniques of large-scale surface mining were developed and appropriately strong machinery was introduced. Output per miner-day, while significantly higher than in underground mines, grew at about the same rate: an average of roughly 1.3 percent per year.

Figure 8.8 Number of Shovels and Draglines in Use by Power Source, 1932–1959 and 1969–1978[a]

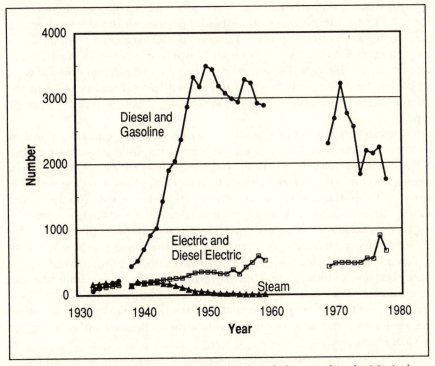

SOURCES: U.S. Bureau of Labor Statistics, *Technological Change and Productivity in the Bituminous Coal Industry, 1920–1960*, Bulletin No. 1305 (Washington, D.C.: GPO, 1961); U.S. Bureau of Mines, "Coal—Bituminous and Lignite," in *Minerals Yearbook* (Washington, D.C.: GPO, annual); U.S. Energy Information Administration, *Energy Data Report, Coal—Bituminous and Lignite in 1976*, DOE/EIA-0118/1(76) (Washington, D.C.: GPO, 1978), and *Bituminous Coal and Lignite Production and Mine Operations—1978*, DOE/EIA-0118(78) (Washington, D.C.: GPO, 1980).

[a] We have been unable to find data for the years 1960 to 1968.

1950 to Date: Growth and Dominance of Surface Mining

The Increasing Scale of Equipment and Use of Electricity

Between 1950 and 1973, surface mine labor productivity grew nearly three times faster than it had prior to mid-century and reached a record high for the bituminous coal-mining industry of 36.7 tons per miner-day (see Figure 8.2). This difference in growth rate is partly attributable to the major increases in the size of excavating and coal-handling equipment that occurred over this period, making economic the mining of western coal seams of thickness and extent far exceeding those in the East.

Between mid-century and the present, the bucket capacity of large shovels and draglines increased from around 40 to well over 100 cubic yards. And for unconsolidated overburden or in the softer shales, a new machine was introduced that turned out to be a better partner to the stripping shovel than the dragline. Electric-powered "bucket wheel excavators" using a series of whirling buckets and conveyors could move large quantities of earth over long distances. Finally, as large power shovels for overburden removal were improved, smaller versions were developed for ripping coal from the exposed seam and loading it into rail cars, into haulage trucks, or on to conveyors.

Most large shovels and draglines are driven by electric motors powered by alternating current delivered via a trailing cable. Indeed, Figure 8.8 hints of a trend away from excavating machines driven by diesel and gasoline engines and toward machines driven by electricity. Also, growing numbers of large off-road haulage trucks and front-end loaders are driven by electric motors. These motors, however, are not energized via cable but rather by on-board diesel-powered ac generators with solid state rectifiers.

Why is electricity used for large mobile machinery when it either requires a mammoth extension cord or additional transformations of energy on the machine itself? The answer to this question is reminiscent of the historical power distribution revolution in manufacturing: when power is transmitted to end-use points via electricity, mechanical power transmission systems (in the present case, shafts, gears, clutches, universal joints, etc.) are eliminated.

On most electric shovels and draglines, a pair of large ac motors operates continuously at full speed. These drive dc generators that energize dc motors, each matched to the desired power and speed range of a particular machine function. For example, a 35 cubic yard dragline (relatively small by present standards) might include two 500-hp hoist motors, two 500-hp drag motors, four 135-hp swing motors, and two 20-hp boom hoist motors, all powered by individual dc generators driven by ac motors rated in total at 2,500-hp.[38] Such an arrangement gives simplicity and flexibility to what would otherwise be an extraordinarily complex and cumbersome mechanical power transmission system. Reliability is higher, lifetime is longer, and maintenance is much simpler with electric power transmission than with mechanical power transmission. In addition, the spare parts inventory is considerably smaller.[39]

On diesel-electric haulage trucks and front-end loaders, each axle or wheel is driven by its own electric motor. Power is transmitted from the engine to the drive train electrically rather than mechanically. According to users, such equipment has higher reliability, lower maintenance costs, and longer life than diesel-mechanical equipment of comparable size.[40] Indeed, due to the limitations of mechanical power transmission, the trend to ever-larger straight-diesel units had run its course by the mid-1970s. Diesel-electric systems, however, have since made still larger vehicles practical. In particular, electric haulage trucks of 170 tons capacity have gained a firm foothold in the surface mining industry, and diesel-electric trucks of even larger capacity are being introduced.

The Onset of Reclamation

Beginning around 1969 a number of states enacted laws that required mining companies to restore mined lands to their approximate original contours and agricultural quality. During the early 1970s these state reclamation laws were more significant than federal ones. Because reclamation involves the use of capital, labor, and energy in activities that do not yield coal, a decrease in output per unit of input is to be expected. But state surface mine reclamation requirements varied in the time of enactment, in the specific provisions, and in the level of enforcement. As a result, the labor productivity record for the nation as a whole appears to falter for several years before entering a period — 1974 to 1978 — of more serious decline (Figure 8.2). A strong association between state reclamation laws and reduced productivity has been demonstrated econometrically by Baker and others.[41]

State reclamation laws were followed on the national level in 1977 by the Federal Surface Mine Control and Reclamation Act. By 1978, reclamation had become almost standard practice. In that year (based on data for 73 percent of the total amount of surface-mined coal) about 106 square miles of land surface were disturbed during mining operations, while 93 square miles were reclaimed.[42]

Despite the onset of land reclamation, surface-mined coal continued to represent a growing portion of total coal production. In each year after 1973, more than half the bituminous coal mined in the United States came from surface mines. Reclamation and the growth of surface mining are reflected in the energy statistics for the coal-mining industry as a whole. Between 1972 and 1977, fuel use by the industry increased by 79 percent. Most of this was probably due to surface mining, since direct fuel use represents a small fraction of the total primary energy used in underground mines (see, for example, Table 8.1). Thus, even though electricity purchases increased by 26 percent over the period, the electric fraction for the industry as a whole fell to its lowest value in the data record (Figure 8.6).

Current Developments

Labor productivity on the surface began to move upward again in 1979, just as it did underground. By this time, reclamation techniques had been refined; larger, improved equipment was being deployed; and a number of new technologies were being developed. The following paragraphs briefly describe several new systems that could appreciably affect surface mining labor productivity.

Computer monitoring. Large surface mining shovels and draglines are being equipped with computer-controlled monitoring systems. Bucket fill, load and cycle times, rope positions and forces, boom or swing angles, and motor field strengths are measured or calculated. Real-time performance data are displayed before the operator in the cab, helping him to reduce errors and to maximize dragline or shovel production. Some dragline monitoring and control systems also determine when the momentum and position of the bucket are such that it could damage the boom or other parts of the machine; the computer then takes control, slowing the swing rate or making other necessary adjustments. Records of production and machine condition are transmitted to the mine data management office, facilitating timely maintenance and decreasing downtime. At a North Dakota mine of the Knife River Coal Company, stripping costs were reduced from 10 percent to 15 percent in the year following the installation of such a computer system.[43]

Remote operation and computer-aided control. The number of possible routes that trucks can take to various loading, dumping, and coal depot areas is large at major surface mines. Several types of automatic truck dispatching systems are in use or under development. For example, one large surface mine utilizes a system known as ATID (Automatic Truck Identification and Dispatching). The system automatically determines and reports the location of all haulage trucks, shovels, and other large equipment in relation to coal seams, spoil banks, and haulage roads. It combines this information with other data and selects optimal routes and destinations based on minimizing the idle time of each piece of equipment. The computer assigns a route to each truck and generates reports on production and productivity. Mine operators using computer dispatch systems have reported production increases of as much as 14 percent.[44]

New mechanized operations. Surface mine conveyor systems are sensitive to large rocks and other materials that can cause blockages and result in damage to the belt or hopper before the system can be shut down. High-capacity, mobile, in-pit crushers have been developed that are able to reduce not only coal, but also large boulders to a conveyable size. Thus, the complete elimination of haulage trucks — with large cost savings — appears feasible. In one mine an electric-powered, crawler-mounted "feeder-breaker" will reduce 1,200 tons of coal per hour from chunks as large as 3.5 feet to 8-inch lumps, and feed it to a crawler-mounted conveyor system. This mine is the first in the United States to rely entirely on primary size reduction and conveyors for coal handling rather than on trucks. At least as far as coal handling is concerned, it is a prototypical "all-electric" surface mine.[45]

Continuous, automated mining. Bucket wheel excavators (BWEs) have been used in Europe for decades in high-volume surface mining operations. The BWE produces small size materials, can feed onto a variety of haulage modes, and can deliver overburden back to a cut for reclamation. In unconsolidated overburden, in the softer shales, or in lignite, BWEs have the lowest operating cost per ton — by far — of any mining method. The main disadvantage of BWEs is their relative lack of mobility. But if a large mine is to be worked for ten or more years — as at Texas Utilities' Big Brown mine — BWEs used in conjunction with conveyor systems can obviate the need for truck haulage or other rehandling of material. Surface mining is then transformed from an intermittent to a continuous operation and from one at least partially based on diesel fuel to one based almost entirely on electricity.[46]

MINING AND MANUFACTURING: SOME COMPARISONS

Prior to the 1880s, underground coal-mining tasks were done by hand. Mechanical power could be transmitted from steam engines only with great difficulty and risk. The electric distribution of power had so many advantages that underground mines adopted electricity almost as soon as it became available. Thus, electricity virtually created mechanized underground mining.

Most of the manufacturing industries, on the other hand, mechanized much earlier — first via water power and then via steam power. Between 1890 and 1930, electric power largely replaced steam power with marked improvement in productive efficiency. But large productivity gains were not immediately achieved; indeed, they came in at least two stages. Small improvements in productive efficiency were realized when large motors were directly substituted for steam engines. Much more significant gains in productivity were associated with the subsequent reorganization of production that was made possible by electric unit drive. Reorganization led to the integration of production — the coordination of related mechanized processes and the adjustment of the tempo of these processes with one another so that they could be carried out in a natural or continuous sequence.

In underground coal mining, productivity improvement also came in two stages. Steady but comparatively slow productivity growth occurred prior to 1950 as coal cutting and hauling were mechanized via electricity. But the room-and-pillar setting and the delayed adoption of mechanical loaders prevented the emergence of mechanically integrated production until the second half of the twentieth century. Then new electric-powered machinery was introduced with which coal could be cut, loaded, and removed from the face in a continuous sequence; much more significant productivity growth accompanied this mechanical integration of production.

During the 1960s manufacturing industries began to employ electronic computer-related technologies. Computers were used for process simulation and design, and electronic feedback control devices were applied in production automation. The underground coal-mining industry, on the other hand, has only recently begun to use electronic information technologies to automate production. There is evidence, however, that use of these electric technologies can have an appreciable impact on productive efficiency.

Large-scale surface mining was not possible before mechanization. Specialized machinery began to be introduced around World War I, but it was mid-century before appropriate machines and techniques were widely employed. Electric power was not integral to the mechanization of surface mining (as it was to underground mining), but electricity has always played an important supporting role. Mechanical power transmission proved to be cumbersome and costly on very large mobile machinery; electric power was often advantageous even though it had to be delivered to the machines via cable or generated on board.

Larger and larger machines were deployed (particularly after 1950), and many of them used electricity provided in these ways. This large machinery facilitated the mining of coal seams of thickness and extent far exceeding those previously mined, and productivity grew rapidly. Better integration of operations became necessary during the 1970s as reclamation was added to the tasks of the surface miner. Electronic equipment and cable- and diesel-electric machinery currently being introduced may improve integration and could even lead to continuous, automated production of coal from some surface mines.

Overall, then, the history of underground coal mining parallels that of manufacturing in its widespread adoption of electromechanical operations and its subsequent two-stage gains in productivity. Surface mining, in contrast, is a more recent facet of the industry in which electricity has played a less central role in production. In both parts of the coal mining industry, electricity is now supporting progress toward automated production systems, just as it is doing in the American manufacturing industries.

Notes

1. Arthur Donovan, "Carboniferous Capitalism: Excess Productive Capacity and Institutional Backwardness in the U.S. Coal Industry," *Materials and Society* 7, no. 3/4 (1983): 270–271; Keith Dix, "Work Relations in the Coal Industry: The Handloading Era, 1880–1930," in *Case Studies in the Labor Process*, ed. Andrew Zimbalist (New York: Monthly Review Press, 1979), 158–163; Howard N. Eavenson, "Seventy-Five Years of Progress in Bituminous Coal Mining," in *Seventy-Five Years of Progress in the Mining Industry, 1871–1946*, ed. A. B. Parsons (New York: American Institute of Mining and Metallurgical Engineers, 1947), 225–226, 229.

2. Dix, "Coal Industry," 162–163.

3. Eavenson, "Bituminous Coal Mining," 227.

4. William H. Patchell, *The Application of Electric Power to Mines and Heavy Industries* (New York: Van Nostrand Co., 1913), 64–65, 214.

5. Ibid., 11.

6. Eavenson, "Bituminous Coal Mining," 227–228.

7. Harold Barger and Sam H. Schurr, *The Mining Industries, 1899–1939: A Study of Output, Employment, and Productivity* (New York: National Bureau of Economic Research, 1944), 173.

8. Dix, "Coal Industry," 164.

9. Ibid., 164, 166; Barger and Schurr, *Mining Industries,* 173, 175.

10. Some nonelectric locomotives were also used. Steam locomotives were employed in Pennsylvania and West Virginia mines around the turn of the century, but their use was discontinued because of the smoke and gases emitted. Compressed air locomotives were introduced in the 1880s, but were replaced within twenty years. A few gasoline locomotives were used between 1904 and 1915, but were also discontinued because of emissions and because they proved to be mechanically unsatisfactory. Robert L. Marovelli and John M. Karhnak, "The Mechanization of Mining," *Scientific American* 247, no. 3 (September 1982): 94, 96; Eavenson, "Bituminous Coal Mining," 229–231; Dix, "Coal Industry," 167.

11. C. L. Christenson, *Economic Redevelopment in Bituminous Coal: The Special Case of Technological Advance in United States Coal Mines, 1930–1960* (Cambridge: Harvard University Press, 1962), 134; Dix, "Coal Industry," 166; Marovelli and Karhnak, "Mechanization," 101; Patchell, *Electric Power,* 191.

12. Marovelli and Karhnak, "Mechanization," 97.

13. Eavenson, "Bituminous Coal Mining," 232.

14. U.S. Bureau of Labor Statistics, *Technological Change and Productivity in the Bituminous Coal Industry 1920–1960*, Bulletin No. 1305 (Washington, D.C.: GPO, 1961), 56; Christenson, *Economic Redevelopment*, 134.

15. Marovelli and Karhnak, "Mechanization," 96; Christenson, *Economic Redevelopment*, 134–135. Roof bolting, in the 1950s and 1960s, was only slightly more electricity intensive than timbering. Currently, electricity is being used more extensively in roof bolting.

16. U.S. Congressional Research Service, *The Coal Industry: Problems and Prospects,* stock number 052-070-04767-4 (Washington, D.C.: GPO, 1978), 25–26; U.S. Bureau of Labor Statistics, *Technological Change,* 16.

17. Christenson, *Economic Redevelopment*, 135; U.S. Energy Information Administration, *Coal Data: A Reference,* DOE/EIA-0064(84) (Washington, D.C.: GPO, 1985), 11.

18. Eavenson, "Bituminous Coal Mining," 234.

19. U.S. Bureau of Labor Statistics, *Technological Change,* 22.

20. U.S. Congressional Research Service, *The Coal Industry,* 26.

21. Hittman Associates, Inc., "Fuel and Energy Consumption in the Coal Industries," FEA-EI-1659 (Columbia, MD: Hittman Associates, 1974), III-60.

22. Hittman Associates, Inc., "Underground Coal Mining: An Assessment of Technology," EPRI AF-219 (Columbia, MD: Hittman Associates, 1976), 5–99. Ironically, current attempts to substitute diesel for electric haulage also add to the demand on the (electric-powered) ventilating system. "Diesels Pull Through Underground," *Coal Age* 92, no. 5 (May 1987): 34.

23. Bernard A. Gelb and Jeffrey Pliskin, *Energy Use in Mining: Patterns and Prospects* (Cambridge, MA: Ballinger Publishing Co., 1979), 132.

24. It is tempting to try to relate this change in trend to the rise of surface mining: if surface mining were more energy intensive than underground mining, there might be good correlation between overall energy intensity and type of mining. But the few reliable data that are internally consistent suggest otherwise. According to Table 8.1, surface mining was less, not more, energy intensive than underground mining in 1972.

25. C. L. Christenson and W. H. Andrews, "Manpower and Technology in Bituminous Coal Mining: 1956–70," OLMA 91166632 (Bloomington; Indiana University,

1970), 6; U. S. Congressional Research Service, *The Coal Industry*, 114–115; Marovelli and Karhnak, "Mechanization," 101–102.

26. Daniel R. Walton and Peter W. Kauffman, "Preliminary Analysis of the Probable Cause of Decreased Coal Mining Productivity (1969–1976)," FE/8960-1 (Reston, VA: Management Engineers, 1977), I2–I3; Joe G. Baker, "Determinants of Coal Mine Labor Productivity Change" (Oak Ridge, TN: Oak Ridge Associated Universities, Manpower Research Programs, 1979), 14–15; U.S. Bureau of Labor Statistics, *Technology, Productivity, and Labor in the Bituminous Coal Industry, 1950–79*, Bulletin 2072 (Washington, D.C.: GPO, 1981), 41.

27. Hittman (1976), "Underground Coal Mining," 2–3 and 5–55.

28. See, for example, W. L. Wearly, "Technological Solutions to Declining Productivity in Underground Mining," *Mining Congress Journal* 60, no. 7 (July 1974); "Stemming the Slide in Productivity is a Job for Both Machinery Manufacturer and Mine Operator," *Coal Age* 81, no. 7 (July 1976); U.S. Congressional Research Service, *The Coal Industry*, 24; Edward F. Denison, "Effects of Selected Changes in the Institutional and Human Environment Upon Output Per Unit of Input," *Survey of Current Business* 58, no. 1 (January 1978): 33–35 and 43.

29. Walton and Kauffman, "Coal Mining Productivity"; Baker, "Mine Labor Productivity"; U.S. Bureau of Labor Statistics, *Technology, Productivity*.

30. Personal communication, Clyde Huffstutler, Division of Industry Productivity and Technology Studies, U.S. Bureau of Labor Statistics, Washington, D.C., November 1985.

31. "Computers Monitor Conditions, Operations at U.S. Mines," *Coal Age* 91, no. 12 (December 1986); "New Technology Boosts Production and Lowers Costs," *Coal Age* 91, no. 7 (July 1986); "Safety and Productivity: The Search Continues at BOM," *Coal Mining* 23, no. 3 (March 1986); "U.S. Underground Mine Monitoring: A Technology Comes of Age," *Coal Mining* 22, no. 12 (December 1985); "Microcomputer Keeps Watch at Emerald Mine," *Coal Mining* 24, no. 4 (April 1987); "Deserado Gains Edge Through Computer Monitoring," *Coal Mining* 24, no. 12 (December 1987).

32. "Out-of-Sight Operator Controls Mining Machine," *Coal Age* 90, no. 2 (February 1985); "Remote Control Promises Safety and Productivity Gains," *Coal Mining* 22, no. 10 (October 1985); "Terry Eagle Coal Fine-Tunes Production Systems," *Coal Mining* 22, no. 12 (December 1985); "New Technology Boosts Production and Lowers Costs," *Coal Age* (July 1986).

33. "Remote Control Promises Safety and Productivity Gains," *Coal Mining* (October 1985); "Mobile Roof Supports Make Pillar Extraction Safer," *Coal Age* 91, no. 5 (May 1986).

34. U.S. Energy Information Administration, *Quarterly Coal Report*, DOE/EIA-0121 (84/2Q) (Washington, D.C.: GPO, 1984); "Longwalls Advance Automatically," *Coal Age* 90, no. 2 (February 1985); "Remote Control Promises Safety and Productivity Gains," *Coal Mining* (October 1985); "Longwall Automation Grows," *Coal Mining* 24, no. 5 (May 1987); "Holton Automates Its Longwall With a Shearer Initiation System," *Coal Age* 92, no. 7 (July 1987); "Longwall Productivity Continues Upward Trend," *Coal* 25, no. 8 (August 1988).

35. U.S. Bureau of Labor Statistics, *Technology, Productivity*, 8; Gelb and Pliskin, *Energy Use*, 143; Christenson, *Economic Redevelopment*, 127; Marovelli and Karhnak, "Mechanization," 99–100; U.S. Congressional Research Service, *The Coal Industry*, 28–29.

36. Barger and Schurr, *Mining Industries*, 139.

37. Christenson, *Economic Redevelopment*, 129.

38. "Wide Variety of Models and Types Offered by Manufacturers of Surface Mining Equipment," *Coal Age* 81, no. 10 (October 1976): 75. In recent years a few

cable-powered electric mining shovels with ac motor drives have been built. The Bucyrus-Erie "acutrol" system allegedly vastly reduces routine maintenance because of the elimination of dc motors with their commutators and carbon brushes. See *World Mining Equipment* 9, no. 8 (August 1985): 4.

39. "Wide Variety of Models and Types," *Coal Age* 81, no. 10 (October 1976): 229.

40. Ibid., 204b.

41. Baker, "Mine Labor Productivity," 15, 76–77; Walton and Kauffman, "Coal Mining Productivity," 3; U.S. Bureau of Labor Statistics, *Technology, Productivity*, 17; Libby Rittenberg and Ernest H. Manuel, Jr., "Sources of Labor Productivity Variation in the U.S. Surface Coal Mining Industry, 1960–1976," *The Energy Journal* 8, no. 1 (January 1987): 87–100.

42. U.S. Energy Information Administration, *Coal Data*, 10.

43. "Freedom Mine Sets Sights on 9.9 Million Tons Per Year," *Coal Mining* 22, no. 7 (July 1985); "Electronics Making Machines Brainy," *Coal Age* 90, no. 8 (August 1985); *Coal Age* 90, no. 10 (October 1985): 59; *Coal Age* 91, no. 2 (February 1986): 70; "Microprocessor-Aided Equipment Proves Productive and Reliable," *Coal Age* 92, no. 7 (July 1987): 55; "It Pays to Squeeze More Out of Truck-Shovel Operations," *Coal Age* 92, no. 12 (December 1987): 64.

44. "Truck Selection in Surface Mining," *World Mining Equipment* 10, no. 1 (January 1986); "Computer Monitors and Controls All Truck-Shovel Operations," *Coal Age* 90, no. 3 (March 1985); "Microprocessor-Aided Equipment," *Coal Age* (July 1987): 53; "Improving Surface Mine Productivity: Truck Dispatching," *Engineering and Mining Journal* 188, no. 8 (August 1987): 69.

45. "In-Pit Conveyors and Crushers Cut Surface Mining Costs," *Coal Age* 90, no. 7 (July 1985); *Engineering and Mining Journal* 187, no. 2 (February 1986): 51; "Feeder-Breaker Speeds Coal Out of Pit by Conveyors," *Coal Age* 92, no. 6 (June 1987); "Improving Surface Mine Productivity: In-Pit Crushing," *Engineering and Mining Journal* 188, no. 8 (August 1987): 71.

46. "Earthmoving Equipment Trends," *World Mining Equipment* 9, no. 11 (November 1985); "Surface Mining: Equipment Selection and Use are the Keys to Profits," *World Mining Equipment* 10, no. 1 (January 1986); "Cross-Pit Mining System Proving Its Capability in Texas," *Coal Age* 92, no. 8 (August 1987).

Works Cited

Joe G. Baker, "Determinants of Coal Mine Labor Productivity Change" (Oak Ridge, TN: Oak Ridge Associated Universities, Manpower Research Programs, 1979).

Harold Barger and Sam H. Schurr, *The Mining Industries, 1899–1939: A Study of Output, Employment, and Productivity* (New York: National Bureau of Economic Research, Inc., 1944).

C. L. Christenson, *Economic Redevelopment in Bituminous Coal: The Special Case of Technological Advance in United States Coal Mines, 1930–1960* (Cambridge: Harvard University Press, 1962).

C. L. Christenson and W. H. Andrews, "Manpower and Technology in Bituminous Coal Mining: 1956–70," OLMA 91166632 (Bloomington: Indiana University, 1970).

"Longwall Productivity Continues Upward Trend," *Coal* 25, no. 8 (August 1988).

Coal Age Magazine
"Computer Monitors and Controls All Truck-Shovel Operations," 90, no. 3 (March 1985).

"Computers Monitor Conditions, Operations at U.S. Mines," 91, no. 12 (December 1986).

"Cross-Pit Mining System Proving Its Capability in Texas," 92, no. 8 (August 1987).

"Diesels Pull Through Underground," 92, no. 5 (May 1987).

"Dragline Productivity — It's Important to Get It Right in the Planning Stage," 81, no. 7 (July 1976).

"Electronics Making Machines Brainy," 90, no. 8 (August 1985); no. 10 (October 1985); 91, no. 2 (February 1986).

"Feeder-Breaker Speeds Coal Out of Pit by Conveyors," 92, no. 6 (June 1987).

"Holton Automates Its Longwall With a Shearer Initiation System," 92, no. 7 (July 1987).

"In-Pit Conveyors and Crushers Cut Surface Mining Costs," 90, no. 7 (July 1985).

"It Pays to Squeeze More Out of Truck-Shovel Operations," 92, no. 12 (December 1987).

"Longwalls Advance Automatically," 90, no. 2 (February 1985).

"Microprocessor-Aided Equipment Proves Productive and Reliable," 92, no. 7 (July 1987).

"Mobile Roof Supports Make Pillar Extraction Safer," 91, no. 5 (May 1986).

"New Technology Boosts Production and Lowers Costs," 91, no. 7 (July 1986).

"Out-of-Sight Operator Controls Mining Machine," 90, no. 2 (February 1985).

"Stemming the Slide in Productivity is a Job for Both Machinery Manufacturer and Mine Operator," 81, no. 7 (July 1976).

"Wide Variety of Models and Types Offered by Manufacturers of Surface Mining Equipment," 81, no. 10 (October 1976).

Coal Mining Magazine

"Deserado Gains Edge Through Computer Monitoring," 24, no. 12 (December 1987).

"Freedom Mine Sets Sights on 9.9 Million Tons Per Year," 22, no. 7 (July 1985).

"Longwall Automation Grows," 24, no. 5 (May 1987).

"Microcomputer Keeps Watch at Emerald Mine," 24, no. 4 (April 1987).

"Remote Control Promises Safety and Productivity Gains," 22, no. 10 (October 1985).

"Safety and Productivity: The Search Continues at BOM," 23, no. 3 (March 1986).

"Terry Eagle Coal Fine-Tunes Production Systems," 22, no. 12 (December 1985).

"U.S. Underground Mine Monitoring: A Technology Comes of Age," 22, no. 12 (December 1985).

Edward F. Denison, "Effects of Selected Changes in the Institutional and Human Environment Upon Output Per Unit of Input," *Survey of Current Business* 58, no. 1 (January 1978).

Keith Dix, "Work Relations in the Coal Industry: The Handloading Era, 1880–1930," in *Case Studies in the Labor Process,* ed. Andrew Zimbalist (New York: Monthly Review Press, 1979).

Arthur Donovan, "Carboniferous Capitalism: Excess Productive Capacity and Institutional Backwardness in the U.S. Coal Industry," *Materials and Society* 7, no. 3/4 (1983).

Warren D. Devine, Jr., "Electricity and Technical Progress: The Bituminous Coal Mining Industry, Mechanization to Automation," ORAU/IEA-87-1(M) (Oak Ridge, TN: Institute for Energy Analysis, Oak Ridge Associated Universities, 1987).

Howard N. Eavenson, "Seventy-Five Years of Progress in Bituminous Coal Mining," in *Seventy-Five Years of Progress in the Mining Industry, 1871–1946*, ed. A. B. Parsons (New York: American Institute of Mining and Metallurgical Engineers, 1947).

Engineering and Mining Journal
   "Improving Surface Mine Productivity: In-Pit Crushing," 188, no. 8 (August 1987).
   "Improving Surface Mine Productivity: Truck Dispatching," 188, no. 8 (August 1987).

Bernard A. Gelb and Jeffrey Pliskin, *Energy Use in Mining: Patterns and Prospects* (Cambridge, MA: Ballinger Publishing, 1979).

Hittman Associates Inc., "Fuel and Energy Consumption in the Coal Industries," FEA-EI-1659 (Columbia, MD: Hittman Associates, 1974).

Hittman Associates Inc., "Underground Coal Mining: An Assessment of Technology," EPRI AF-219 (Columbia, MD: Hittman Associates, 1976).

Personal communication, Clyde Huffstutler, Division of Industry Productivity and Technology Studies, U.S. Bureau of Labor Statistics, Washington, D.C., November 1985.

Robert L. Marovelli and John M. Karhnak, "The Mechanization of Mining," *Scientific American* 247, no. 3 (September 1982).

H. F. McDuffie, Kenneth Warren, and Harvey S. Leff, "Mining," in *Industrial Energy Use Data Book*, ed. Lisa Carroll (Oak Ridge, TN: Oak Ridge Associated Universities, 1980).

William H. Patchell, *The Application of Electric Power to Mines and Heavy Industries* (New York: Van Nostrand Co., 1913).

Libby Rittenberg and Ernest H. Manuel, Jr., "Sources of Labor Productivity Variation in the U.S. Surface Coal Mining Industry, 1960–1976," *The Energy Journal* 8, no. 1 (January 1987).

U.S. Bureau of Labor Statistics, *Technological Change and Productivity in the Bituminous Coal Industry 1920–1960*, Bulletin No. 1305 (Washington, D.C.: GPO, 1961).

U.S. Bureau of Labor Statistics, *Technology, Productivity and Labor in the Bituminous Coal Industry, 1950–79*, Bulletin 2072 (Washington, D.C.: GPO, 1981).

U.S. Bureau of Mines, *Minerals Yearbook* (Washington, D.C.: GPO, 1969).

U.S. Congressional Research Service, *The Coal Industry: Problems and Prospects*, stock number 052-070-04767-4 (Washington, D.C.: GPO, 1978).

U.S. Energy Information Administration, *Annual Outlook for U.S. Coal 1985*, DOE/EIA-0333(85) (Washington, D.C.: GPO, 1985).

U.S. Energy Information Administration, *Coal Data: A Reference*, DOE/EIA-0064(84) (Washington, D.C.: GPO, 1985).

U.S. Energy Information Administration, *Quarterly Coal Report*, DOE/EIA-0121 (84/2Q) (Washington, D.C.: GPO, 1984).

Daniel R. Walton and Peter W. Kauffman, "Preliminary Analysis of the Probable Cause of Decreased Coal Mining Productivity (1969–1976)," FE/8960-1 (Reston, VA: Management Engineers, Inc., 1977).

W. L. Wearly, "Technological Solutions to Declining Productivity in Underground Mining," *Mining Congress Journal* 60, no. 7 (July 1974).

World Mining Equipment Magazine

"Earthmoving Equipment Trends," 9, no. 11 (November 1985).

"Surface Mining: Equipment Selection and Use are the Keys to Profits," 10, no. 1 (January 1986).

"Truck Selection in Surface Mining," 10, no. 1 (January 1986).

9 | Transportation: Electricity's Changing Importance Over Time

Calvin C. Burwell

The advantage of flexibility in delivering electric energy by wire to stationary work locations becomes a major disadvantage when mobility is required. As we have seen for underground coal mining, the disadvantage of providing electricity for mobile machinery can be overcome if conditions warrant, but this has not been the case for the majority of transportation services. Thus, while the major theme of this book concerns the strong ties between electricity and technological progress, in mobile applications energy's ties with technological progress are manifested mainly in liquid fuels combined with the internal combustion engine — not in electricity combined with the electric motor.

Even so, the ties between electricity and mobility are not completely absent. In transportation they existed at an early date, particularly in streetcars and other urban transit systems, but then the electricity ties weakened, essentially because the internal combustion engine was more compatible with the requirements of mobility. In the future, the links between technological progress in mobile applications and electricity may again become strong, but this depends upon technologies that are still in a developmental phase, particularly those for electric automobiles.

ELECTRIFIED URBAN RAIL SYSTEMS

Before 1830 America was largely rural. But in New York, the nation's largest city with a population of about 200,000 people, one of the nation's first public transportation conveyances, a horse-drawn stagecoach, was established in 1827.

Within a few years the efficiency of wheels on rails had been demonstrated and this spurred the development of horse-drawn street railway systems. In 1832, service began on the nation's first street railway in New York City. By 1880, more than 400 street railways had been established in urban centers.

The expansion of cities eventually caused the use of horsecars to run into limits. Manure in the streets became a growing nuisance. In addition, with the need to maintain five to ten horses for each streetcar, many cities incurred heavy costs in the care and feeding of several thousand horses devoted to streetcar service. A great epidemic claiming thousands of horses in 1872 forced cities to look to mechanization of the streetcars.[1]

San Francisco placed the first cable car system into operation in 1873 powered by a stationary steam engine in a central power station. The cable ran under the track between the rails through a system of pulleys. Within a few years, a number of other cities realized the advantages of a cable car system. By the 1890s, when cable railways reached their zenith, two dozen American cities had cable systems in operation. Not only did the cable railway eliminate the many expenses associated with animal traction, it also allowed speeds of 9 to 10 mph, significantly faster than any horse-pulled streetcar.

Although the cable cars had their advantages — they were relatively quiet and clean, and fairly reliable in operation — they also had their shortcomings. Cable systems required a high amount of maintenance. In addition, considerable energy was expended in keeping heavy cables several miles long in continuous motion whether or not the cars were moving.

To overcome these and other difficulties, attention turned to the electric motor as a means of powering streetcars. Thus, in 1885 a two-mile passenger railway in Baltimore that had been using horsepower was electrified, and by 1888 at least twenty street railways in various cities covering nearly 100 miles were operating with electric traction.

Although the use of electricity to provide motive power for streetcars began to take hold, numerous problems hindered its widespread acceptance. Two principal problems were (a) designing a way to couple the motor to the drive wheels, and (b) discovering an effective way to transmit electric current from a central powerhouse to the cars.[2] Solutions to these and other problems were incorporated in the Richmond, Virginia, trolley system electrified in 1888, and this convinced those planning trolley systems in other cities that electricity was superior to other forms of motive power. Within a three-year period following 1888, over 100 cities adopted the use of electricity for their trolley systems.

City transit systems also spawned interurban railways in the early part of this century. Many of the interurban railways of New England were actually extensions of existing city streetcar lines along country roads. However, in many other areas of the nation, interurban lines were built to more substantial standards than most streetcar lines. The building of interurban lines began on a widespread scale in 1900 and was substantially completed by 1915.

The widespread use of private automobiles after World War I led to the demise of the interurban railway industry. By 1940 less than 20 percent of the interurban system, which at its peak covered more than 15,000 miles, was still operating.[3] Eventually, the electric streetcar and interurban rail systems gave way to fuel-powered buses and private automobiles.

Changes Over Time in Electrified Urban Transit

Rapid Transit Systems

In 1867, New York became the first city in the United States to operate a rapid transit system. Chicago followed with its own system in 1892. Boston and Philadelphia built rapid transit systems shortly after the turn of the century. These transit systems were initially powered by steam but were later converted to electric operation with the development of electric motors and remote control equipment.

As can be seen in Table 9.1, the few rapid transit systems that were in place in the late nineteenth and early twentieth century continued to be the only ones until after mid-century when new systems came into operation in several cities. The table lists cities having rapid transit systems in operation as of 1988 and those planned to start up in subsequent years. Start-up dates are somewhat ambiguous because sections of a system are often in operation long before the full system is completed.

Mileage data are also ambiguous because the full mileage originally planned may never be completed or the system may eventually be extended beyond original plans. As the table shows, the mileage for the post-1984 startups is small compared to earlier systems, except for Dallas. And there is now serious doubt whether the Dallas system will ever be built.[4]

Indeed, there is currently little prospect for major new systems beyond those listed, although a study by the Urban Mass Transportation Administration[5] has identified nine other cities that should be prime candidates for rail rapid transit: Cincinnati, Columbus, Denver, Honolulu, Houston, Kansas City, Louisville, Minneapolis, and Seattle. The study also identified desirable route extensions of transit systems in cities that presently have them. Total mileage of the new transit systems plus extensions of the existing rapid transit lines would be about 350 miles. The total construction cost for these projects is an estimated $19 billion (1982 dollars). The major benefit claimed for the new transit lines would be the displacement of private automobiles and the associated major savings in oil, in addition to the environmental benefits of reduced traffic congestion and improved air quality.

Although there is extensive debate concerning the relative costs and benefits of rapid transit systems, there is no question about the form of energy most suited for their propulsion. Electricity produces no fumes or soot, and it is the only practical form of energy to power systems in which so much of the trackage is in tunnels underground.

Table 9.1   Rapid Transit Systems in the United States

	Year of startup	Miles in operation	Additional miles under construction or approved
New York	1867	258	10
Chicago	1892	98	10
Boston	1901	42	
Philadelphia	1907	38	
Cleveland	1955	19	
San Francisco	1972	72	2
Washington, D.C.	1976	70	31
Atlanta	1979	25	28
Baltimore	1983	14	2
Miami	1984	21	24
Buffalo	1985	6	
Portland (light rail)[a]	1986	15	
San Diego (light rail)	1986	6	11
Sacramento (light rail)	1987	18	
Pittsburgh (light rail)	1987	11	1
San Jose (light rail)	1988	8	12
St. Louis (light rail)	1992 (est.)		18
Los Angeles	1992 (est.)		18
Dallas	1999 (est.)		52

SOURCES: "LRT Guide," *Mass Transit* (March 1987); "World Metro Guide," *Mass Transit* (October 1987); and *Transit Fact Book,* American Public Transit Association (1985).

[a] "Light rail" is a type of electric transit vehicle railway with a "light volume" traffic capacity compared to "heavy rail." Light rail may be on exclusive or shared rights-of-way, high or low platform loading, multicar trains or single cars, automated or manually operated. In generic usage, light rail includes street cars, trolley cars, and tramways; in specific usage, light rail refers to very modern and more sophisticated developments of these older rail modes. The other entries in this table refer to heavy rail. Heavy rail is a type of transit railway with the capacity for a "heavy volume" of traffic and characterized by exclusive rights-of-way, multicar trains, high speed and rapid acceleration, sophisticated signaling, and high platform loading. Also known as subway, elevated (railway), or metropolitan railway (metro).

Trolley Buses

Trolley buses, sometimes referred to as trolley coaches, are electrically driven transit buses that receive their power from a pair of overhead trolley wires. They operate over the roadway on pneumatic tires and are in some ways like diesel-driven transit buses.

Trolley buses were first used in this country in 1910 in a suburban Los Angeles community. The steep grades on this route made other forms of public transportation, with their less reliable traction, impractical. Trolley buses were soon employed in other cities where, for example, streetcar tracks could interfere with mainline railroad tracks. Philadelphia installed its trolley bus system in 1923 for this reason.

In the 1930s, trolley bus systems achieved widespread acceptance, and they were installed in cities throughout the United States. Street railway systems were deteriorating and major investments were required to repair tracks and rebuild the equipment. In addition, the introduction of the automobile had caused ridership on the streetcars to decline, thereby rendering certain streetcar routes uneconomical. In many instances, the electrical distribution system was in good order and conversion of a streetcar line to a trolley bus line was relatively inexpensive.

By the early 1950s, when trolley bus use reached its peak, 54 cities had systems which totaled 6,500 vehicles nationwide. Many of the streetcar systems had been phased out by this time and motor buses, except for lesser traveled routes, were usually not as economical to operate as trolley buses.[6]

After 1950, trolley coach systems went into a period of decline. The overall decline in the urban transit industry was a major cause, but there were numerous other factors. New diesel buses that were competitive with the trolley coach in terms of both performance and operating costs became available. Highways were improved as automobile use increased. Many streets were converted to one-way thoroughfares, thus requiring extensive replacement of the overhead wiring if trolley coaches were to continue operating. The move to suburbia meant extending routes to provide service to areas having lower population densities.

As of 1970, only five cities in the United States — Boston, Philadelphia, Dayton, Seattle, and San Francisco — still operated electric trolley buses, but many buses were in need of replacement. Each of these cities had its own reasons for keeping the trolley buses running. Boston, for example, had a partial underground operation that would have required mechanical ventilation had motor buses been used through the tunnel. In Dayton, the president of the bus company, a strong advocate of the trolley bus system, turned the national decline in trolley bus use into a benefit for his own company by buying used equipment at salvage prices from other city transit systems. San Francisco found the trolley buses more suitable for the hilly terrain along many of the routes in the city, and public support for the trolley bus was strong.

In the late 1960s, Toronto, Canada, completed a study of its trolley bus system and found that continued operation of its electric buses would cost 28 percent less per mile to operate than motor buses. A manufacturer was found to produce a large order of new buses for that system, and the overhead wiring system was improved. This action led other cities to reconsider their systems and served to renew interest in the trolley bus industry. Since 1969, the five cities in the United States that operate trolley systems have all purchased new electric trolley buses and have planned extensive upgrading of their electrical distribution systems.[7]

It is difficult to generalize about the costs of trolley bus operation compared with motor bus operation because of the unique situations that are encountered in each city and because the motor and trolley buses are usually subjected to different types of service demand. Motor buses are more likely to serve suburban areas with fewer stops and higher average speeds, whereas trolley buses are assigned the more demanding urban areas with more frequent stops and slower average speeds. Thus, historical cost data may show a lower cost per mile for motor bus operation than the cost that would have resulted if motor buses had been operated under the same service conditions as trolley buses.[8]

There are also environmental factors to be considered. Whereas electric bus service requires overhead wiring (a form of visual pollution), motor buses generate far more noise and produce thermal and exhaust pollution.

Conservation of petroleum-based fuels is another consideration. There is a potential market for over 9,000 trolley buses in this country which, if put in place, would replace nearly 90 million gallons of diesel fuel per year.[9] However, expansion of the electric trolley system in the United States would require that cities, perhaps in conjunction with private business interests, develop financial plans to provide for the initial investment in fixed facilities.[10]

ELECTRIC AUTOMOBILES

In 1899, despite their limited performance and a price twice that of their gasoline-engine counterparts, 1,575 electric passenger cars were produced in the United States compared to 936 gasoline-powered vehicles. The electrics had many advantages: quietness, reliability, easy starting, low maintenance and vibration, freedom from odor and heat, and freedom from the use of oil and gasoline. For the time, their relative performance was good; in 1900 an electric automobile won a 50-mile race on Long Island, averaging more than 24 mph.[11]

It seems strange now that the battery-powered personal automobile was favored briefly so early in history. However, the distances traveled were short, and the alternatives of horse-drawn, steam and "infernal" combustion carriages were not that attractive. All were expensive to own and to care for, and initially none offered an advantage in speed, comfort, or convenience. However, continued improvement in the performance of internal combustion engines, their increasing public acceptance, and the development of a supporting infrastructure (places to buy gasoline and oil) while electric vehicle technology lagged, led to the dominance of the gasoline-powered car. By 1904 more than 85 percent of the cars produced in the United States had gasoline engines.[12]

Current Developments

Recent years have seen a renewed interest in the electric automobile in response to rising concerns about the availability, price, and security of imported petroleum supplies and the problems of environmental pollution connected with the use of internal combustion engines.

Natural resources of petroleum are relatively limited, but, more important, global distribution of the remaining reserves is such that the United States has grown increasingly dependent upon oil supplies originating in insecure regions of the world. Total automotive transportation (including trucks and buses) today accounts for 50 percent of total oil consumption in the United States — an amount that exceeds the nation's total oil imports. This situation has produced adverse effects on the U.S. balance of trade and poses an overhanging threat to national security.

If electric automobiles were to become an important transportation mode, they could not only relieve the oil import problem, but could also help to reduce the environmental pollution associated with the use of the internal combustion engine. Electric vehicles are essentially nonpolluting at the point of use. There are pollutants associated with the central station generation of electricity, but they can be more easily controlled at a central location than those produced by millions of individual internal combustion vehicles.

Electric vehicles using today's technology already find acceptance for special purposes such as golf carts and forklift trucks and in enclosed or geographically confined settings such as airport terminals. Some retirement communities are being designed so that electric vehicles serve all community requirements for shopping, golfing, and so on; thus, the primary vehicle in these communities is electric, with a second conventional automobile used only for special purposes, such as long trips. Interestingly enough, according to the 1977 Nationwide Personal Transportation Study, an electric vehicle with a range of 100 miles could electrify more than 60 percent of all of the travel of multicar households.[13] Such electrification and the consequent use of electricity for home battery recharging could increase residential electricity consumption by as much as 30 percent.[14]

The operating cost of electric automobiles is competitive. For example, small passenger vehicles consume 0.3 to 0.6 kWh per mile including all inefficiencies after the electric meter. At 0.5 kWh per mile and 6¢ per kWh, the energy cost of operation is 3¢ per mile. This is equivalent to a gasoline-fueled vehicle capable of delivering 30 miles per gallon in city driving at a gasoline price of 90¢ per gallon.

Electricity costs for battery recharge are likely to be low because the recharging would usually occur during off-peak hours. Many electric utilities encourage the use of off-peak power by selling electricity during off-peak periods at prices much lower (e.g., 50 percent lower) than at other times. Given mass production economics, the first cost of electric vehicles should also compare favorably with today's gasoline-fueled automobile in many situations.[15]

Although the operating costs of electric vehicles are competitive, higher initial costs and lower levels of performance and comfort (i.e., auxiliaries) result in their not being competitive with current motor vehicles. The primary performance issue is the limited range between battery recharges. Moreover, because the power source is the limiting feature, other power equipment, including air conditioning, can be provided only at the expense of decreased range. Current strategy calls for continuing research and development as well as the introduction of electric vehicles where their disadvantages are of less concern.

The most favorable near-term market is for fleet service vans for activities such as mail delivery and telephone service on urban routes where distances are short, the climate is moderate, operations are mainly during the working day, and routine service and maintenance is available. In the United States since 1985, twenty-two companies including eleven utilities have been testing the GM Griffon van in their service fleets. Through August of 1987 these vehicles had been driven 320,257 miles, and worldwide similar vehicles have logged over 5 million miles.[16]

Despite the potentially favorable impacts of their use, there are major problems connected with the development of commercially successful electric vehicles. The biggest problem is that state-of-the-art lead-acid batteries are presently limited to less than 100 miles between stops for recharging. Moreover, the battery pack is expensive and may need to be replaced once or twice during the life of the vehicle.

Batteries based on iron-nickel cells are being developed that are capable of delivering about twice the energy of lead-acid batteries, and have the potential to perform favorably when compared to internal combustion engines in many situations.[17] Development of even higher performance cells with the potential for automobile use in intercity travel is also under way. If one of these advanced concepts — such as the high-temperature sodium-sulfur cell, which has about four times the capacity of the lead-acid cell — is successfully developed, the performance and range of electric vehicles could be on a par with that of today's gasoline-fueled automobiles.[18]

Unless there are unexpected technological breakthroughs or severe oil or pollution crises, the chances are that electric vehicles will only gradually work their way into the transportation picture. They will first capture market niches for which they are particularly adapted — for example, urban fleets. Then as electric vehicle technology improves and manufacturing costs decline, they should penetrate other strata of the market.

RAILROAD MOTIVE POWER

Early Electrification

The prevailing technology in long-distance railroads at the turn of the century relied upon coal-steam locomotives as the source of motive power. However, even within this technology, an essential auxiliary role for the use of locomotives powered by electricity arose from the need for a nonpolluting source of power for moving trains through tunnels and densely populated urban centers. The first important installation occurred in 1895 in a Baltimore & Ohio Railroad tunnel in the city of Baltimore.

Another major instance of electrification as a means of improving air quality occurred after the turn of the century at the Grand Central Station in New York City. In this case large capital expenditures to replace otherwise adequate existing trackage were apparently not justified in strictly commercial terms. But the approximately 700 steam trains entering or leaving the station daily created an intolerable pollution problem. Not only did the smoke and cinders foul the air, they also severely restricted vision in the two-mile Park Avenue tunnel that led trains into Grand Central. Several serious accidents occurred because signals were impossible to read. One particular accident created a public furor that resulted in the state legislature passing an act prohibiting steam operation into the station after July 1908.[19]

Although electrification of Grand Central Station had been mandated by the state legislature for health and safety reasons, the economics of the new electric system also proved to be quite favorable in some respects. Electric locomotives consumed power only when they were moving trains, whereas steam locomotives consumed power at all times unless they were completely shut down. Electric locomotives were available for service much more of the time with lower maintenance costs than steam locomotives. The electric locomotives also eliminated smoke clean-up problems, an added economic incentive. These direct benefits of electrification in turn allowed the eventual construction of buildings over the tracks leading to the terminal, an important indirect economic benefit given the high cost of land in New York City.

The New Haven Railroad used Grand Central Station and was therefore also obligated to electrify its operations in New York City.[20] The first regular trains to operate with electric power on the New Haven left Grand Central Station in July 1907. Extension of the electrified system continued until 1927, at which time over 600 miles of the New Haven line were electrified.[21]

Several other railroads also made decisions to electrify a portion of their lines because of tunnels. Another early tunnel electrification was that of the St. Clair Tunnel between Sarnia, Ontario, and Port Huron, Michigan, completed in 1908. Two years later the Michigan Central, a subsidiary of the

New York Central, opened a tunnel under the Detroit River. In this case, primary power was supplied by Detroit Edison, one of the first times a railroad had relied on an electric utility company rather than generating its own electricity. Several other tunnel railroads were electrified in the subsequent years.

Two western railroads made decisions to electrify their lines in the early 1900s solely for economic reasons. The first of these was the Butte, Anaconda, and Pacific Railway in Montana, which was built primarily to serve the Anaconda Copper Mining Company near Butte, Montana.[22] Electrified operations began in 1913 and continued until 1967, by which time a change in mining operations had greatly reduced traffic over the line to the point where electrified operations were no longer economic, and diesels were substituted.

The second western railroad to electrify its lines was the Chicago, Milwaukee, and St. Paul Railway. Electrification was installed first on a section of line running through Montana in 1915. That year, severe weather conditions revealed the clear-cut advantages of electric locomotives over steam when temperatures dropped as low as $-40°F$, severely reducing steam locomotive efficiency.[23]

The greatest electrification project of any railroad in the United States was that of the Pennsylvania Railroad. The Pennsylvania had experience with electric railroading as early as 1895 when it electrified a seven-mile branch line in New Jersey. In 1910, the Pennsylvania built its own terminal in New York City. Between 1913 and 1928, the Pennsylvania installed electricity on several of its suburban lines in and around Philadelphia. Then in 1928, the railroad announced plans to electrify 325 route-miles between New York, New York; and Wilmington, Delaware; and between Trenton, New Jersey; and Columbia, Pennsylvania; that were based upon the use of steam locomotives. These lines carried some of the heaviest traffic in the nation. Even though many of the route-miles had four main tracks, the capacity of the railroad had been reaching its limit along these busy lines.[24] Within a year, it was announced that the initial program would be expanded to include track all the way to Washington, D.C.

In 1935, passenger service to Washington began using electric locomotives. Electric freight service into Washington began the following year. In 1938, the electrified system was extended to Harrisburg, Pennsylvania, the last major expansion of electrification on the Pennsylvania line. Over 650 route-miles and over 2,100 track-miles were eventually electrified, most of which is still in operation today.

The overall size of the Pennsylvania's electrified system made its operations more economic than was the case for railroad systems making only limited use of electricity.[25] Thus, when the advent of the diesel locomotive led many other railroads to decide to abandon their electrified systems, the

Pennsylvania continued to upgrade its existing electrified lines and equipment.

The success of early electrifications appeared to ensure the widespread use of electricity by railroads in the United States. While at first the electric locomotive was limited to situations in which smoke and soot from steam locomotives had created a nuisance, the economic advantages of electric locomotive operations compared with steam engines soon became evident. In addition to lower maintenance costs and the fuel savings in terminal areas, electrification allowed the capacity of heavily traversed mainlines to be increased without laying additional track. (For more explanation, see note 24 at the end of this chapter.) And fuel savings, even in mainline railroad service, were substantial when electric locomotives were used.[26] Although the electric locomotives themselves were generally more expensive than steam locomotives of comparable power, the greater availability of the electrics and their much longer service lives could outweigh the initial cost considerations. At the same time, on heavily traveled lines the savings in operating costs could offset the large capital outlay for line electrification.

Greater operational flexibility was also achieved. Whereas steam locomotives were designed for one forward direction, the symmetrically designed electric locomotives permitted travel in either direction without having to turn the locomotives around. Operation of the electric locomotives was much simpler, since engineers and firemen did not have a boiler to tend, and fueling stops were unnecessary.

Recent Trends in Railroad Motive Power

The high expectations for railroad electrification were not realized, however. Demand for freight and passenger services declined or leveled off during the 1920s and 1930s, until stimulated once again by wartime demands in the early 1940s. Without growth and the revenue created by high load factors for use of railroad capacity, plans for further railroad electrification were shelved. The electrification completed in 1938 by the Pennsylvania Railroad turned out to be the last major electrification project by any U.S. railroad.

Dieselization after World War II

The nation emerged from World War II with the need to modernize its railroad equipment. Diesel engine technology, proven in military tanks and submarines, offered an attractive alternative to both steam locomotives and railroad electrification and avoided the major disadvantages of each. (In a diesel-electric locomotive, a diesel engine drives an electric generator whose output is fed to the motors that turn the wheels.)

The diesel-electric locomotive eliminated the smoke and soot that precluded the use of coal-steam engines in terminals and tunnels. It also removed the need to shovel coal and trail a coal car. It eliminated the need to stop for water and avoided freezing problems associated with the use of water and the exhaust of steam in cold weather. With its electric drive and control system, the diesel-electric was comparable to the all-electric locomotive in its operating flexibility. It could operate in either direction, achieve easy startup from a cold shutdown condition, and operate multiple locomotives from one locomotive.

The overall energy efficiency of the diesel-electric locomotive was typically more than four times that of its coal-steam counterpart and was nearly comparable to that of the all-electric system.[27] The energy efficiency of the diesel-electric meant that the energy cost advantage that formerly favored the all-electric now favored the diesel-electric as well. So without an operating cost advantage compared to diesel-electric locomotives, there was no economic justification for incurring the high capital cost of electrifying trackage for all-electric trains.

Moreover, the volume of railroad passenger and freight services, stimulated briefly by wartime conditions, soon began again to level off or decline. The energy cost and other advantages of operating the diesel-electric locomotive resulted, after World War II, in the rapid displacement of steam engines (and later in the displacement of some small, isolated or underutilized electric systems). The diesel-electric locomotive became the railroad standard.

Energy Cost and Availability and Current Prospects

The Arab oil embargo in 1973 and subsequent events in the Middle East have raised concerns about the future cost and availability of liquid fuels. From 1970 to 1980, the cost of oil more than tripled relative to the cost of electricity for railroad operations.[28] As a result, the energy cost advantage in railroad operations once again began to favor the all-electric system.

A study mandated by Congress and undertaken for the Federal Railroad Administration in 1977 concluded that the 14 percent of the railroads that carried about 50 percent of the freight hauled enough freight each year over their 26,000 miles of track (i.e., in excess of 30 million tons) to justify the capital costs of electrification.[29] In a study published in 1982, capital costs (in 1980 dollars) were estimated at $15.7 billion, and the expected annual savings of $1.5 billion implied nearly a 10 percent rate of return on the investment (Table 9.2).

More detailed studies for particular rail segments followed soon after.[30] But as yet, modern electric railways for freight movement remain in the study stage, and the nation's railroad lines remain less than 1 percent electrified. In the 1980s, increased competition in the transportation industry, financial difficulties on the part of many carriers, and low diesel oil costs in the past few years, have largely precluded serious consideration of the major investments required for railroad electrification.

Table 9.2 Capital and Annual Costs and Credits for Electrification of 26,000 Route-Miles of Railroad ($1980, in Millions)

Investment item	Cost (credit)
Catenary[a]	6,040
Substations	3,000
Utility connections	390
Signaling and communications	4,400
Civil reconstruction	1,660
Electric locomotives	5,240
Diesel locomotives	(5,060)
Net investment	15,670

Operation item	Annual operating costs (credits)
Diesel locomotive replacement	(281)
Diesel energy	(1,500)
Electric energy	1,110
Diesel locomotive maintenance	(1,360)
Electric locomotive maintenance	409
Catenary maintenance	114
Net annual savings	1,508

SOURCE: W. H. Sutton, "Electrified Rail and Rapid Transit Applications and Outlook," in *Proceedings of the Atomic Industrial Forum Workshop on the Electric Imperative*, Monterey, CA (Washington, D.C.: AIF Inc., March 1982).

[a] The term "catenary" in the context of electric railroads refers to overhead power distribution equipment.

RAILROAD ELECTRIFICATION IN OTHER COUNTRIES

Although there is great potential for using electricity to power railroads, the trend over the past forty years in the U.S. has been to reduce the electrification already in place. Less than half the route-mileage of steam railroads that was once electrified still remains electrified.[31] This situation is in sharp contrast to that prevailing in most other industrialized countries, where electrification of railroads has been undertaken on a vast scale (see Table 9.3). Several countries, especially the USSR, West Germany, Poland, and India, have installed significant electrified mileage during the past twenty years.[32] A substantial portion of railroad movement in other countries is now done with electric power.

In many countries the total number of electrified route miles has not necessarily increased, and in a few it has decreased by significant amounts over the past two decades, as a result of reductions in overall railroad mileage. For instance, France has reduced its total mileage by over 2,400 miles, the United Kingdom by about 7,000 miles, and West Germany by 1,600 miles. These highly industrialized countries developed their railroad networks before the advent of modern highway trucks and motor buses, so

Table 9.3   Electrification of Railroads in Selected Countries

Country[a]	Percentage electrified 1980	Country[a]	Percentage electrified 1980
Japan	100	West Germany	34
Switzerland	99	USSR	30
Sweden	63	France	27
Netherlands	59	Poland	26
Norway	58	Czechoslovakia	25
Austria	52	United Kingdom	19
Italy	51	Yugoslavia	19
Brazil	36	India	15
Spain	36	United States	1

SOURCES: *Jane's World Railways, 1961–1962, 1970–1971,* and *1980–1981* (London: annual); and *Association of American Railroads 1981 Fact Book* (New York: annual).

[a] Only countries with 1,000 or more route-miles of electrified railroads in 1980 are included (excludes narrow-gauge railroads—less than 4'8½").

some of their railroad lines have been abandoned for economic reasons. Yet, of the countries that have abandoned considerable trackage, all but the United States have progressed significantly with the electrification of the remaining railroads.

Most of the countries included in Table 9.3 have maintained a relatively stable railroad network. The exceptions are Brazil, which has developed 1,500 miles of new lines to reach its interior points; India, which has increased its mileage by 2,200 miles by both the construction of new lines and the conversion of narrow-gauge lines,[33] and has built 800 miles of high-speed track devoted primarily to express trains; Yugoslavia, which has increased its mileage by 1,900 miles; and the USSR, which now has an additional 10,000 miles of lines as the result of a newly constructed transcontinental railroad and other recently constructed lines.

A number of countries not included in Table 9.3 are progressing with railroad electrification projects but at present have less than 1,000 miles of electrified lines. Australia has recently electrified a number of its suburban rail lines. Tentative approval has been given to electrify the 600 miles of rail line between Sydney and Melbourne, one of Australia's busiest trunk lines. The Australian government has indicated that, in principle, electrification of numerous other lines would be desirable, but a more urgent need in the country is the conversion of some of its busiest rail lines to standard gauge.

Belgium and Bulgaria will soon have over 1,000 miles of electrified rail routes each. Both of these countries have about 2,500 miles of total rail routes. Finland, East Germany, Hungary, and Mexico have major rail electrification projects underway, although the total electrified route-miles for each of these countries will remain below 1,000 miles after completion of these projects. In China, 87 percent of all trains were still powered by steam as late as 1978. With approximately 30,000 miles of railroad routes, China now has or is in the process of installing nearly 1,000 miles of electrification on rail lines converging on Beijing.[34]

Various reasons exist for electrification in different countries. For instance, railroad electrification allows Australia to make use of its large coal reserves rather than continuing to depend on diesel fuel. Austria, with abundant hydroelectric resources, cites independence from foreign coal supplies as a reason for electrification. (Until fairly recently, coal-fired steam locomotives were still operating in Austria.) Numerous countries besides Austria have continued to operate steam locomotives well into the 1970s: Czechoslovakia, East Germany, India, Bulgaria, China, and South Africa. For these countries, electric power has economic as well as operating advantages over steam power. Inducements to convert to electric power (or diesel) from steam are much greater than those to convert from diesel power.

The fact that railroads in other countries are nationally owned has made it easier for foreign railroads to finance the initial construction of electrified systems. In the United States, railroad prosperity seriously declined in the years following World War II, leaving very limited resources for any capital improvements, let alone major electrifications. In addition, after World War II the railroads of the United States remained intact, but the railroads of Japan and many European countries were extensively damaged. Since complete rebuilding of many foreign railroad systems was necessary (which meant new tracks, locomotives, and rolling stock), electrification of these "new" railroads could be planned at the outset.

Differences in the economies of the various countries have also led to patterns of transportation that are quite dissimilar. Most U.S. citizens have been able to afford automobiles, and new public highways have been constructed for their use. In many other countries, only the wealthy can afford private transportation, so the demand for government-financed modern, public transportation has been much greater.

ELECTRICITY'S ROLE IN REDUCING TRANSPORT NEEDS

Electrotechnology and its products play a substantial though indirect role in reducing the need for transportation services. They do so by decreasing the weight that is transported, by decreasing the transportation distance that is required, and by transporting materials and energy in fixed conduits.

Weight Reduction

The weight to be transported can be reduced by preconcentration of the materials to be shipped and by the use of lightweight materials in vehicles and packaging. The preconcentration of mined materials such as metal ores and coal is often accomplished through mechanical operations powered by electricity. Some food products such as milk powder, frozen orange juice, and other frozen fruits and vegetables have some of the natural moisture removed before shipment. In addition, the frozen condition preserves the product and eliminates the requirement for special processing and packaging. For food products the moisture is often removed by evaporation under vacuum in order to retain the natural flavor and appearance as much as possible. Vacuum evaporation and food freezing again depend upon mechanical operations and the related use of electricity for motor drive.

The use of lightweight materials is illustrated not only by electroprocess products such as aluminum, but also by the growing use of plastic materials. Most of the energy for the production of plastics comes directly from fuels but, even so, two to three times as much electricity is used per pound for the production of plastic as is used per pound for the production of the materials they displace, such as glass, steel, and paper products.

Aluminum is used in automobiles to reduce their weight and thereby improve their fuel economy. The use of aluminum grew from 80 pounds per automobile in 1975 to 140 pounds in 1983.[35] Aluminum is also displacing steel in utility truck bodies and railroad hopper and gondola cars. Trucks equipped with aluminum service bodies are 400 to 500 pounds lighter than their steel counterparts. The reduced weight means that a three-quarter-ton vehicle with two rear wheels can provide the same payload formerly provided by a one-ton vehicle with four rear wheels.

Aluminum gondola and hopper rail cars use up to 5 tons of aluminum to displace 12 tons of steel. They reduce diesel fuel use by 0.06 to 0.08 gallons per car-mile and increase the locomotive payload about 6 percent.[36] Finally, aluminum is now used for more than one-third of all food and beverage containers — nearly 50 billion units a year.[37] A 12-fluid ounce aluminum can weighs 0.043 pounds, less than one-tenth as much as a glass bottle.

The use of plastics in vehicles and containers to reduce weight is analogous to the use of aluminum but the specific applications differ. The use of plastics has grown from 155 pounds per automobile in 1975 to 220 pounds in 1984. Plastics are used for body parts and trim where the requirement for strength is secondary to those of appearance and corrosion resistance, whereas aluminum displaces steel applications in parts where strength and wear resistance are important, such as the engine head.

Plastics displace glass, steel, and paper containers for milk, wine, motor oil, and soft drinks. Plastic bags, trays, and film, which are much lighter in weight, displace paper meat trays and shopping bags. For example, polystyrene meat trays are only one-third as heavy as their cardboard counter-parts.[38]

Distance Reduction

Electric transmission systems have all but eliminated siting constraints imposed by the need for power in manufacturing. As a result, manufacturers can disperse fabrication and assembly operations as needed for close proximity to labor and product markets. To the extent that metropolitan areas generate sufficient quantities of obsolete scrap materials, they can be recycled locally by electric processes to supply local markets. This mechanism has already been responsible for substantially restructuring the steel industry. Local recycling of other materials such as paper, copper, and aluminum is also important today, and the recycling of rubber and glass are likely to be of growing importance in the future.

Electric processes dominate recycling operations because of their special characteristics. Examples are the high energy efficiency of arc melting steel scrap, the high quality of rubber devulcanized with electromagnetic radiation, the reduction in aluminum loss to oxidation when the aluminum is remelted electrically in a controlled atmosphere, and the high purity of copper metal recycled electrolytically. In addition, materials recycling tends to be decentralized and small scale compared to primary production, and capital costs tend to favor electric processing at low levels of production.

Transportation in Fixed Conduits

The high initial cost of providing a pipeline system precludes transport in this way for all but a few high-volume commodities such as oil; natural gas; industrial gases such as oxygen, nitrogen, and hydrogen in heavily industrialized areas; and, prospectively, major coal slurry lines. Pipelines are an energy-efficient way to move high-volume bulk commodities at low speed because no energy is required to move containers (including returning them empty) and because electric motors are efficient in providing the motive power. By far the most important application is in transporting and distributing water. More than 100 billion kWh of electricity is used annually to transport water for urban and rural uses. The energy required to produce this amount of electricity is equal to 5 percent of all the energy used for transportation services.

Energy distributed by the power transmission network eliminates the need for over-the-road transport of fuels as well as the weight of the transport vehicle and of parasitic tares: the water and ash content of wood and coal, or the heavy steel bottle needed to contain pressurized or liquefied fuel gases. For example, 25 million households now use electricity for part or all of the fuel otherwise needed for home heating. Each household using electricity instead of oil for water heating and space heating eliminates the annual distribution of about three tons of oil — more if coal, wood, or bottled gas is displaced. Of course, fuel must still be shipped to the power plant, but that is centralized in contrast to the need for distribution to millions of homes.

AN UNCERTAIN FUTURE

Transportation in the United States today depends, with few exceptions, upon equipment that is tied to the use of liquid fuels. This was not always the case, nor is it as much the case in many other countries. Electricity was very important early in the twentieth century when, having taken over from horses and steam power, electric motors powered the trolley cars and trains that made up the nation's urban transport system. Even electric automobiles were comparatively important in the infancy of the motor car industry, but the internal combustion engine became the standard automobile power plant early on and continues to be to this day. Electricity also powered many U.S. railroad lines during the first half of the century, and for a time it looked to be the wave of the future in replacing steam locomotives, but diesel-electric engines became the dominant technology following World War II.

The electrification of transportation in the United States is far less prevalent than in many other countries where institutional, geographic, and demographic factors have made electrified railroads economically feasible. Nevertheless, technologies based upon the use of electricity have indirectly eased the burden on American transportation systems by making possible innovative methods for weight reduction, distance reduction, and movement via fixed conduits such as pipelines.

Whether the direct electrification of transportation will become of major importance in this country will depend largely on technological progress in battery research. Improved batteries could make electric automobiles a feasible economic alternative in achieving driving ranges and other operating features comparable to those of the conventional internal combustion engine. Concerns about protecting environmental air quality and reducing the dependence on insecure supplies of imported oil provide an added impetus for developing such batteries. Nevertheless, the outlook for electric automobiles remains uncertain.

Notes

1. Several means of mechanical traction were tested on streetcars. Small steam locomotives were tried as early as 1860, but their smoke, cinders, and noise frightened horses along the way and were generally unappealing to the public. Ammonia locomotives were also tried. Ammonia under pressure in tanks was used to drive pistons, then collected in water in another tank. The ammonia could be recycled by

heating the water. Another "fireless" locomotive used a soda motor: caustic soda was allowed to react with water to produce steam. Compressed air was still another method tested to propel streetcars. Although these means of motive power had their merits, none of them proved satisfactory under continuous operating conditions.

2. Frank Sprague is credited with the design and development in 1885 of a satisfactory method for mounting electric traction motors directly on an axle. Previously, motors had been mounted on the car body and connected to the driving axles with belts, cables, or other means of flexible connection. These driving mechanisms had been a source of endless trouble for streetcar operators. The new mounting method devised by Sprague permitted exact alignment of the motor with the driving axles, regardless of how uneven the track might be. Sprague incorporated this design into the streetcars that he built for the Richmond system. Notwithstanding this major improvement, however, the Richmond railway allowed Sprague to experiment in many ways to improve the streetcar and electrical distribution system. Armature brushes, originally made of copper, brass, or bronze, sometimes had to be replaced daily due to the rapid wearing of the soft material. Carbon brushes, which are much more durable, were finally adopted. Problems with gear breakage and chipping were eliminated with proper lubrication. Problems with the overhead trolley system were diminished when a satisfactory current collection device was used.

3. Electrical technology for the early interurban railways did not differ much from that used on street railways. Until 1904, nearly all interurban lines used low-voltage dc, since there was no effective alternative. Because of the extended distances involved on the interurban lines, however, a method had to be devised to compensate for the loss in line voltage at places far removed from the power supply point. Rotary converters, devices that can change the voltage level, were employed in substations along the interurban routes to remedy this problem. High-voltage current could be transmitted over great distances with relatively little line loss. At substations, the voltage was reduced by the rotary converters to a level that could be used in the interurban cars. The rotary converters could use ac electricity as input power, and as ac generation became more popular, most later interurbans adopted the use of ac to transmit electric power. Some of the first interurban lines were eventually converted to ac operations. Transformers were employed at a later date to supply ac directly to the interurban cars. This required the use of ac motors, some of them very similar to the dc motor.

Many of the rural areas served by the interurban routes did not have commercial power; for this reason many interurban companies also developed into power companies. A few became quite sizable and are still operating today.

4. At present the prospects for new mass transit systems are not good. Harvard economics professor John F. Kain discusses the reasons in some detail in his article, "Choosing the Wrong Technology: Or How to Spend Billions and Reduce Transit Use" in *The Journal of Advanced Transportation* 21 (Winter 1988): 197ff.

5. Boris S. Pushbarev, Jeffrey M. Zupan, and Robert S. Cumella, *Urban Rail in America: An Exploration of Criteria for Fixed Guideway Transit* (Bloomington: Indiana University Press, 1982); see also, "Rail Transit: Fourteen Forgotten Cities," *Railway Age* (June 14, 1982), 28–33.

6. The economics of trolley buses was generally more favorable than that of early diesel buses. Early motor buses required daily servicing for fueling and oil had to be housed in heated facilities in cold weather to prevent freezing of the cooling system, and required more maintenance and a large spare parts inventory. In addition, the trolley bus was superior in terms of acceleration and had a longer economic lifetime.

7. Ned L. Treat, *Electric Railroads and Trolley Systems: Past, Present and Future*, ORAU/IEA-83-123(M) (Oak Ridge, TN: Oak Ridge Associated Universities, Institute for Energy Analysis, March 1984), 37–38.

8. The results of a study in Vancouver, British Columbia, to determine whether an existing electric trolley bus operation should be refurbished or replaced with a motor fleet are summarized below. For this particular city, retaining the trolley bus system was clearly the best economic decision.

The costs would be significantly different for a city that did not have an existing overhead trolley system in place. Assuming that $400,000 is the average cost per mile of overhead trolley, 100 miles of overhead trolley would cost $40 million. Unless other trolley bus operation costs decline, the capital investment in overhead wiring appears to make the trolley bus uneconomical. However, mass production of trolley buses, as is done for motor buses, could significantly reduce total trolley system costs and perhaps make some new trolley bus systems economically advantageous.

Technical improvements in the control and propulsion systems of electric buses may result in additional advantages for trolley systems. Currently, most electric buses are equipped with resistor switch controls for accelerating and braking, and dc motors are now generally used to propel the buses. In the future, solid-state chopper controls may convert the dc electricity to three-phase, variable-frequency ac. This in turn would power an induction motor, which has no electrical connection either via slip rings or contact brushes, to the rotating armature. Such a motor requires far less maintenance. Because energy-wasting resistors would not be employed to control motor speed, less energy would be required to operate the vehicle.

9. J. D. Wilkens, A. Schwartz, and T. E. Parkinson, *The Trolley Coach: Potential Market, Capital and Operating Costs, Impacts, and Barriers: Task 2 Report for the*

Comparison of Trolley Bus and Motor Bus Costs in Vancouver, 1978

	Trolley bus (thousands of dollars)	Motor bus (thousands of dollars)
Vehicle purchases:		
1980–1983	36,500	24,440
1995–1998		13,850
Sale of present buses		−800
Special capital costs:		
Trolley bus overhead	570	
Diesel garage		2,230
Trolley bus substations	1,060	1,060
Maintenance and servicing:		
Trolley bus overhead	11,470	
Buses	22,990	33,490
Energy costs:		
Trolley bus electricity	19,230	
Diesel fuel		29,430
	91,820	102,640

SOURCE: John D. Wilkens, Arthur Schwartz, and Tom E. Parkinson, *The Trolley Coach Development and State of the Art: Task 1 Report for the Electric Trolley Bus Feasibility Study*, UMTA-IT-06-0193-79-1, prepared for the U.S. Department of Transportation by Chase, Rosen, and Wallace, Inc., 1979. Available from NTIS, Springfield, VA.

Electric Trolley Bus Feasibility Study, UMTA-II-06-0193-79-2. Prepared for the U.S. Department of Transportation by Chase, Rosen, and Wallace, Inc., 1980. Available from NTIS, Springfield, VA, 1980.

10. Small trolley fleets operate in numerous countries. Switzerland, with an abundance of hydroelectric power, has 99 percent of its railroad system electrified, and has several trolley bus systems operating within its cities. The USSR, which has the largest amount of electrified railroad trackage of any country in the world, has about 20,000 trolley buses in daily operation.

11. C. C. Burwell, *Roles for Electricity in Transportation*, ORAU/IEA-86-2(M) (Oak Ridge, TN: Oak Ridge Associated Universities, Institute for Energy Analysis, April 1986), 9.

12. J. L. Hartman, E. J. Cavins, and E. H. Hietbrinck, "Electric Vehicle Challenges Battery Technology," in *Proceedings of the Fifth Energy Technology Conference* (Washington, D.C.: Government Institutes, February 1978), 387; R. S. Kirk and P. W. Davis, "A View of the Future Potential of Electric and Hybrid Vehicles," in *Proceedings of the Seventh Energy Technology Conference* (Washington, D.C.: Government Institutes, June 1980), 582.

13. Michael M. Collins, "Impact of Electric Vehicles on Electric Utilities," in *Proceedings of the Atomic Industrial Forum on the Electric Imperative*, Monterey, CA (Washington, D.C.: AIF, March 1982).

14. Assumes 840 billion household vehicle miles by household cars, station wagons, and vans (see U.S. Bureau of the Census, *Statistical Abstract of the United States, 1987* [Washington, D.C.: GPO, 1987], Table 1033) at 63 percent electrification at 0.5 kWh per mile.

15. J. G. Asbury, J. G. Seay, and W. J. Walsh, *The Role of Electric Vehicles in the Nation's Energy Future*, ANL/SPG-26 (Argonne, IL: Argonne National Laboratory, November 1984), 20, 22.

16. Electric Power Research Institute, *Building the Electric Vehicle Future: EPRI's Vehicle Development Activities*, EMU3017 (Palo Alto, CA: Electric Power Research Institute, November 1987), 2 and 4.

17. The West German chemical company, BASF, was, in 1987, reported to be in the final stages of developing an automotive battery based on combining a lithium electrode with a newly developed conductive plastic (polypyrrol) that already matches the power density of nickel-cadmium batteries. It should be inexpensive to produce in quantity. See *Business Week* (August 17, 1987), 81.

18. Jacob Jorne, "Flow Batteries," *American Scientist* 71 (September–October 1983): 507–13.

19. Electrification was required by law over only a limited amount of track leading to Grand Central Station, but operating considerations led the railroad to decide to extend electrification to include two major suburban commuter lines. Legal barriers concerning the use of high-voltage ac in New York City and clearance restrictions that prevented the use of an overhead catenary system, plus the success with dc on the Baltimore and Ohio and the elevated railroad lines in Chicago and New York City, led to the selection of a dc third-rail system on the railroad.

20. Previous to the legislative requirement to electrify in New York City, the New Haven had installed dc power on several of its branch lines. This experience and the fact that the New Haven would have to share twelve miles of dc trackage leading in to Grand Central Terminal with the New York Central would seemingly make dc electrification the favored option. However, New Haven decided instead to use ac. Part of the reason was that technological advances had made ac more practical for some applications. One important device was the transformer, which allowed volt-

age step-ups and step-downs and in turn permitted more efficient transmission of electricity over long distances.

21. Although the railroad announced plans to electrify additional major portions of the New York–Boston mainline, which brought about speculation that this entire route would someday be electrified, less than half of this line was eventually electrified (prior to its being taken over by Amtrak). Nevertheless, only two other American railroads surpassed the electrified mileage of the New Haven, and much of the New Haven is still electrified today.

22. Although most railroads that were contemplating electrification after the turn of the century were choosing to do so with ac electricity because of its long-distance transmission advantages, the Butte, Anaconda, and Pacific chose to do so with dc electricity. Some success had already been achieved with the use of higher voltage dc on several interurban lines — 1,200 volts rather than 600 volts. General Electric, an advocate of dc, wanted an opportunity to test its favored current on heavy railroad equipment and thus developed a high voltage dc system for the Butte, Anaconda, and Pacific using 2,400 volts.

23. Electric operations continued until 1974 when inefficiencies associated with operating two disconnected sections outweighed the justifications for another rehabilitation of the equipment and electrical distribution system.

24. Locomotive pulling power is limited by wheel-rail adhesion and traction motor amperage capacity at lower speed ranges but is limited to the capacity of the prime power source at higher speeds. Since the prime power source is the electrical system supplying the railroad, electric locomotives essentially have a prime power source with unlimited capacity relative to their needs. Therefore over relatively flat terrain, the electric locomotive can pull the train faster than its steam-driven counterpart.

25. With its large electric system, the Pennsylvania Railroad could take full advantage of the electric locomotive's advantages in speed and reliability while avoiding the costs and delays associated with starting, stopping, and changing between electric and steam locomotive portions of systems not completely electrified.

26. Steam locomotives were inefficient in the use of coal (i.e., less than 8 percent of the thermal energy was converted to mechanical energy) whereas the thermal efficiency of electric power generation was increasing steadily (e.g., from 10 percent on average in 1920 to 24 percent on average in 1940). This meant the railroad company used less coal in generating power for its electric traction operations per unit of power than for its steam traction operations. S. S. Penner and L. Icuman, *Energy 1, Demands, Resources, Impact, Technology and Policy* (Reading, MA: Addison-Wesley Publishing Co., 1974), 237.

27. Ibid., 237.

28. Edison Electric Institute, *Statistical Yearbook of the Electric Utility Industry* (Washington, D.C.: EEI, various years). The cost of oil to railroads for these calculations has been assumed to be essentially the same as the cost of oil to electric utilities.

29. R. J. Buck, H. B. H. Cooper, P. Elliott, J. N. Martin, and A. Purcell, Unified Industries, *A Report of U.S. Railroad Electrification*, Report FRA/ORD-778-67 (Springfield, VA: Federal Railroad Administration, October 1977).

30. See, for example, K. L. Lawson, J. H. Wujek, and K. J. Ingram, "Effects of Energy-Cost Variation on Feasibility of Electrifying the Cincinnati-Atlanta Main Line of the Southern Railway System," *Transportation Research Record 694*, Railroad Track and Electrification Studies, Transportation Research Board (Washington, D.C.: National Academy of Sciences, 1978).

31. Treat, *Electric Railroads*, 27.

32. It should be noted that the electrified portions of railroad networks generally have the highest traffic densities. Thus, a country with only 30 percent of its railroad route-miles electrified may actually be moving over 80 percent of its rail ton-miles via electrical power.

33. Generally, narrow-gauge (less than 4'8½" between rails) track is built to lesser standards than standard-gauge track and is therefore generally much less costly to build and less expensive to maintain. Increased curvatures are also possible with narrow-gauge, thus making it easier to provide rail linkups to remote areas in rugged terrain, such as mining locations. Operating speeds and load capacities are usually more restricted on narrow-gauge railroads than on standard-gauge lines due to the reduced stability and limited size of the rolling stock that operates on narrow-gauge. High-density rail traffic is therefore generally confined to standard-gauge or broad-gauge railroads. For this reason, narrow-gauge railroad lines were excluded from the table.

However, several countries have a significant amount of narrow-gauge track that is of importance. India has over 18,000 miles of railroad lines less than 3'4" in gauge, about equal to the amount of broad-gauge lines that exist in that country. Switzerland has approximately 900 miles of narrow-gauge railroad line, much of which is electrified. France has over 300 miles of narrow-gauge line, and most other European countries have some narrow-gauge railroads. It is reasonable to expect that these narrow-gauge lines transport proportionately far less traffic than the standard-gauge lines. Narrow-gauge railroads in a few countries, however, probably do transport an important share of the total rail traffic. For example, Brazil has about 15,000 miles of narrow-gauge railroad lines, over 80 percent of its total rail mileage. The Japanese railroad system is primarily narrow-gauge. About 14,000 miles, over 90 percent of the total Japanese rail mileage, is 3'6" in gauge. Undoubtedly, some of these particular narrow-gauge lines have a high traffic density. South Africa, with only narrow-gauge railroads, has over 14,000 miles of rail lines, about 40 percent of which are electrified. Most of the world's rail traffic movements, however, are accommodated on standard- or broad-gauge railroad lines.

34. Foreign data not specifically sourced are generally from the same sources as Table 9.3.

35. U.S. Bureau of Mines, *Mineral Commodities Profiles 1983 — Aluminum* (Washington, D.C.: GPO, 1983), 16.

36. Michael G. McGraw, "Aluminum Service Bodies Offer Many Advantages," *Electrical World* (February 1984), 72–73; and "Streamlining Adds to Aluminum Car's Potential," *Electrical World* (August 1984), 98–103.

37. U.S. Bureau of Industrial Economics, *1983 U.S. Industrial Outlook* (Washington, D.C.: GPO, January 1983), 6–1.

38. National Research Council, *Reference Materials System, A Source for Renewable Materials Assessment*, PB264-494 (Springfield, VA: National Technical Information Service, 1976), 32.

10 | *Agriculture: Mobile Machinery and Other Energy Applications*

Calvin C. Burwell

As late as the beginning of the 1930s, only about 10 percent of the nation's farms were served by central station electricity.[1] The farms that received electrical service were, in the main, either on the edge of urban communities and therefore close to the existing electrical network, or they were specialized dairy or poultry operations that required substantial amounts of motor power for feed and water handling, milking machines, and milk and egg refrigeration.

The chief economic obstacle to electrification was the high cost of extending electric lines to serve isolated farm homes, many of which would, moreover, be able to afford to consume only very small quantities of electricity. A test project by the Alabama Power Company showed, for example, usage by the average cotton farm household of only 240 kWh per year, far below the 600 to 1,000 kWh needed, even at high electricity prices, to support an estimated line cost of $2,000 or so per mile.[2]

Despite its poor economic justification, early proponents of rural electrification argued that its benefits would "result in a higher standard of living which cannot be figured in dollars and cents or in kilowatts."[3] Elsewhere in the world, perhaps because of more densely populated rural areas, industrialized nations were well along in the process of farm electrification, even having resorted to heavy farm subsidies to bring it about. Germany claimed 60 percent rural electrification in 1927. France had achieved 70 percent by 1930. Other countries far ahead of the United States included Finland with 40 percent, Sweden and Denmark with 50 percent, and Czechoslovakia with 70 percent in the 1930s.

ELECTRICITY AND THE QUALITY OF RURAL LIFE

Inevitably, pressures built in the United States to provide the rural dweller "an even chance with his city brother in the comforts of life."[4] After the election of Franklin Roosevelt in 1932, rural electrification, along with other initiatives affecting the production and use of electric power, became an element in the New Deal program to reactivate and reform the American economy. In 1935, the President by executive order set up the Rural Electrification Administration (REA), followed by legislation in 1936 that established the REA on a permanent basis and set up the terms and conditions for its activities.

What soon emerged was a system in which farm cooperatives were chartered to purchase bulk power, in some cases from government hydropower projects like the Tennessee Valley Authority (TVA), and to assist their members in building distribution systems and in wiring homes at low cost. In addition, families needing money for wiring and appliances could apply for small individual loans. In 1936, the government interest rate was 3 percent, lower than the yield on utility securities which in 1935 and 1936 varied from 3.25 percent to 3.88 percent.[5]

For impoverished southern farmers the REA devised the "Arkansas plan." The plan included membership — $5; house wiring — $10; an iron — $3; and a radio — $7. All could be paid for on the installment plan at 89¢ per month which, along with a $1 minimum monthly power purchase for the first 11 kWh, made up the total monthly bill of $1.89. Penniless farmers worked as laborers on REA crews to provide payment and permit those at the bottom of the agricultural ladder some benefits of modern living. Electricity went first for lights, an iron, and a radio. Running water was more expensive and came later. For expensive appliances, farmers in the South gave preference to refrigerators, whereas Northern farmers opted for washing machines.[6]

By 1939, the REA had 417 cooperatives serving 268,000 households, a small fraction of the 5 million farms yet to be served. While well begun, the REA program moved slowly in the early 1940s because of higher federal priorities imposed by World War II and related shortages of manpower, fuel, and copper. By 1944, 55 percent of the farms were still without service, but the program was popular and appropriations were expanded after the war, with the highest annual appropriation being ten times higher than prewar funding rates. By 1953, over 2.5 million farms were connected to REA and for all practical purposes, all American farmers had electricity, whether through REA or otherwise.[7]

Efforts to achieve farm electrification were based largely on a desire to improve the quality of rural life. However, applications of electricity within the farm home also resulted in indirect increases in agricultural production and productivity — the main subject of this chapter — through their beneficial effects on members of the farm population.

As in homes generally (see Chapter 11), the availability of electricity made it possible for farm families to modernize their homes with electrical equipment and to reap the benefits of their use. Electric lighting is credited with reducing home accidents, offering numerous psychological benefits, and providing additional working hours equivalent to 91 eight-hour days per year. Indoor plumbing and the electric washing machine are said to have promoted cleanliness and reduced farm sickness caused by polluted water, while refrigerators are credited with having improved the farm diet and having reduced the high incidence of food spoilage and staphylococcus food poisoning on farms.[8] The radio brought information on agricultural opportunities, weather, and marketing. But more importantly, "With the aid of radio, rural inhabitants became more aware of the ebb and flow of life in the world, and it enabled them to play a less passive role in an industrial society."[9]

ENERGY USE AND TECHNOLOGICAL CHANGE IN AGRICULTURAL PRODUCTION

There is a tendency to think of energy applications in agriculture, particularly as they interact with long-run technological progress, in terms of the tractors and other farm machinery that replaced farm animals and human labor. This perspective is correct so far as field work requiring mobility is concerned; as with general transportation, mobility in agriculture has usually been based on the use of liquid fuels. Such mobile field operations are practically 100 percent nonelectric (see Table 10.1).

However, there are many other uses of energy in agriculture that are not so strongly oriented toward liquid fuels. Table 10.1 shows that electricity, although a secondary source of energy for agriculture as a whole, accounts for a large percentage of the energy consumed by such field applications as irrigation and for quite large percentages in many aspects of livestock production. Moreover, the percentage of electricity has been rising — from 17 percent to 22 percent for all agricultural operations between 1974 and 1981, and from 37 percent to almost 50 percent of energy use in the production of livestock.

Although of secondary rank from an overall quantitative standpoint, electricity has also been of great importance to technological progress in particular aspects of agricultural practice. This can be seen most clearly by examining trends in agricultural technology in relation to energy use.

Mobile Applications

Prior to the development of the internal combustion engine and its commercial use in vehicles, the work of agriculture was provided through human and animal labor. Steam-driven tractors and stationary engines (i.e., for central threshing operations) were available. Even though they were expensive and unreliable, they were sometimes used on large farms. Simple farm machinery (plow, mowing machine, rake) had been used from earliest times

Table 10.1 Energy Use in Agriculture, 1974 and 1981

	1974		1981	
	Primary energy[a] *(10^12 Btu)*	*Percentage electric*	*Primary energy[a]* *(10^12 Btu)*	*Percentage electric*
Field crops				
Field operations	660.0	1	687.0	1
Agricultural chemicals[b]	728.0	12	907.0	12
Irrigation[c]	359.0	47	474.0	49
Crop drying	73.0	8	76.0	15
Frost protection	41.0	5	18.0	14
Grain handling	2.0	5	2.0	15
Other	32.0	56	56.0	69
Subtotal	1,895.0	15	2,220.0	18
Livestock				
Lighting	17.5	100	30.6	100
Feed handling	61.3	19	76.5	21
Waste disposal	21.7	6	29.2	8
Water	19.7	6	34.7	95
Livestock handling	2.9	0	2.8	0
Space heating	7.2	25	9.0	39
Ventilation	21.3	100	42.3	100
Water heating	17.1	60	24.5	74
Milking	8.7	100	14.1	100
Milk cooling	14.2	100	23.5	100
Egg handling	0.3	100	0.5	100
Brooding	24.1	0	36.1	0
Farm vehicle	49.5	0	45.5	0
Other	17.2	13	15.7	23
Subtotal	284.0	37	385.0	49
Total	2,180.0	17	2,605.0	22

SOURCES: Federal Energy Administration and U.S. Department of Agriculture, *Energy and U.S. Agriculture, 1974 Data Base,* FEA/D-76/459 (Washington, D.C.: GPO, September 1976); B. A. Stout, J. L. Butler, and E. E. Garett, "Energy Use and Management in U.S. Agriculture," in *Energy Use and Agriculture,* ed. G. Stanhill (New York: Springer-Verlag, 1984), 175–176; John Hostetler and Gordon Sloggett, "Energy and Irrigation," in *Inputs,* U.S. Department of Agriculture, Economic Research Service, Report No. 10S-8 (Washingt   D.C.: GPO, August 1985), 28.

[a] Primary energy includes purchased fuels and electricity, with electricity counted at 10,600 Btu/kWh.

[b] Energy data include fuel and electricity used by the chemical industry in the manufacture of agricultural chemicals.

[c] Does not include electricity used for delivering surface irrigation water to the farm boundary. One estimate places electricity use by the U.S. Bureau of Reclamation for that purpose at 3.6 billion kWh for 1974, or approximately 38 trillion input Btu.

NOTE: Many of the statistics have been rounded.

and was gradually improved but was usually limited in size to the capability of one or two draft animals. While some commercial fertilizer was used (primarily directly mined materials), fertility of the soil was sustained principally through the use of manure and crop rotation (including fallow periods and legumes in rotation).

The internal combustion engine changed all this. Beginning about 1910, farmers first bought automobiles, then trucks and tractors (Figure 10.1). With inanimate power, the human and the work animal population on farms began to decline. The decline was slow at first because the tractors and trucks were low powered and because many farms could not afford them. The process accelerated in the 1940s as farm labor lost to military operations was replaced by machinery. As an increasing number of farms adopted more and larger tractors, larger farm machines were developed and performed multiple operations simultaneously. Thus, the binder became a combine and eliminated the separate threshing operation, and a corn picker was developed that could also shell the grain.

These changes injected an enormous input of machine power into agriculture, although larger machine sizes slowed the growth in, or even reduced, the actual number of machines. The 20 million (work-animal) horsepower on farms in 1900 had by 1980 grown to 360 million (inanimate) horsepower using 7 billion gallons of liquid fuel annually.

Is there a future for electricity use in these mobile field applications? In the preceding chapter on transportation, we speculated briefly on the prospects for battery-powered automobile operation. Compared with automobiles, tractors' acceleration, braking, speed, and mileage requirements are modest, and it is not as necessary to provide heat or air conditioning for operator comfort. In addition, dc motors deliver high torque at low speed, and electric systems are well suited to provide various power take-off functions that are now driven by expensive and maintenance-prone hydraulic and mechanical power transmission equipment. Also, battery weight would add to vehicle traction and eliminate the need to purchase weights for this purpose when added traction is needed. Since the farmer must transport and store fuels for tractor use, energy supplied by overnight battery recharge would be an added convenience.

Farmers have been found to be receptive to electric vehicles.[10] Among those surveyed twice as much interest was expressed in the "quietness" of operation as in engine efficiency and such optional accessories as power steering and air conditioning. A prototype electric tractor has been developed and evaluated at South Dakota State University. In a 1984 paper, its developers reported that even with lead-acid batteries, the electric tractor could economically perform 50 percent of field operations and 100 percent of chore tasks.[11] However, the lower fuel costs in recent years compared to the early 1980s might result in some modification of these findings.

Figure 10.1   Trends in the Use of Engine-Driven Vehicles in Agriculture

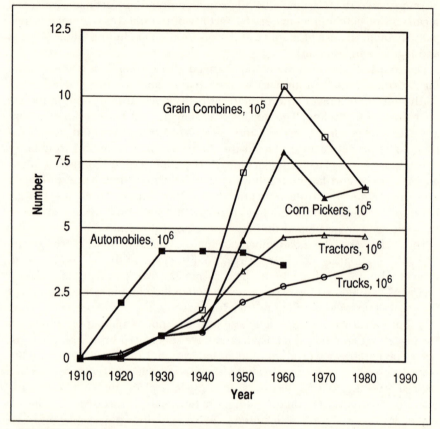

SOURCES: U.S. Bureau of the Census, *Statistical Abstract of the United States, 1984* (Washington, D.C.: GPO, 1984); and *Historical Statistics of the United States: Colonial Times to 1970* (Washington, D.C.: GPO, 1975).

Cultivation Practices

Operations on the land require liquid fuels for mobile machinery. However, the energy used in field operations represents a shrinking share (now about 30 percent) of the total requirement for growing crops. The other energy inputs are for agricultural chemicals, irrigation, and a variety of chores not requiring mobility (see Table 10.1). The growth in these other inputs reflects changing agricultural practices that are linked to the growing use of electricity.

Changes in cultivation practices have been a major factor leading to the expansion of U.S. agricultural output. Less cropland was in use in 1978 (369

Table 10.2 Trends in Average Yield per Harvested Acre for
Large Acreage Crops

Commodity	1940	1950	1960	1970	1982
Wheat (bushel)	15.3	16.5	26.1	31.0	35.5
Corn (bushel)	28.4	37.6	54.7	72.4	113.2
Cotton (pound)	253.0	269.0	446.0	438.0	590.0
Soybeans (bushel)	16.2	21.7	23.5	26.7	31.5
Hay (ton)	1.31	1.38	1.76	2.07	2.50

SOURCE: U.S. Department of Agriculture, *Agricultural Statistics* (Washington, D.C.: GPO, various years).

million acres) than in 1930 (382 million acres) despite the fact that in 1978 some 50 million acres were used to produce corn, wheat, and soybeans for export.[12] During this period, increases in production to satisfy population growth, gains in levels of nutrition, and growth in export demand have depended upon gains in land productivity. (For example, an additional 32 million acres would have been required to produce 1981 corn and soybean crops at 1971 yields.) Improvement in crop yield has been most successful with corn, but the yields of other major crops have also been improved as the result of technological progress (Table 10.2).

Increases in yields have resulted from the development of plant varieties with high yield potential but, more importantly, have depended upon controlling the growing environment so that the yield potential could be approached under field conditions. The major methods for controlling the growing environment, in addition to conventional machine operations (plowing, cultivating), have involved the use of agricultural chemicals (fertilizer, pesticides, herbicides) and water management (field drainage, irrigation). The use of chemicals and irrigation, in turn, depends on inputs of electricity and other sources of energy (see Table 10.1).

Commercial Fertilizer

Commercial fertilizers now provide most of the nitrogen, phosphorus, and potassium needed by growing plants. Of these elements, nitrogen is the largest both in tonnage (50 percent more than the other two elements combined)[13] and in energy required for its production (90 percent of all energy used in fertilizer manufacture). The trend in the use of commercial nitrogen is shown in Figure 10.2. Application of nitrogen per acre, and in total, has more than doubled in the past twenty years, and applications of commercial potassium and phosphorus have followed similar patterns. About one-fourth of all the energy required for fertilizer production is electricity.

The benefits of the intensive use of fertilizer go beyond the immediate gain in yield. For example, fertilizer-intensive crops such as corn can be

Figure 10.2   Use of Commercial Nitrogen Fertilizer, 1960–1983

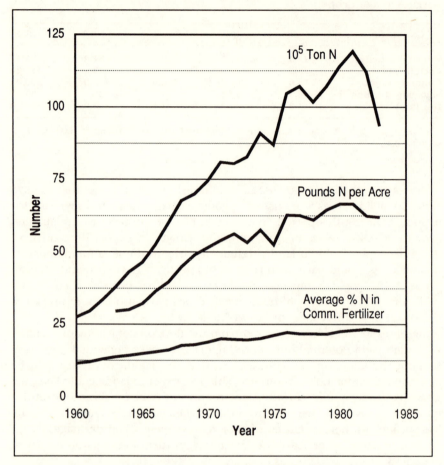

SOURCE: U.S. Department of Agriculture, *Agricultural Statistics* (Washington, D.C.: GPO, various years).

grown year after year without crop rotation. High yield also builds carbon into the soil (because of the increased stalk cellulose produced), which enhances yield potential for the future.[14]

Chemicals Instead of Tillage

Instead of field cultivation, herbicides are being used increasingly to control weeds — a practice called "minimum tillage." A further step in this direction is "no-tillage" agriculture in which a ground cover crop such as rye grass is planted, allowed to grow to the point where it shades the ground (thereby discouraging growth of weeds), and then killed by herbicides that do not affect the crop. The desired crop is planted in the vegetative cover without plowing the ground.

Proponents of no-tillage agriculture point to benefits such as reduced soil loss through erosion, reduced water loss by evaporation from the soil surface, lower labor requirements, and in some cases better scheduling potential for optimum plant yield. Moreover, several tractor-powered field operations (plowing, discing, cultivating) are eliminated. As a result, the energy requirement for no-tillage practice can be substantially less than that for conventional practice (see Table 10.3).

About 25 percent of the primary energy used to produce herbicides and pesticides is for electricity generation. In this way a small amount of energy, some of it electricity, indirectly displaces a much larger amount of diesel fuel that would otherwise be required if cultivation were done mechanically (i.e., by tractors and associated equipment).

Minimum and no-tillage field practices received a strong impetus from the sharp increase in costs of tractor fuel following the international oil dislocations in 1973, and again in 1979. The acreage under no-tillage practice increased from 3 million in 1972 to nearly 17 million in 1985 — an average annual growth of 14 percent.[15] Since the benefits of no-tillage practice go beyond the cost of fuel, it seems likely that once adopted and successful, minimum tillage practices will spread.

Irrigation

The greatest single use of electricity in agriculture is for pumping water. Large-scale irrigation began in the West with the completion of large federal dams on three rivers: the Colorado (Hoover Dam, 1936); the Missouri

Table 10.3 Energy Requirements for Conventional and No-Tillage Production of Corn and Soybeans in Virginia

Crop	Operations	Energy requirement (gallons of diesel fuel per acre)	
		Conventional	No-Tillage
Corn	Tillage and planting	6.35	2.47
	Herbicide/insecticide	—	1.05[a]
	Total	6.35	3.52
Soybean	Tillage and planting	5.63	0.75
	Herbicide	—	0.75[a]
	Total	5.63	1.50

SOURCE: D. H. Vaughn, E. S. Smith, and H. A. Hughes, "Energy Requirements of Reduced Tillage Practices for Corn and Soybean Production in Virginia," in *Agriculture and Energy,* ed. William Lockeretz (New York: Academic Press, 1977), 114.

[a] Includes energy used to manufacture the chemicals, about 25 percent of which is used in the form of electricity.

(Ft. Peck Dam, 1940); and the Columbia (Grand Coulee Dam, 1942). These federal projects brought low-cost water and electric power to land suitable otherwise only for limited grazing.

The shift of cotton growing from the Southeast to the Southwest and of potatoes from Maine to Idaho are indicators of the early successes of irrigation of arid western land. In 1939, the eleven most western states of the contiguous forty-eight contained nearly 90 percent of the irrigation area; otherwise, only the rice states of Louisiana and Arkansas and the citrus crop in Florida used irrigation. Today, the total cropland area is less than it was in 1935, but the amount of land with irrigation capability has increased from 18 million acres in 1939 to over 60 million acres in 1982 — an average annual increase of 2.8 percent.

Irrigation is a way to control the environment of the growing crop and thereby increase productivity. This is accomplished not only by adding water to dry land but also by allowing the use of the best climate and soils for a particular crop in areas that would be unsuitable without irrigation and by multiple cropping if the growing season is long. Even in naturally watered farming areas, irrigation is usually beneficial.

During the 1970s, when demand for export grains increased rapidly, the area under irrigation grew in states like Nebraska, Missouri, Georgia, and Michigan. In dry years, the capability to irrigate is the difference between low yield or crop failure and a predictable high yield. In Nebraska, each inch of water shortage in July reduces corn yield about 25 percent (i.e., 40 bu/acre); a one-inch shortage in August reduces corn yield about 20 bu/acre.[16] Also, when crop yields are low, the crop value is likely to be high so that drought years become highly profitable years for those farms with irrigation capability.

There has been a shift away from irrigation by surface flooding toward sprinkler irrigation, despite its associated requirement for pressurized water. Total irrigated acreage increased by 5.5 million (6 percent) from 1976 to 1984, while acreage using pressurized irrigation techniques increased some 7 million acres (47 percent).[17]

Sprinkler irrigation techniques reduce water use by 30 percent to 50 percent for a given yield, but require ten to twenty times as much energy per gallon of water distributed as does field flooding.[18] Electricity is the most energy-efficient and reliable source of energy for pumping water and is now used for half of the power requirements for irrigation (Table 10.1). Annual farm use of electricity for irrigation grew from 16 billion kWh in 1974 to 22 billion kWh in 1981, and it now approaches 60 percent of all electricity used for crop production.

The benefits of pressurized irrigation systems must go beyond those associated with surface flooding to justify their greater first cost and extra energy use. Besides saving water, the pressurized system can deliver metered quantities of fertilizer and chemicals uniformly to the growing crop and displace the tractor-powered field operations otherwise required. The control available in this way reduces the fertilizer requirement by 30 percent to 40 percent. Uniform water application permits a relatively constant moisture level to be maintained in the soil, which often results in a more uniform, higher quality crop. Without surface flooding, field operations can proceed even as the crop is receiving sprinkler water, and one person can manage ten times as much sprinkler-irrigated acreage as surface-irrigated acreage.[19]

Sprinkler operation is an application for mechanical energy that can come from diesel or gas engines or electric motors. If electric motors are chosen, the irrigation application brings electricity to the field where it can also be used for other purposes. For example, electricity can provide the means of moving the sprinkler systems as well as providing power for pumping. Large center-pivot sprinkler irrigation systems that gradually rotate through the field can serve an entire section of land (640 acres) from a single pumping point. The half-mile-long sprinkler arms comprise more than 20 sections of sprinkler pipe, each over 100 feet long. These sections are connected at towers, each of which is driven by a variable-speed electric motor mounted directly on the drive wheels. The towers move in a circle and can be kept in alignment via a laser beam.[20]

Additional uses for electricity in the field go beyond pumping. High-dosage ultraviolet radiation successfully destroys the DNA in the bacteria that feed on the iron content of the irrigation water and cause clogging of drip irrigation holes.[21] And electrostatic charging of pesticide droplets has proved to be three to four times more effective than conventional spray techniques.[22] The charged pesticide adheres to the underside of the foliage where pests often reside, but where it is difficult to reach with uncharged overhead spraying.

Finally, motorized irrigation equipment is now being adapted to sow seeds as well. Because the irrigation equipment is elevated above the growing crop, the seeding can take place prior to harvesting the crop it replaces. In this way enough time can be saved to permit two crops to be grown each growing season. In the future, the electrical distribution system for irrigation might be adapted also to serve vehicular motors for field operations, especially if those operations are combined with no-tillage cultivation practices.

Livestock and Poultry Production

While the major use of electricity in agriculture is for pumping water, electric technology also finds numerous applications in animal production because of the many small labor-oriented chores of raising animals. Electricity is normally available in animal shelters for lighting purposes, and electric appliances can be used without complications of fuel supply. Moreover, the fumes from the use of fuel in internal combustion engines would be objectionable, fire hazards would be greater, and the noise of internal combustion engines would, in some cases, be disruptive. As a result, electricity has played an important role in increasing the productivity of animal raising in the postwar period. Poultry production is the primary example.

At the end of World War II, poultry production was still largely a farm sideline. Flocks numbering dozens or at most a few thousand birds provided for on-farm consumption and a small supplementary source of income.[23] Today's poultry production is factory like by comparison. Flocks may number 100,000 birds.

The birds are confined in shelters where climate, photoperiod, feeding, and sanitation are carefully controlled and monitored. Ventilation fans, heaters, and evaporative coolers control temperature and humidity. Mechanical feeders and waterers control the quantity and timing of feeding. For laying flocks, machines also control the photoperiod and collect, grade, and pack the eggs. Except for heating, environmental control is almost entirely electric.

In the postwar period, research and development in genetics, nutrition, and disease control produced major advances. The management of the growing environment made possible with electric technology enabled producers to capitalize on these scientific advances and to build the larger flocks necessary to make vertical integration profitable. Table 10.4 illustrates the result. From 1950 until 1980, while beef and consumer prices tripled, poultry prices rose less than 20 percent. Largely as a result of falling real prices, annual per capita poultry consumption rose from 25 to 61 pounds over this time period. Chicken was converted from a Sunday treat to a bargain staple.

With the shift to commercial-sized flocks, liquid petroleum gas largely displaced the use of electricity for chick brooder heat.[24] But more recently, the high cost of this gas combined with a brooder house design that permits off-peak use of electricity is encouraging a return to the use of electricity.

Table 10.4   Selected Consumer Prices, 1950–1980

	1950	1960	1970	1980
Chicken, whole (cents per pound)	60	43	41	71
Beef, chuck roast (cents per pound)	62	62	73	182
Consumer Price Index, all items (1967 = 100)	72	89	116	247

SOURCES: Food prices: U.S. Bureau of Labor Statistics, unpublished data; Consumer price index: U.S. Bureau of the Census, *Statistical Abstract of the United States, 1984* (Washington, D.C.: GPO, 1984) and *Historical Statistics of the United States: Colonial Times to 1970* (Washington, D.C.: GPO, 1975).

The building design incorporates a concrete floor with embedded electric heaters or pipes for hot water. The concrete floor stores and releases heat gradually, and since most of the heat is needed during the cold nighttime hours, almost all of the electricity required can be provided during off-peak periods.

Electricity provides all the energy for many animal-raising needs such as lighting, ventilation, milking, and milk cooling. Even for space and water heating, electricity is used for a large and growing share of the energy requirements for animal raising: 74 percent versus 60 percent of water heating requirements and 39 percent versus 25 percent of space heating requirements in 1981 compared with 1974. New techniques have clearly contributed to the economic use of electric heat.

In dairy operations, heat rejected from the refrigeration system for milk cooling can be used to heat water for washing the electric milking machines and associated milk piping system. If water storage capacity is adequate, most of the water heating requirement can be met in this way, and the rest of the electricity for water heating can be used during off-peak periods.[25]

ELECTRICITY'S CONTRIBUTION TO FARM TECHNOLOGY

Technological progress in farming is often identified with the large improvements that have occurred in the efficiency of growing field crops through the expanding use of farm machinery powered by liquid fuels. The long-term displacement of farm animals by machines and the massive flow of labor from agriculture into manufacturing in the course of the twentieth century can be largely accounted for in this way.

But this is not the whole story. Technologies involving growth in the use of irrigation, fertilizers, and other agricultural chemicals have also made major contributions to the phenomenal growth in the productivity of American agriculture. All of these are strategic inputs into agriculture, and they have required very large amounts of energy in their own production, including a sizable fraction in the form of electricity. Electricity thus accounts for some 20 percent of all energy inputs required in producing field crops, although in particular operations, such as irrigation, the electricity fraction is in the neighborhood of 50 percent and growing.

Apart from field crops, a wide range of farming activity is concerned with livestock and poultry production. Here, the 50 percent energy share accounted for by electricity is substantially larger than in the production of crops. Electrification has been particularly important in achieving a total transformation of production in some of these industries, most conspicuously in the case of poultry operations, which are now almost factory-like in their organization. This thoroughgoing change in technology based on electrical techniques is clearly reflected in the highly favorable recent history of poultry prices to the consumer.

Although leadership of the twentieth-century revolution in agriculture operations belongs to liquid fuels and the internal combustion engine, the importance of electricity continues to grow. A significant boost in electricity's importance could also occur in the future if advances in battery technology were to permit the electrification of tractors and other equipment used in field operations.

Notes

1. D. C. Brown, *Electricity for Rural America* (Westport, CT.: Greenwood Press, 1980), 10. Another 4.8 percent of the farms had home generators.

2. Ibid., 9.

3. Ibid., 9.

4. Ibid., 11.

5. Ibid., 66.

6. Ibid., 70.

7. Ibid., 113.

8. Information in this paragraph is from Brown, *Electricity*, 116–118. Farm household electrification also released farm women for operating trucks and tractors and for other production activities. In an analogous fashion, electric appliances in the urban home have made it easier for women to take jobs in commerce and industry (see Chapter 11).

9. Brown, *Electricity*, 118.

10. Peter H. Calkins, "Electric Farm Vehicles: How Much Oil Can Be Replaced?" Paper presented at the 28th Annual Conference of the Food and Energy Council, Williamsburg, VA, September 22–24, 1981.

11. Les Christianson, Ralph Alcock, and Lowell Endahl, "The Electric Farm Tractor Is Here." Paper presented at the Food and Energy Council Annual Conference, Kansas City, MO, July 30–August 1, 1984.

12. U.S. Department of Agriculture, *Agricultural Statistics* (Washington, D.C.: GPO, various years).

13. Ibid.

14. R. E. Lucas, J. B. Holtman, and L. J. Connor, "Soil Carbon Dynamics and Cropping Practices," in *Agriculture and Energy*, ed. William Lockeretz (New York: Academic Press, 1977), 345–347.

15. "A Look at Changing Tillage Practices," *No Till Farmer* (June 1985), 8.

16. A. Sanghi and R. Klepper, "A Method for the Economic Analysis of Irrigated Farming With Diminishing Ground Water Reserves," in Lockeretz, ed., *Agriculture and Energy*, 178.

17. "1984 Summary — U.S. Irrigated Acreage," *Irrigation Journal* (March–April 1985), 30.

18. G. Sloggett, "Energy Used for Pumping Irrigation Water in the United States, 1974," in Lockeretz, ed., *Agriculture and Energy*, 114.

19. "Production Advantages Evident With New Irrigation System," *Irrigation Journal* (July–August 1984), 13–17; Sam Tobey, "Lo-flo Drip Irrigation," *Irrigation Journal* (September–October 1984), 26–33.

20. Lockwood Corporation, "Best in Irrigation" sales brochure, Gering, NB.

21. Robert E. Flatow, "Iron Bacteria — The Invisible Threat to Drip Irrigation Systems," *Irrigation Journal* (May–June 1985), 30–32.

22. "Chementator," *Chemical Engineering* (October 11, 1976), 67.

23. The description in this and the following paragraph follows Gordon Sawyer, *The Agribusiness Poultry Industry: A History of Its Development* (Jericho, NY: Exposition Press, 1971); and various issues of *Poultry Tribune and Poultry Digest.*

24. Electric Power Research Institute, "Electric Brooding — Poultry," *Technical Brief* EMU.95.4.87 (Palo Alto, CA: Electric Power Research Institute, 1987).

25. Electric Power Research Institute, "Water Heating for Production Agriculture," *Technical Brief* EMU.86.8.86 (Palo Alto, CA: Electric Power Research Institute, 1986).

11 | The Home: Evolving Technologies for Satisfying Human Wants

Calvin C. Burwell and Blair G. Swezey

This chapter on electricity in the home departs briefly from the general theme of electricity in industrial production. It is included primarily because household electric applications provide a good case study for examining the interaction of electricity use and technological change.

The residential sector has over time become the largest component of electricity utility sales, having grown from a mere 15 percent of utility sales in 1930 to almost 35 percent in 1986. Industrial sales, which represented over half (55%) of utility sales in 1930, have fallen to a 35 percent share in 1986, equal to the residential share (see Figure 11.1).

The greatest increase in residential electricity use occurred between the end of World War II and the beginning of the 1970s. During this period, innovations in household electric-based technologies, coupled with declining real electricity prices, led to the rapid adoption of many electrical appliances and, consequently, to a greater than threefold increase in electricity use per household (Figure 11.2). While electricity prices have trended upward since 1973, leading to electricity conservation and end-use efficiency improvements, electricity use in the home has continued to increase, although at a much slower pace. New household technologies have continued to enlarge the scope of household electricity applications.

THE EVOLUTION OF HOUSEHOLD APPLICATIONS

Homes were wired initially for electric lighting. Prior to the advent of electric lighting, candles, and gas and oil lamps were used for inside illumination. Although electric arc lighting was in use, it was suitable only for outside areas because of the noise, fumes, and large size associated with the equipment of the time — characteristics just the opposite of those now identified with electricity in the home.

The use of electricity in the home resulted from the development of two technologies pioneered by Thomas Edison — one for supplying electricity

Figure 11.1   U.S. Electric Utility Sales by Sector, 1930–1988

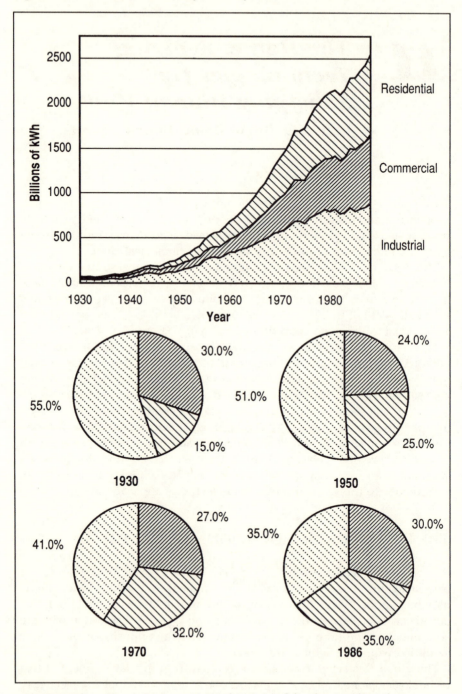

SOURCES: Edison Electric Institute, *Historical Statistics of the Electric Utility Industry Through 1970* (New York: EEI, 1974) and *Statistical Yearbook of the Electric Utility Industry* (Washington, D.C.: EEI, annual).

Figure 11.2   Residential Electricity Use versus Price, 1950–1988

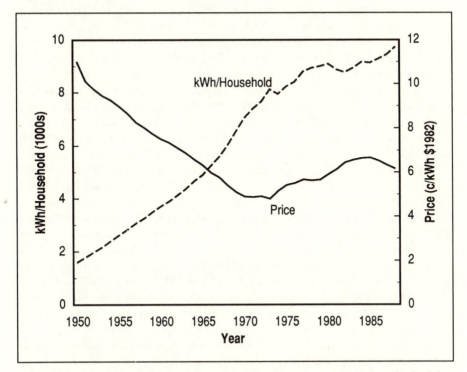

SOURCES: Kilowatthour and price data: Edison Electric Institute, *Statistical Yearbook of the Electric Utility Industry* (Washington, D.C.: EEI, annual).

GNP deflator: U.S. Council of Economic Advisers, *Economic Report of the President* (Washington, D.C.: GPO, February 1988), Table B-3.

Households: U.S. Bureau of the Census, *Current Population Reports*, Series P-20, no. 382 (Washington, D.C.: GPO), and *Statistical Abstract of the United States 1988* (Washington, D.C.: GPO, 1988), Table 56.

and the other for using it. It was Edison's genius to link central station electric power generation (his Pearl Street station began operation in 1882) with his end-use technology of the totally enclosed electric lamp. Edison's persistence resulted in the development of a filament that allowed the electric lamp's output to be specified by adjusting the combination of the filament diameter and length. Despite its initial high cost, electric lighting was rapidly adopted because of its cleanliness, convenience, and safety.

Lighting, the initial use in the home, continued to be the major use until nearly 1930,[1] but today it accounts for less than one-tenth of total residential electricity consumption. The substantial growth of electricity in the home is essentially a story of applications for purposes other than lighting, resulting from advances in home technologies.

The major growth in electric appliance use began during the 1920s, a decade that witnessed the introduction of the tube radio, the refrigerator, and the electric water heater. By the early 1930s, during the Depression, almost all urban homes were wired and had electric irons; 70 percent had radios, and 20 percent to 50 percent had electric refrigerators, washing machines, toasters, vacuum cleaners, and coffee makers.[2]

Each subsequent decade witnessed the introduction and adoption of new and improved electrical applications. The 1950s were a period of particularly rapid change, perhaps because of technological developments in the 1930s and 1940s that could not then be marketed (or even manufactured) due first to the Depression and then to the war. Major appliances included the refrigerator-freezer, television, clothes dryer, automatic washing machine, and room air conditioner along with a variety of small appliances, such as the steam-spray iron, electric blanket, and electric frying pan.

Major appliances and technologies introduced in the 1960s included the color television, dishwasher, central air conditioning and space heating, frost-free refrigerator-freezer, and waste disposal. Those penetrating the household market in the 1970s included the microwave oven, heat pump, trash compactor, and food processor. In the 1980s, new products for the home have included the home computer and its accessories, large screen television, videocassette recorder, compact disc player, home satellite receiver, heat pump water heater, photocell-operated mercury and sodium vapor lamps for outside areas, and the side-by-side refrigerator-freezer with automatic ice and cold water dispensers.

The profusion of household electrical applications that have developed over time can be attributed generally to the unique qualities of electricity as an energy form. Its inherent flexibility, convenience, and cleanliness make it appropriate for use with a great variety of technological mechanisms. Actual rates of appliance adoption, on the other hand, have been conditioned by other factors such as the capital and operating costs of the equipment, the competition from other fuels to provide the same energy service, the physical ease of introduction into the home, household income levels, and demographic trends.

For instance, the historical trend of decline in the price of electrical appliances has provided a major impetus to their adoption. This trend is shown in Figure 11.3, which tracks change in the prices of selected major household appliances (expressed in constant dollars) over the past thirty years. If adjustments were made for appliance quality, that is, the size and added features, the downward price trend would be much more pronounced. For example, the 7-cubic-foot refrigerator of the 1950s is now a 19-cubic-foot refrigerator-freezer with automatic defrosting and ice-making features. The 13-inch color television of the 1970s is now larger and can be remotely controlled. Today's electric range has a self-cleaning oven. And the energy efficiency of all electric appliances has been improved.

Figure 11.3 Price Trends of Household Appliances, 1955–1987

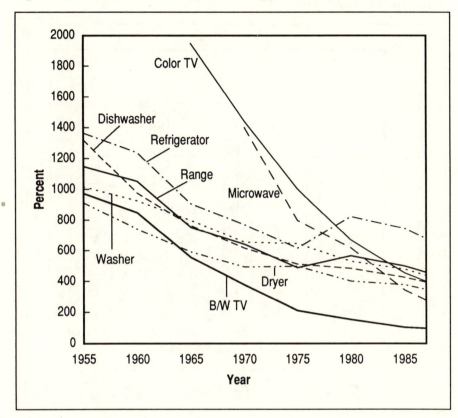

SOURCES: Appliance prices: U.S. Bureau of the Census, *Statistical Abstract of the United States, 1969* (Washington, D.C.: GPO, 1969), Table 1157; *1982–1983,* Table 1434; *1988,* Table 1297.

GNP deflator: U.S. Council of Economic Advisers, *Economic Report of the President* (Washington, D.C.: GPO, February 1988), Table B-3.

The rise in income levels over time has also played an important part in the growth and proliferation of household electrical applications. Figure 11.4 illustrates the connection between income level and appliance ownership patterns. Looking at a cross section of family income groups, we can see that ownership of a broad array of major appliances increases as family income level rises. Long-term increases in family income (Figure 11.5) have made these and other appliances accessible to a growing number of households.

Postwar demographic changes, particularly trends in female labor force participation rates and household size (Figure 11.6), have also had a profound influence on the adoption of household electrical appliances. As more

Figure 11.4 Electrical Appliance Penetration Rate as a Function of Family Income Level, 1984

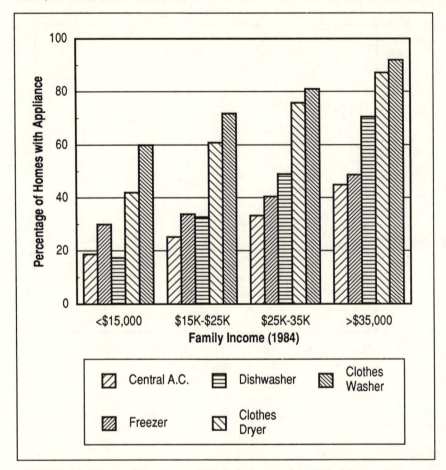

SOURCE: U.S. Bureau of the Census, *Statistical Abstract of the United States, 1989* (Washington, D.C.: GPO, 1989), Table 1249.

women, traditionally the homemakers, have entered the work force, time (both for labor and leisure activities) has become a more valuable commodity, promoting the increased adoption of household labor-saving devices. The two wage-earner family has also created a level of affluence that makes labor-saving appliances more affordable to the household.

Improvements in household technology (with electrification frequently at its core) have also served to facilitate the growth of female participation in the paid labor force. In other words, the expanded variety and use of electrical appliances has been both a cause and an effect of broad social

Figure 11.5   Trends in U.S. Family Income, 1950–1985

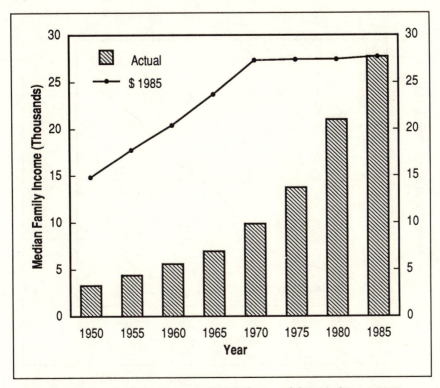

SOURCE: U.S. Bureau of the Census, *Statistical Abstract of the United States, 1987* (Washington, D.C.: GPO, 1987), Table 732.

trends influencing the role of women in the economy. As a leading student of the subject has put it "there is a dynamic interaction between the social changes that married women were experiencing and the technological changes that were occurring in their homes."[3]

GROWING DIVERSITY OF ELECTRICAL SERVICES IN THE HOME

A useful perspective on the evolution of household electrical applications can be obtained by considering the growing variety of services that electrical technologies provide. Figures 11.7, 11.8, and 11.9 chart the course of the key technologies and electrical services that have transformed the American home during the twentieth century, for the most part in the years since World War II.

Figure 11.6   Postwar Demographic Trends, 1950–1987

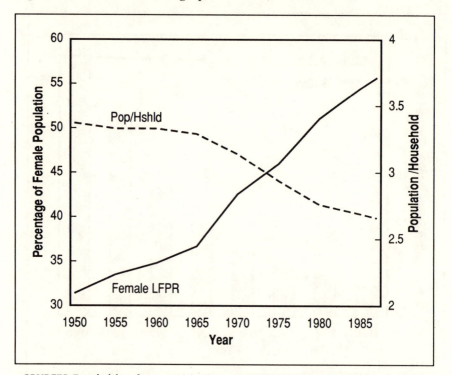

SOURCES: Female labor force participation rate (LFPR): U.S. Bureau of Labor Statistics, *Handbook of Labor Statistics*, Bulletin 2217 (Washington, D.C.: GPO, June 1985), Table 5; U.S. Bureau of the Census, *Statistical Abstract of the United States, 1988* (Washington, D.C.: GPO, 1988), Table 607.

Population/Household: U.S. Bureau of the Census, *Current Population Reports*, Series P-20, No. 382 (Washington, D.C.: GPO) and *Statistical Abstract of the United States, 1988* (Washington, D.C.: GPO, 1988), Table 56.

Our system of classification distinguishes between two broad classes of electrical technologies that underlie the services provided: (1) mechanical and electronic; and (2) thermal. This distinction serves to separate those household services (in the first category) that are tied to electricity as the essential energy source from others (mainly the large-heating technologies) where there continues to be cost competition between electricity and other energy forms.

Mechanical and Electronic

Household services based upon mechanical and electronic technologies can be classified into three (somewhat overlapping) broad categories that are charted in Figures 11.7, 11.8, and 11.9: refrigeration and air conditioning;

Figure 11.7 Trends in the Adoption of Refrigeration and Air Conditioning, 1933–1987

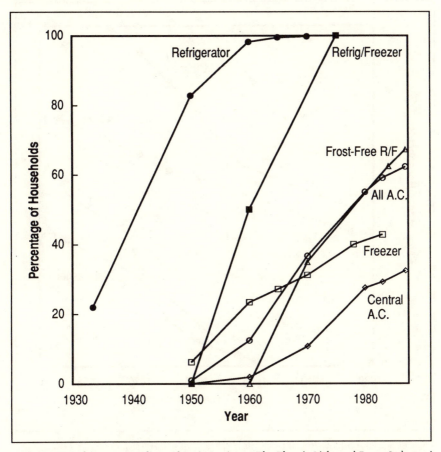

SOURCES: Refrigerators: Edison Electric Institute, *The Electric Light and Power Industry in 1933* (New York: EEI, October 1934) 1; John Tansil, *Residential Consumption of Electricity*, ORNL/NSFEP-51 (Oak Ridge, TN: Oak Ridge National Laboratory, July 1973); U.S. Bureau of the Census, *Statistical Abstract of the United States, 1971* (Washington, D.C.: GPO, 1971), Table 1098; *1981*, Table 1384; *1988*, Table 1227.

Freezers: John Tansil, *Residential Consumption of Electricity*, ORNL/NSFEP-51 (Oak Ridge, TN: Oak Ridge National Laboratory, July 1973); U.S. Bureau of the Census, *Statistical Abstract of the United States, 1971* (Washington, D.C.: GPO, 1971), Table 1098; *1981*, Table 1384; Electric Power Research Institute, *Trends in the Energy Efficiency of Residential Electric Appliances*, EM-4539 (Palo Alto, CA: EPRI, April 1986), Table 1-2.

Air conditioning: John Tansil, *Residential Consumption of Electricity*, ORNL/NSFEP-51 (Oak Ridge, TN: Oak Ridge National Laboratory, July 1973); U.S. Bureau of the Census, *Statistical Abstract of the United States, 1988* (Washington, D.C.: GPO, 1988), Tables 1221 and 1225.

Figure 11.8   Trends in the Adoption of Major Labor-Saving Appliances,
1933–1987

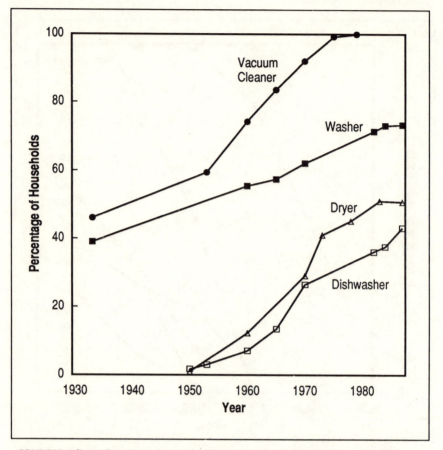

SOURCES: Edison Electric Institute, *The Electric Light and Power Industry in 1933* (New York: EEI, October 1934); John Tansil *Residential Consumption of Electricity,* ORNL/NSFEP-51 (Oak Ridge, TN: Oak Ridge National Laboratory, July 1973); U.S. Bureau of the Census, *Statistical Abstract of the United States, 1971* (Washington, D.C.: GPO, 1971), Table 1098; *1981,* Table 1384; *1985,* Table 1320; *1988,* Table 1227; Electric Power Research Institute, *Trends in the Energy Efficiency of Residential Electric Appliances,* EM-4539 (Palo Alto, CA: EPRI, April 1986), Table 1-2.

labor saving; and entertainment and information. What is striking about most of these services is the rapidity of their adoption in a large percentage of U.S. homes. Some of them are, indeed, virtually indispensable to full participation in twentieth-century American life.

What is equally striking is the great variety of applications within each of the broad categories and the growing proliferation of applications over

Figure 11.9 Trends in the Adoption of Entertainment Appliances, 1933–1989

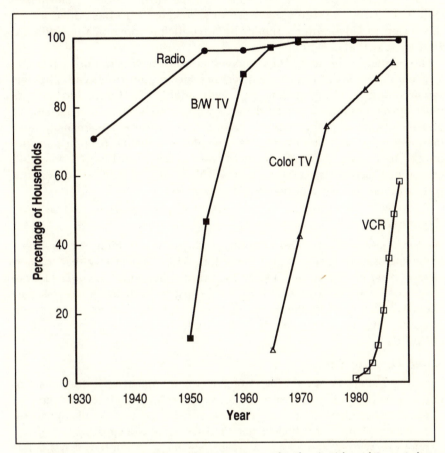

SOURCES: Radio, television: Edison Electric Institute, *The Electric Light and Power Industry in 1933* (New York: EEI, October 1934); John Tansil, *Residential Consumption of Electricity,* ORNL/NSFEP-51 (Oak Ridge, TN: Oak Ridge National Laboratory, July 1973); U.S. Bureau of the Census, *Statistical Abstract of the United States, 1971* (Washington, D.C.: GPO, 1971), Table 1098; *1981,* Table 1384; *1988,* Table 1227; *1989,* Table 900.

VCRs: Lee Schipper, "Approaching Long-Term Energy Demand: Energy Demand for Production or Pleasure?," Lawrence Berkeley Laboratory, May 1988 draft.

time. The entertainment and information category, in particular, clearly illustrates a category of service that continues in unabated fashion to offer a widening variety of choices for use in the home.

First came the tube radios, which provided a new mechanism for mass communication both in terms of news and other information and radio shows for entertainment. As electronics technology evolved, visual image was added to the audio medium via television. And the black and white picture tube was soon improved upon with the introduction of color television. In this decade, the videocassette recorder (VCR) has added a new dimension to video entertainment by introducing enormous flexibility in viewing selection and timing.

More recently, the advent of semiconductor electronics has made possible a whole new class of information technologies within the home. The personal computer, although not yet widely adopted by households, makes interactive communications available as well as the electronic processing of household data and information. Computers also offer a new form of entertainment capability. The interactive capabilities available through semiconductor technology will also lead to the future adoption of household computer-based control systems that will allow the ultimate in convenience, comfort, and flexibility in household operation.

Thermal

We turn now to thermal applications, a category in which electric-based technologies have rarely, if ever, been the only practical way that the services can be provided. Small-scale heating applications, considered first, have become increasingly electrical. But in large-scale heating services, particularly space heating, there continues to be a lively competition between electricity and the direct use of fuels, in which comparative prices play a critical part.

Small Heating Services

Various possibilities existed for electric heating appliances at the outset of home electrification in the early 1900s. However, at that time, the high cost of electricity made heating applications expensive even for small appliances. The electric iron, in which electricity had strong advantages, was an exception and its widespread adoption came very early. Not only was it a labor-saving device, but, without electricity, a stove had to be used to heat two or more irons that were used alternately. The irons were heavy and had minimal heat capacity; any dirt or smudge transferred from stove to iron ended up on clean clothing; the stove heat was greatly in excess of the amount transferred to the irons and kept the kitchen hot on ironing days; the ironing temperature was uncontrollable other than through careful attention and experience; and the low heat capacity of the iron seriously constrained the use of water as an aid to ironing.

The cost of electricity long ago ceased to be a constraint on its use in other small heating appliances. Moreover, improvements in these appliances, primarily in controlling their operations, have enhanced their convenience well beyond the level possible with nonelectric alternatives. Manual toasters evolved into the automatic pop-up variety that has since been superseded by the countertop toaster oven, which can accommodate a variety of warming and cooking chores. The early heavy electric iron later gained automatic temperature control and is now lightweight and includes automatic steam and water-spray capability. Similarly, the electric percolator gained automatic temperature control that now has separate electric systems for making filtered coffee and keeping it warm. Virtually all households now have electric toasters and coffee makers, and to a lesser extent, a variety of other small heating devices (see Figure 11.10).

The microwave oven is the most recent example of an electric heating device that carries important advantages over traditional methods. It constitutes a completely new way to cook that has no fuel-using alternative. Microwave energy is deposited volumetrically in the food, leading to a substantial savings in the time required for preparation. Neither the oven nor the container is heated, and cooking can be accomplished in the serving dish, which eliminates the food loss and clean-up requirements associated with separate cooking vessels.

Microwave cooking is particularly well suited to defrosting and heating the frozen foods that are now offered commercially in packaging appropriate for microwave preparation and table use. Such products and technologies, designed around the needs of the growing number of households with two working adults, have evolved to fit social trends.

Large Heating Services

Adoption of electricity for cooking, water heating, and space heating is a relatively recent development. It has occurred on a significant scale only since about 1950 for cooking and water heating and since about the 1960s for space heating (Figure 11.11). Several factors account for the delayed penetration of these applications: costs were often not competitive with fuel-using technologies; turnover rates on such equipment in existing housing were low; and large-scale electric heating technologies usually required separate wiring, often calling for the provision of a new power supply to older homes.

More recently, in a substantial fraction of new housing these large electric heating services have been incorporated as original equipment. Table 11.1 shows the percentage of housing units using electricity as the main source of energy for cooking, water heating, and space heating as of November 1984. The table shows the relatively high addition rate for electric service in homes built since 1970 and also implies a substantial shift to electric services even in homes constructed prior to 1970 (when compared with the historical data presented in Figure 11.11).

Prior to the 1970s, adoption of electric space heating occurred primarily in regions having access to low-cost hydropower, such as the Pacific Northwest and the Tennessee River Valley. With the decline in the cost of electricity relative to oil and gas that occurred in the 1970s (Figure 11.12) as well as the growing percentage of housing being constructed in Sunbelt regions

Figure 11.10 Trends in the Adoption of Small Heating Appliances, 1933–1987

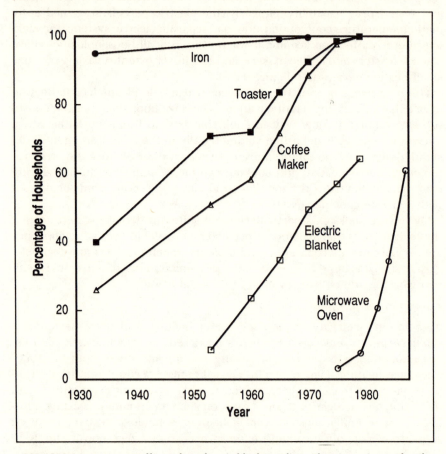

SOURCES: Iron, toaster, coffee maker, electric blanket: Edison Electric Institute, *The Electric Light and Power Industry in 1933* (New York: EEI, October 1934); U.S. Bureau of the Census, *Statistical Abstract of the United States, 1971* (Washington, D.C.: GPO, 1971), Table 1098; *1981*, Table 1384.

Microwave oven: U.S. Bureau of the Census, *Statistical Abstract of the United States, 1981* (Washington, D.C.: GPO, 1981), Table 1384; *1985*, Table 1320; *1988*, Table 1227.

with lower than average heating loads, electric space heating began to be adopted more generally. Recently it accounted for about 40 percent of the space heating equipment in new housing construction.

The shift to electric space heating has been encouraged by the maturation of heat pump technology. The heat pump delivers two to three times more heat than the thermal equivalent of the electricity it uses. In addition, the heat pump, now available either as a central system or as a modular unit

Figure 11.11   Trends in the Adoption of Large Heating Appliances, 1933–1987

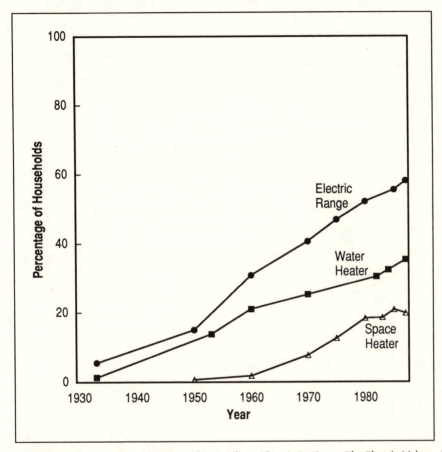

SOURCES: Electric range (electric cooking): Edison Electric Institute, *The Electric Light and Power Industry in 1933* (New York: EEI, October 1934); U.S. Bureau of the Census, *Statistical Abstract of the United States, 1988* (Washington, D.C.: GPO, 1988), Tables 1221 and 1225.

Water heat: Edison Electric Institute, *The Electric Light and Power Industry in 1933* (New York: EEI, October 1934); John Tansil, *Residential Consumption of Electricity*, ORNL/NSFEP-51 (Oak Ridge, TN: Oak Ridge National Laboratory, July 1973); U.S. Bureau of the Census, *Statistical Abstract of the United States, 1971* (Washington, D.C.: GPO, 1971), Table 1098; *1985*, Table 1320; *1988*, Table 1227.

Electric space heating: U.S. Bureau of the Census, *Statistical Abstract of the United States, 1988* (Washington, D.C.: GPO, 1988), Tables 1221–1225.

installed through a wall or window, serves both heating and air condition-ing loads. So the migration trend to warmer regions and the subsequent adoption of air conditioning reinforces adoption of electric heating via the versatile heat pump.

Table 11.1 Adoption of Large Electric Heating Services; Percentage of Housing Units Using Electricity as the Primary Source of Heat for Cooking, Water Heating, and Space Heating, as of November 1984

Heating service	Period of housing construction								Total, as of Nov. 1984
	Pre-1939	1940–1949	1950–1959	1960–1964	1965–1969	1970–1974	1975–1979	1980–1984	
Cooking	36	48	56	55	56	67	79	80	55
Water heating	21	31	27	33	33	47	52	50	34
Space heating	4	10	8	13	20	30	40	39	17

SOURCE: U.S. Energy Information Administration, *Residential Energy Consumption Survey: Housing Characteristics, 1984* (Washington, D.C.: GPO, October 1986), 87.

Figure 11.12   Electricity-Fuels Price Ratios in Residential Use, 1960–1987

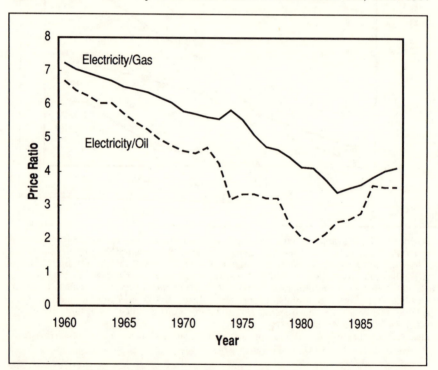

SOURCES: Electricity price: Edison Electric Institute, *Statistical Yearbook of the Electric Utility Industry* (Washington, D.C.: EEI, annual).
   Natural gas price: American Gas Association, *Gas Facts* (Washington, D.C.: AGA, annual).
   Heating oil price: American Petroleum Institute, *Basic Petroleum Data Book* (Washington, D.C.: API, annually).
   NOTE: The electricity-fuels price ratios are calculated as a ratio of dollars per Btu of electricity delivered to dollars per Btu of gas delivered.

FUTURE OUTLOOK FOR HOUSEHOLD ELECTRICAL SERVICES

Typical requirements for electricity in the home as of the early 1980s are shown in Table 11.2 by various services. These service requirements multiplied by the fraction of homes having each service yields an average residential use of about 9,400 kWh per year. Such information suggests that around 1980 electric service in the residential sector had reached about 40 percent of its potential based on the then existing housing stock, level of occupancy, types of electric services provided, and the amount of electricity consumed by the average appliance.

It is to be expected, of course, that such forces as technological innovation, social and demographic change, and greater affluence will, as in the past, create new uses for electricity in the home. But, on the other hand, many new models of existing electrical appliances already offer improvements of up to 50 percent in energy use efficiency, and some of the new electrical household services will very likely be associated with relatively lower intensities of power use than in the past.

Rather than try to predict which applications will be widely adopted and the quantities of electricity they will consume, we will briefly identify a partial catalog of possibilities that are now evident in various categories of

Table 11.2   Residential Electric Services, Circa 1980

Service	Annual usage, kWh	Fraction adopted	Approximate average annual use per household, kWh
Space heating	9,000	0.18[a]	1,600
Water heating	4,000	0.32	1,280
Air conditioning	2,500	0.57	1,420
Refrigeration	1,700	1.00	1,700
Freezer	1,300	0.38	500
Lighting	1,000	1.00	1,000
Clothes dryer	1,000	0.47	470
Range	800	0.54[b]	430
Other	1,000	1.00	1,000
Totals	22,300		9,400

SOURCES: Annual usage: Howard S. Geller, *Energy Efficient Appliances* (Washington, D.C.: American Council for an Energy-Efficient Economy, June 1983). Typical requirements for the various services in homes in which they were used as of the early 1980s.

Fraction of homes having each service: U.S. Energy Information Administration, *Residential Energy Consumption Survey: Housing Characteristics 1980* (Washington, D.C.: GPO, June 1982).

[a] Excludes the additional 11.3 percent that use electricity as a secondary fuel for space heating.

[b] Defined as those that cook with electricity.

use, with no effort to quantify their ultimate effects on electricity consumption.

Health, Comfort, and Convenience

In-home services for an aging and health conscious population are likely to expand. Already available and in limited use are air-to-air heat exchangers, automatic humidity control systems, and electrostatic precipitators for home dust control. For example, manufacturer shipments of electronic air cleaners have averaged about 200,000 units annually in the 1980s.[4] Like the initial need for space cooling, the markets for these appliances are now primarily associated with residences where there are unusual health problems that can be alleviated through their use. Other potential health care amenities include systems for monitoring vital signs and performing health-related diagnostic tests in the home. The air-to-air heat exchanger also offers a potential to save energy for space conditioning and to reduce the natural radon level in afflicted houses.

For large homes, there is a trend toward modular space and water heating systems. These modular systems (e.g., multiple heat pump and water heating systems for different parts of the home) eliminate long ducting and piping runs, thereby adding space to the home and allowing individual control of energy service for different purposes and locations in the home. They also improve reliability and reduce the requirements for energy, water, and sewer service. Modular appliances tend to be electric because the extra cost for fuel supply and flue gas venting at multiple locations makes the direct use of fuel less practical.

Entertainment and Information

The adoption of technology to provide home entertainment continues to expand rapidly. Growth in shipments of video cassette recorders (VCRs) has exploded in recent years as the technology has improved and costs have fallen (Figure 11.13). Other video-related technologies, such as television projection and satellite receiver systems, have also seen growth. In audio entertainment, compact disc (CD) technology is just beginning to make inroads (Figure 11.14). Other applications envisioned for CD technology, such as the storage and retrieval of vast amounts of information, imply greatly expanded use in the future.

Electronic office equipment has made important inroads into the home appliance market. Personal computers for the home had seen dramatic growth up to 1984, but they then declined somewhat in what looks to have been a temporary retreat (Figure 11.15). The personal computer performs a variety of tasks, such as correspondence and record keeping, and fulfills many creative and leisure needs as well. The decreasing cost of home copiers and telefax equipment ensures increased adoption for home office

Figure 11.13 Manufacturer Shipments of Videocassette Recorders, 1981–1987

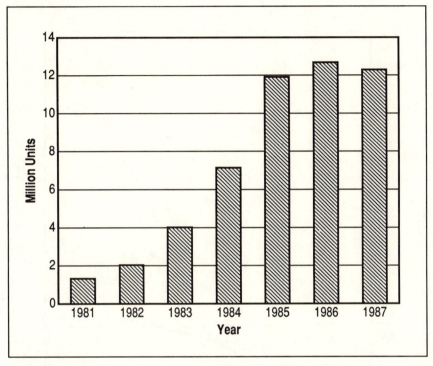

SOURCE: U.S. Bureau of the Census, *Statistical Abstract of the United States, 1986* (Washington, D.C.: GPO, 1986), Table 1384; *1988*, Table 1297; *1989*, Table 1321.

use. Several manufacturers have recently introduced "personal" machines that combine telefax, telephone, answering machine, and copier.[5]

Leisure

The home swimming pool and hot tub are no longer considered luxuries limited to affluent households. In addition to their popularity among all age groups, such facilities are increasingly identified with health-related decisions to improve the physical fitness and relieve the discomfort of an aging population. The swimming pool requires motors for water recirculation and filtering and automatic (robot) cleaning, as well as power for nighttime lighting. In the future, ozone generators and ultraviolet light at the pool may displace the use of chemicals for virus and bacteria control in pool water.

Figure 11.14   Manufacturer Shipments of Selected Video and Audio
Home Appliances, 1981–1987

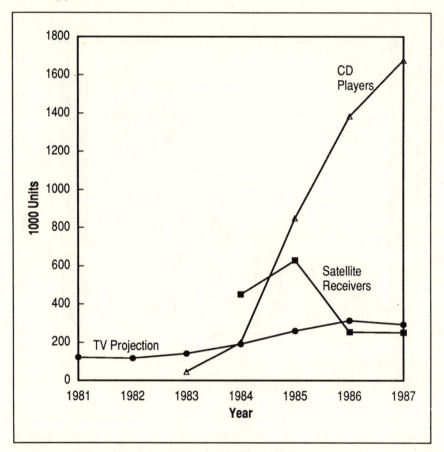

SOURCE: U.S. Bureau of the Census, *Statistical Abstract of the United States, 1986* (Washington, D.C.: GPO, 1986), Table 1384; *1988*, Table 1297; *1989*, Table 1321.

Transportation

If research efforts now underway to develop higher performance batteries
for automotive use prove successful, it seems likely that electric vehicles will
be adopted for some fraction of automobile use. Electricity requirements
for automobile batteries are potentially a very large increment to household
electricity use (see Chapter 9).

Figure 11.15   Manufacturer Shipments of Selected Home Office
Equipment, 1981–1987

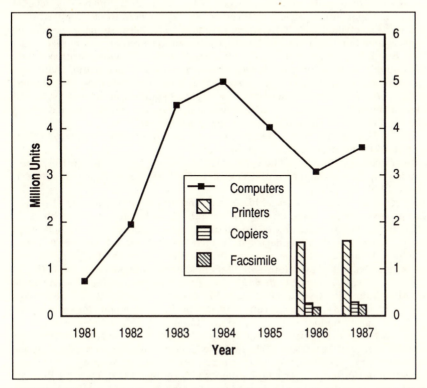

SOURCE: U.S. Bureau of the Census, *Statistical Abstract of the United States, 1984* (Washington, D.C.: GPO, 1984), Table 1422; *1988*, Table 1297; *1989*, Table 1321.

ELECTRICITY'S IMPRINT

Electricity as an agent of technological progress has left as strong an imprint on our daily lives at home as on the performance of work in the industrial sector. Further, no end is in sight, given the rapid proliferation of electronic, mechanical, and thermal applications of electricity in the home.

By enhancing the efficiency of household operations, electric-based technologies have supported such broad demographic and social trends as greater female participation in the paid labor force and the growth in family income and living standards. Greater affluence, in turn, has led to an ever-growing demand for the new and improved services that electrical appliances in the home provide. Dependence on electricity for innovative residential applications is well established historically, and this pattern is expected to continue in the future.

Notes

1. It is necessary to distinguish between urban and rural households because there were important historical differences in their access to electricity. Urban homes were electrified largely between 1910 and 1930. Because urban residences were close together, electrical distribution was economic even for residential usages of a few hundred kWh per year. On the other hand, service to widely scattered farm homes could be justified only for levels of household usage much higher than those being estimated for average farm consumption. Rural electrification was delayed, therefore, until federal assistance became available, beginning in the mid-1930s. Full residential access to electricity — urban and rural — was not achieved until the early 1950s. See Chapter 10 for more on rural electricity usage.

2. Edison Electric Institute, *The Electric Light and Power Industry in 1933* (New York: EEI, October 1934), 13.

3. Ruth Schwartz Cowan, "The Industrial Revolution in the Home: Household Technology and Social Change in the Twentieth Century," *Technology and Culture* 17 (January 1976), 22.

Much has been written by social historians, and others, on the impact of techno-logical progress in the home on women's role in the labor force and at home. In addition to Cowan's writings, other useful discussions include: Martha Moore Tres-cott, ed., *Dynamos and Virgins Revisited: Women and Technological Change in History* (Metuchen, NJ and London: Scarecrow Press, 1979); Christine E. Bose, Philip L. Bereano, and Mary Malloy, "Household Technology and the Social Construction of Housework," *Technology and Culture* 25 (January 1984): 53–82; Rosalind Williams, "The Other Industrial Revolution: Lessons for Business from the Home," *Technology Review* (July 1984), 31–40; Sarah Fenstermaker Berk, *Women and Household Labor* (Beverly Hills and London: Sage Publications, 1980), including articles by Susan M. Strasser and John P. Robinson on the relationship between household technology and household work; Charles A. Thrall, "The Conservative Use of Modern House-hold Technology," *Technology and Culture* 23 (April 1982): 175–195; Joann Vanek, "Time Spent in Housework," *Scientific American* (November 1974), 116–120. Relevant Ph.D. dissertations by Susan M. Strasser and Joann Vanek are available through University Microfilms, International, Ann Arbor, Michigan.

4. U.S. Bureau of the Census, *Statistical Abstract of the United States, 1982–1983* (Washington, D.C.: GPO, 1983), Table 1434, and *1988*, Table 1297.

5. Susan M. Gelfond, "Will there be a Fax in Every Foyer?" *Business Week* (August 3, 1987), 82.

Part IV
Long-Term Quantitative Trends:
Electricity Use, Productivity Growth,
and Energy Conservation

Introduction

Parts I and II dealing with manufacturing contained much industrial detail to document the book's general theme that energy in the form of electricity has had a special connection with technological progress in twentieth-century America. In this part of the book, we deal with the same broad theme, but we shift the main angle of vision to the quantitative analysis of long-term trends for manufacturing as a whole, rather than the description of specific industries.

Chapter 12 analyzes quantitative indicators of the connections between the electrification of production operations and the overall growth of productivity in the manufacturing sector. Chapter 13 follows with a quantitative assessment of the relationship between the substantial growth in the electricity fraction of total energy consumption over the long term and the large improvements in energy efficiency that were achieved. To anticipate the major conclusion: we find that the common source of these two sets of historical trends (electricity up, productivity up; and electricity up, energy intensity down) is electricity's role as an agent of technological progress.

The starting point for the analysis in Chapter 12 is the importance of productivity gains as a component of long-term economic growth in the United States. For the full period covered by our statistics — 1899 to 1985 — manufacturing output grew at an annual average rate of 3.7 percent, of which more than half was accounted for by improvements in multifactor productivity. This single statistic, which measures the amount by which the rate of output growth exceeded that for the combined inputs of labor and capital, shows the enormous importance of improvements in productivity to overall economic growth, and to the phenomenal long-term increases in living standards that have been achieved in the United States.

The main purpose of the chapter is to quantify electricity's role in supporting this growth in productivity. To do this we focus on the changing mix over time of inputs of capital, labor, electricity, and nonelectric energy into manufacturing production, with an emphasis on growth in the use of electricity relative to the use of other inputs.

We assign special importance to growth in electricity use relative to growth in capital inputs (i.e., plant and equipment) because the latter provide the major means whereby new and improved technologies are brought into operation. The quantitative analysis shows that in the course of the twentieth century growth in the use of electricity in manufacturing has far outstripped growth of all the other inputs that we measured. Electricity grew very rapidly compared with capital, about two and a half times as fast, while nonelectric energy increased at only half the rate of growth in the inputs of capital — a comparison that, we believe, signifies the strong affinity of long-term technological progress in manufacturing for energy in the form of electricity.

After an introductory scan of the comparative growth rates of the various inputs over the full eighty-five-year period covered by our statistics, the bulk of the analysis in Chapter 12 focuses on a detailed examination of data relating to three long subperiods (1899–1920; 1920–1948; 1948–1985). This approach allows us to relate quantitative indicators of the growth in electricity use to electricity's changing role in the manufacturing technology that characterized these different time periods. Technological developments differed markedly among the three periods, as did the growth in electricity's use relative to the use of other inputs.

In Chapter 13, we turn from productivity growth, as such, to an analysis of the impact of the growth in the consumption of electricity on the overall efficiency of energy use. This analysis begins by displaying a striking but little-known aspect of the history of energy use in the United States: the coexistence throughout much of the twentieth century of a rising electricity share of total energy consumption and a declining intensity of overall energy use relative to output. These two trends coexisted not only in manufacturing, but also in the total U.S. economy and the private domestic business economy.

Considering the thermal energy losses sustained in converting raw fuels into electric power, it is puzzling to discover that a strong rise in the electricity fraction of total energy has coexisted with a persistent rise in the overall efficiency of energy use. The chapter documents, however, that this is what has happened, and it explains why this outcome is, in fact, a consequence of the unique attributes of electricity as a production factor.

Chapter 13 compares two time periods — 1920 to 1929 and 1973 to 1985 — which enable us to highlight some important historical differences in the impacts of electrification on productivity growth and energy conservation. During the 1920s there were particularly high rates of both productivity growth and energy efficiency improvement, and there were also very strong increases in the electrification of plant and equipment. From the historical evidence in Chapters 1 and 12, we know that these results reflected the electrification of machine drive and an accompanying transformation in the overall organization of factory production, two interrelated developments that constituted a major technological thrust in manufacturing during this period.

From 1973 to 1985, on the other hand, there were high rates of energy efficiency improvement but only slow improvements in productivity. There was also a substantial slowdown in the growth of electricity relative to new increments of capital. Chapter 13 examines the sources and implications of the comparatively slow growth in electricity use and rapid growth of capital that occurred after 1973. This analysis points not to a decline in the importance of electricity as the energy form required by new technology, but to the very different electricity use requirements that exist between electrification of machinery as the major technological thrust (as in the 1920s), and the current technological thrust in automatic control and information technologies. The new technologies simply require less energy. Counterweighting this shift toward reduced electricity consumption by the newer electro-technologies is still another current technological development: the increasing electrification of the heavily energy-using processing industries that were previously strongly dominated by fuel-based technologies.

This final section of the book closes by speculating about the effects that such technological changes may have on the future course of the three parallel trends — increases in the electricity fraction of total energy use, the growth in economic productivity, and the improvements in the overall efficiency of energy use — that have, to date, characterized the long-term historical record in the United States.

12 | Electrification and Productivity Growth in Manufacturing

Sidney Sonenblum

QUANTITATIVE OVERVIEW

In this chapter we present evidence in support of the hypothesis that technological progress in manufacturing during much of the twentieth century has been related to the adoption and spread of production processes and modes of organization that are dependent on the use of electricity. Our analysis is based on an examination of long-term quantitative trends in output, productivity, and the use of labor, capital, electricity, and nonelectric energy inputs for manufacturing as a whole.[1] These quantitative findings are evaluated against a background of historical information describing evolutionary changes in technology, particularly as they relate to the organization of manufacturing production.

Figure 12.1 shows that manufacturing output during the twentieth century has grown at an annual average rate of 3.7 percent, with more than half of this growth attributable to improvements in multifactor productivity (the ratio of output to the combined inputs of capital and labor).[2] Over the same period, the use of electricity in manufacturing grew far more rapidly than output or any of the other individual inputs — at an average annual rate of more than 8 percent compared to 1.5 percent for nonelectric energy; 1.3 percent for labor; and just over 3 percent for the capital used in manufacturing.

Clearly, the long-term growth in capital stock has been geared to a substantially faster increase in electric than in nonelectric energy. This finding is of particular interest because to the extent that additions to plant and equipment incorporate technological progress, a faster rate of growth for electricity relative to capital than for nonelectric energy relative to capital signifies that, over time, technological progress in manufacturing has tended to show an affinity for energy in the form of electricity.[3]

For analytic purposes, we have divided the entire span of years covered by the statistics into three major periods: 1899 to 1920, 1920 to 1948, and 1948 to 1985.[4] The dividing years — 1920 and 1948 — not only mark the begin-

Figure 12.1   Output, Productivity, and Inputs, 1899–1985

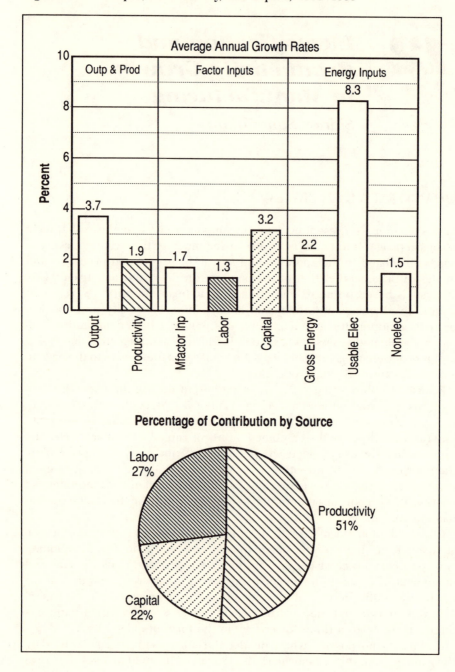

ning of the two postwar eras, but also the start of periods of significant shift in electricity's role in manufacturing technology. This is the case particularly in connection with technologies that have led to fundamental changes in the overall organization of production systems, resulting in across-the-board impacts in practically all manufacturing industries.

Several major themes dominate these three periods. Here is a preview:

1899–1920: The first period carried over into the twentieth century a pre-electrical organization of production in manufacturing. The primary emphasis in manufacturing continued to be, as it had been in the late nineteenth century, on increasing the scale of production as a way of meeting the demands of the growing mass market for goods. Electricity began to be used as an input into productive operations, but it was generally looked upon as little more than another source of heat and power that could serve as a supplement to coal and other traditional energy sources. With only a few exceptions (particularly in the electroprocessing industries discussed in Chapter 3), electricity was not yet being exploited for its ability to change the way in which production was carried out.

1920–1948: During this period the focus in the organization of production changed from a primary emphasis on achieving expansion of scale to a growing attention to the opportunity to accelerate the flow of throughput within the factory. Major advances came about as a result of the introduction of electrical systems for distributing power within the plant in place of earlier steam-based systems of power distribution. Electric unit drive, in particular, was of crucial importance because, as indicated in Chapter 1, the growing use of power delivered by wire to individual machines eliminated obstacles to the movement of work and restrictions on the design of plants that had been imposed by mechanical systems based on the use of shafts and belting. This change opened the way for altogether new approaches to laying out the factory and sequencing the flow of work.

1948–1985: The organization of production during these years began to shift from a primary focus on accelerating the flow of throughput to include a new and still evolving emphasis on achieving greater flexibility in manufacturing operations. The automatic control of production has been a key ingredient in achieving productivity gains and flexibility during these years. Electricity has become absolutely indispensable in this phase of technological evolution because the equipment required can be run by no other energy form.

Figure 12.2 shows that productivity did not grow at similar rates during the three periods. It grew most rapidly in the second period — 1920 to 1948 — when it accounted for fully three-fourths of the growth in total manufacturing output. This performance stands in sharp contrast to the period before 1920 and that following 1948. The contrast with pre-1920, when productivity improvement accounted for only one-fifth of overall output growth, is particularly striking. Although productivity growth continued to be the most important factor during the post-1948 period, output growth in

Figure 12.2 Output, Productivity, and Multifactor Inputs, 1899–1920, 1920–1948, 1948–1985

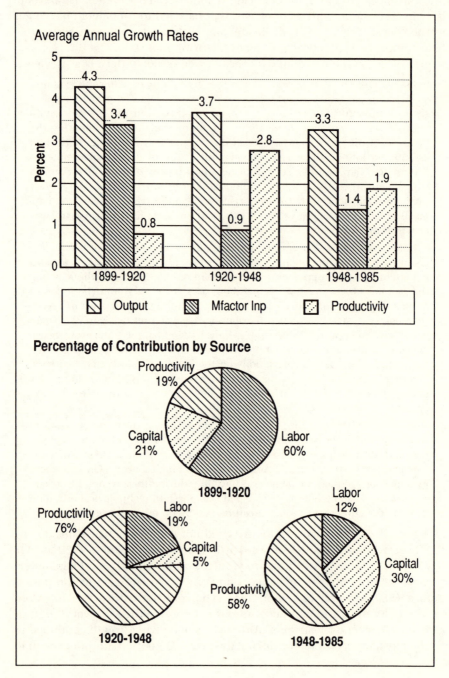

these years has depended on more nearly equal contributions from productivity gains and the growth in combined labor and capital inputs.[5]

Just as productivity growth rates varied sharply over time, so too did the growth rate of electricity as compared to the other inputs. As shown in Figure 12.3, electricity's fastest rate of growth occurred between 1899 and 1920. One reason is that its growth during this period began from a negligible base. Another reason is that all of the other inputs also experienced their fastest rates of growth in these years. Rapid growth of all inputs is, indeed, the hallmark of this early period, reflecting the fact that the high rate of growth in manufacturing output during these years was brought about largely through the expansion of inputs rather than through productivity growth.

During the period from 1920 to 1948 all of the input growth rates slowed down. The slowdown was mild for electricity as compared to the other inputs, resulting in a sharp increase in the rate of electricity growth relative to the growth of other inputs. During this period electricity grew at about 7 percent annually compared to about 0.5 percent for nonelectric energy,

Figure 12.3   Energy and Factor Inputs, 1899–1920, 1920–1948, 1948–1985, Average Annual Growth Rates

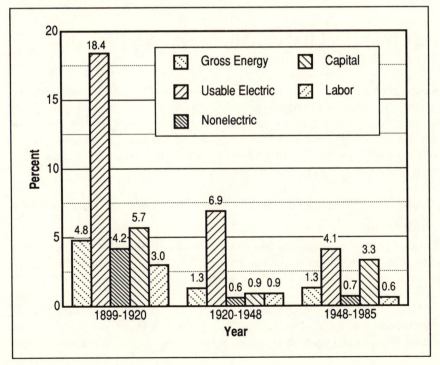

and almost 1 percent each for labor and capital — a comparative growth rate for electricity that is far higher than during the other two periods. We believe, as explained later, that the relatively rapid growth in multifactor productivity and the rapid increase in the use of electricity relative to other inputs during the period from 1920 to 1948 are interrelated.

During the 1948 to 1985 period, electricity again grew more rapidly than any of the other inputs, but by a greatly reduced margin. This was particularly true in relation to capital, whose growth rate of 3.3 percent drew close to electricity's 4.1 percent annual rate.

In the remainder of this chapter, we consider each of these three periods in fuller detail. We will be particularly interested in understanding the reasons for the continuous shift from nonelectric to electric energy use. Because, as noted earlier, much of technological progress is embodied in additions to plant and equipment, changes in the ratio of the growth in electricity and other energy use to the growth in capital stock should provide a useful indicator of energy-technology linkages. To what extent do variations in the rate and direction of these changes reflect the changing role of electricity and other energy forms in the evolution of manufacturing technology over time? As we weave together the descriptive and the statistical sides of our analysis for the three separate periods, we will be giving much attention to this question.[6]

ELECTRICITY'S CHANGING ROLE DURING THE TWENTIETH CENTURY

As the technological basis for the organization of manufacturing evolved during the twentieth century, so did electricity's role. We turn now to an examination of these changes by time periods. For ease in presentation, the periods are designated as Stages I, II, and III, despite substantial overlap. What we are trying to capture is the direction and strength of change using both historical descriptions and quantitative measures.

Stage I (1899–1920): Enlarging the Scale of Operations

The Setting[7]

Because Stage I carried into the early twentieth-century production practices that had developed at an earlier time, it is helpful to refer back to the latter half of the nineteenth century when discussing them. Until the middle of the nineteenth century, American manufacturers depended almost exclusively on wind, wood, and water as their sources of heat and power.[8] The relative scarcity of these energy forms and uncertainty as to their availability in sufficient amounts when and where they were needed put severe constraints on industrial location, technological advance, and output growth.[9]

The widespread use of coal to produce steam for power and heat came to the United States in the second half of the nineteenth century. Coal had particular qualities that made it more efficient and more reliable as a source of heat and mechanical power than were the earlier energy forms; and it was also more generally available. These qualities affected technological progress in significant ways, but the most important was that coal constituted a vastly greater and more mobile source of energy than could be provided through the use of wind, wood, and water — a factor of enormous importance in an economy that was moving toward mass production. Coal's role in loosening energy supply constraints reinforced an emerging attitude toward the organization of production that emphasized the importance of scaling up the capacity of plants in order to expand manufacturing output. This philosophy carried over into the early twentieth century.

Along with an emphasis on enlarging the scale of production went a strong trend toward consolidation of all the operations required to produce a complete product. Such consolidation created very large plants that could be responsive to the growth of mass markets because the materials, parts, products and labor needed for production were readily accessible and subject to unified control.[10]

Within the fabricating industries, production was typically based on the assembly of interchangeable components.[11] Normally the cluster of equipment, labor, and materials needed to produce a particular component "was located in any convenient space [within the plant] without regard to being adjacent to the next operation in the production sequence."[12] Each cluster required convenient access to mechanical power. Thus, as described in Chapter 1, the machines were located close to the line shafts of the steam-based power distribution system, and the production activities were arranged linearly in order to conform with that system.

An analogous situation prevailed in the materials processing industries — those producing homogeneous liquids, gases, or solids. Although produced in bulk, rather than assembled from separate components, each specific type of material was made in a separate area of the plant containing the equipment, labor, and feedstock needed to produce that material. As noted in Chapter 3, only in the later years of Stage I was it recognized that many of the same basic processes were required for different products, enabling production to be organized around types of operations and processes rather than types of products. However, whether organized according to products or processes, the spatial organization of production in these industries was oriented around the coal-burning furnaces used to provide process heat and constrained by the steam powered pumps, compressors, and conveyors used for bulk materials transport.

The Role of Electricity[13]

Manufacturing was using very little electricity at the turn of the century even though it was widely understood that electricity had properties that were in many ways superior to steam. In particular, electricity made it possible for industry to develop power from a variety of fuel and hydro sources, to obtain power at low cost over what were then great distances, to use power efficiently in small shops, to operate machines intermittently, to design equipment with greatly enlarged capacity, to construct more efficient prime movers, to be more flexible in the operation of production machinery, and to create essentially new types of industrial materials.

In a few instances, the use of electricity increased during Stage I partly because of its special qualities. For example, in Chapter 3 we saw that wholly new industries arose, based on products like aluminum and silicon carbide whose manufacture was unique to electricity. However, in most of the cases where electricity was introduced during Stage I, it was not because of its unique qualities as an energy form, but because it provided a convenient, and sometimes less costly, supplement to other energy sources.

Electricity's restricted role during Stage I was to an extent dictated by the existing infrastructure for delivering mechanical power within the plant. As described in Chapter 1, the pre-electric factory used a line drive system in which (a) power generated by prime movers based on waterwheels or steam engines was (b) transferred to belts and pulleys that (c) turned line shafts, which (d) rotated countershafts that were (e) belted to the production equipment and (f) provided the mechanical power for operating the equipment. This power distribution system imposed a restrictive pattern on plant design, but that fact would not be appreciated until the advantages of an electric-based power distribution system were better understood.

When line drive was initially electrified, it was modified only to the extent that electric motors, rather than steam engines, provided the power to run the belts and pulleys. Applied in this way, electricity did not significantly alter the direct line drive distribution system, but it could still be beneficial. It offered advantages when only small amounts of power were needed, or when the preferred location for a factory was distant from a fuel site, or when an industry (such as clothing, textiles, and printing) wanted to minimize dirt and dust within the factory.

The first major step away from traditional line drive distribution came with the emergence of electric group drive, described in Chapter 1. Group drive modified the pre-electric distribution system by supplying power to a group of machines performing similar operations without the use of long line shafts or even, in some cases, countershafts. A single electric motor turning a relatively short line shaft did the job. This considerably reduced the physical constraints on plant layout and work flow, and it also reduced the power lost to friction in turning shafts.

Both line drive and group drive power distribution systems were designed to function as an entity. Either everything operated or nothing did. With line drive the plant's entire network of shafts and countershafts was in continuous rotation, except when something broke down, in which case the entire network stopped operating.

Group drive offered a considerable advantage over line drive because a particular group of machines could be operated independently of the rest of the plant. With group drive, too, the entire network of lines and shafts operating a particular group of machines was either completely shut down or in continuous operation. But the evolution toward group drive, plus the growing use of unit drive (which was to become the main thrust during Stage II), represented transitional steps from the pre-electric system to the uniquely electrical mechanical power distribution system that was soon to become pervasive.

Selected Quantitative Findings

The Growth of Electricity Use (Figure 12.4). Usable electricity (the caloric equivalent of the kilowatthours of electricity actually consumed) was a comparatively unimportant component of total energy use in manufacturing during Stage I. Even though its annual rate of growth was very high, usable electricity accounted for less than 2 percent of the total growth of energy in Stage I and constituted only 1 percent of the total energy consumed in manufacturing in 1920.[14] However, the usable electricity share of total energy underestimates the relative importance of electricity in production because practically all of the electric Btus are put to use, whereas there are significant losses when nonelectric energy is employed.

Electric Energy Services (Figure 12.5). The purposes for which electricity was used in manufacturing changed sharply in the first two decades of the twentieth century, pointing the way toward major transformations yet to come. At the beginning of Stage I, .30 percent of the electricity used was devoted to machine drive and 70 percent to materials conversion and supporting services such as lighting. By the period's end there was a complete reversal of this ratio with 71 percent going into machine drive. While in 1920 the 29 percent of electricity used for materials conversion and supporting services was almost equally divided between these two classes of use, support services (essentially lighting) had been six times as important as materials conversion in 1899. Increasingly, therefore, electricity was being used directly in the manufacturing process, either for motors or in materials conversion, rather than just as a supporting service.

Another significant aspect of electricity use is that its growth was due mainly to the increase in electricity's share of all energy used in machine drive — less than 4 percent in 1899 to over 50 percent by 1920 (measured in terms of thermal inputs into generation). In materials conversion and

Figure 12.4 Electricity and Energy Use During Stage I, 1899–1920

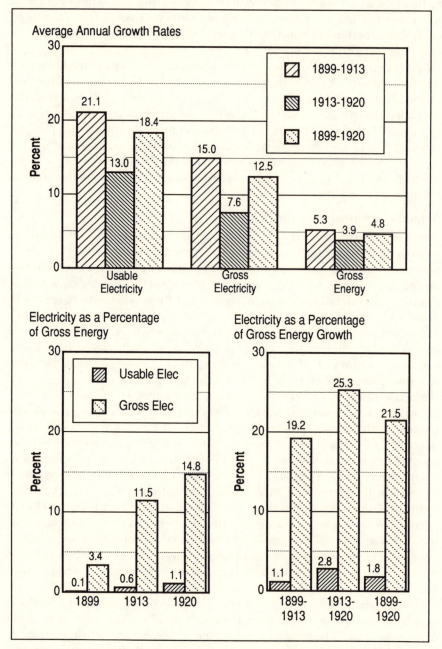

NOTE: "Usable electricity" represents the calorific value of the kilowatthours of electricity used, while "gross electricity" represents the calorific equivalent of the raw energy consumed in generating this quantity of electricity.

Figure 12.5 Energy Services During Stage I, 1899–1920

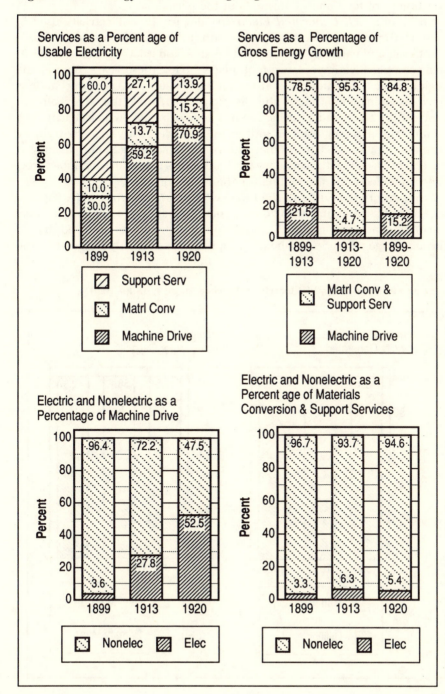

support services, on the other hand, electricity continued to be a minor constituent of total energy throughout the period.

Utility and Self-Generated Electricity (Figure 12.6). Electricity used in manufacturing was produced partly within the manufacturing establishment itself and partly by electric utility plants, and relatively little change in their comparative importance took place during these years. Both at the beginning of the century and at the end of Stage I central power stations were providing about half of the electricity used, while the other half was provided by the manufacturing sector's generating its own electricity.

Similarly, primary electric motors which use utility-provided power accounted for somewhat less than half of manufacturing motor capacity at the end of Stage I, with secondary electric motors that utilize manufacturer-generated electricity accounting for the remainder.[15]

Productivity Growth and the Energy/Capital/Labor Input Mix (Figures 12.7 and 12.8). Manufacturing output increased at over 4 percent per year during Stage I. This relatively rapid rate of growth was achieved mainly through large increases in labor and capital inputs. Together, growth in the

Figure 12.6   Sources of Electricity During Stage I, 1899–1920

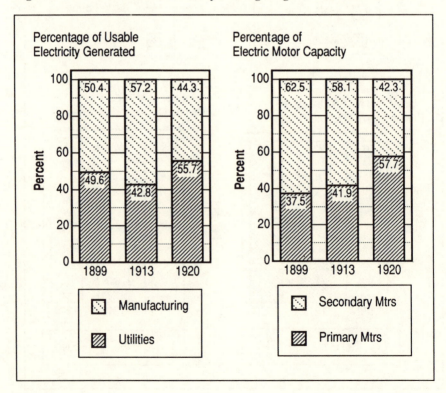

Figure 12.7 Output, Productivity, and Multifactor Input During
Stage I, 1899–1920

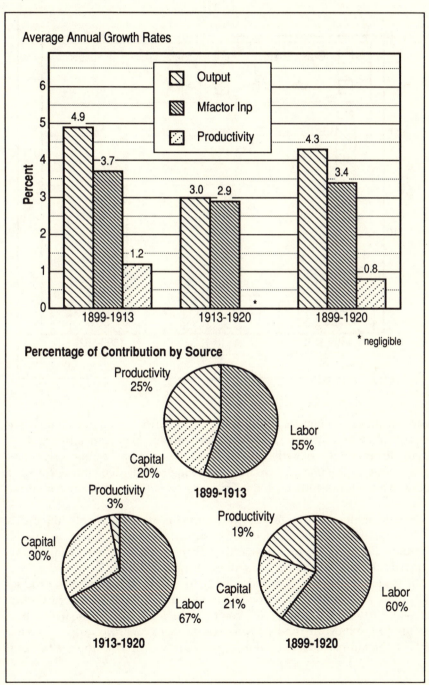

Figure 12.8 Energy and Factor Inputs During Stage I, Average Annual
Growth Rates

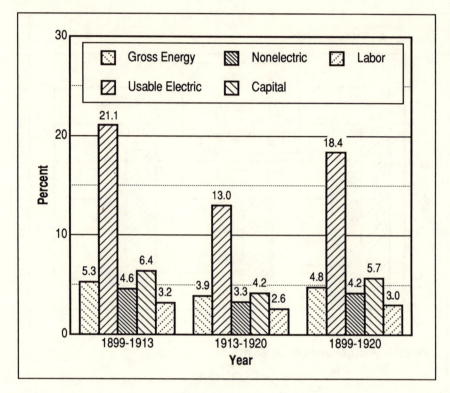

inputs of these two factors accounted for four-fifths of the overall increase
in output, far and away larger than their relative importance in the later
stages. Productivity growth, on the other hand, made a relatively small
contribution to the increase in manufacturing output, accounting for only
one-fifth of total output growth. This was far smaller than in any other
stage.

The rate of increase in total energy use during Stage I was only four-fifths
of the capital increase, but electricity grew three times as fast as capital.
This increase in electricity use relative to capital during Stage I came about
primarily because of the electrification of machine drive.[16]

The overall trends for Stage I also apply to each of its subperiods. Productivity gains were slow (relatively modest up to 1913 and nonexistent
between 1913 and 1920), while growth in factor inputs, primarily labor,
accounted for most of the relatively substantial growth in output. During
each of the subperiods, capital and total energy grew at about the same

rates, while electricity increased much faster than capital and nonelectric energy.

Stage II (1920–1948): Accelerating the Rate of Throughput

The Setting[17]

During these years the focus of managerial attention shifted from enlarging the scale of operations to increasing operating efficiency by speeding up the rate of throughput in the plant. High priority was assigned to modifications of factory design and layout in order to better integrate worker and machine tasks. Advances in the electrification of machine drive were indispensable to the realization of these new objectives and may, indeed, have served to stimulate the new managerial perspectives that emerged.

Speeding up the work flow depended on improving the coordination of workers and machines. This was made possible in both the fabricating and materials processing industries by organizing production as a precise sequence of tasks that moved materials and parts according to the logic of the work process itself. Machinery was arranged spatially so that products moved from one operation to the next with little or no waiting, and machine capacities were balanced so that the production rate at each workplace could match that of its neighbors.[18] Unlike Stage I, where production activities were rooted to a fixed mechanical power distribution system, during Stage II they were linked to each other so as to assure a continuous and speedy flow of materials and parts through the various steps in production.[19]

At the same time, task performance was becoming increasingly routinized; machines were designed and labor was trained to perform specific, highly detailed tasks; production lines and even multiple production lines were used to move the work in process and to reduce bottlenecks; and, long production runs were adopted to minimize disruptions associated with frequent setting up.

Critical to the advances that were achieved in the 1920s was the vision of production as a continuous process. The availability of electric power was a key precondition for achieving that vision. The contribution of management was to recognize the new potentials that had been created and to make the deliberate effort to capitalize on them.[20] It required a new understanding to realize that the output growth needed to satisfy the rapidly expanding markets for manufactured goods no longer had to depend primarily on the traditional approach of increasing labor and capital inputs, but could come to a major extent from improving the work flow to raise efficiency.

The Role of Electricity[21]

The electric line and group drive systems that took hold during the early 1900s were not significantly different from steam-based systems in the rigid infrastructure they required for power distribution. As a result, they carried little potential for reorganizing factory production systems. Unit drive, on the other hand, by consolidating on one frame an individual electric motor (fed by wire) and the machine it operated, effectively negated the distance between the source and use of power. As recounted in Chapter 1, space that had previously been cluttered with power distribution equipment could now be freed up and utilized to design a more effective arrangement of the production process.

Since under unit drive machines do not have to be placed in relation to line shafts, the various fabricating operations — forming, joining, finishing, and assembling — could now be treated as a consecutive sequence in the production process. The logic of this functional sequence could be spatially followed so that the handling of materials and parts could be minimized and the continuity of their movement maintained. Since unit drive could also be used to drive conveyors and other transportation equipment, it was also instrumental in improving efficiency of the plant's transporting equipment, which was a major step toward linking disparate operations in manufacturing production.

Unit drive made fundamental contributions to a more efficient organization of manufacturing production by allowing a more effective use of space in the factory, by speeding up the movement of materials and parts, by reducing the time required between successive production steps, and enhancing the ability of workers and machines to specialize in particular tasks. In addition to these fundamental contributions, Chapter 1 notes other improvements attributable to unit drive that brought additional benefits. The elimination of shafting made unobstructed overhead space available for traveling cranes, which not only facilitated continuous movement but also improved illumination, ventilation, and cleanliness. Also, because the effective size of the production facility was no longer constrained by the power distribution system, factory capacity could be easily increased.

These innovations required the joint use of electricity as the essential energy form and the electric motor as the device that put that electricity to use. The fundamental changes that have been described required the two together. This combination, by providing an economically feasible alternative to ever-increasing size in the scale of manufacturing operations, also paved the way for efficient small-scale production, permitting manufacturers to decentralize their locations and take advantage of opportunities afforded by proximity to markets or to particular supply sources of labor and materials.

Selected Quantitative Findings

The Growth of Electricity Use (Figure 12.9). The shift toward electricity use in manufacturing that began in Stage I accelerated during Stage II. Usable electricity accounted for about one-seventh of the growth in total energy during Stage II as compared with one-fiftieth during Stage I. The shift to electricity was particularly pronounced during the decade of the 1920s, when nonelectric energy actually declined while usable electricity increased at an annual rate of over 9 percent and accounted for three-fourths of the growth in total energy. This trend continued during the 1930s and 1940s, although at a reduced pace.

Electric Energy Services (Figure 12.10). Although productivity and output increased at a relatively rapid rate during Stage II, the energy used for machine drive did not. In particular, during the 1920s, when productivity gains averaged more than 5 percent per year, the quantity of gross energy used for machine drive declined by 2 percent per year. This happened not because machine drive became a less important production activity, but because a shift to electric machine drive permitted a large reduction in the use of nonelectric energy.

Machine drive accounted for most of the electricity used throughout Stage II. Electrification of machine drive became almost total. At the start of Stage II, about half of machine drive was electric and, by the end, nine-tenths was. The quantity of nonelectric energy used for machine drive actually declined during Stage II. Electricity not only accounted for the entire growth in machine drive, but it was also used to replace the loss of nonelectric drive.

Throughout Stage II, 85 percent to 95 percent of the energy used in materials conversion and support services continued to be nonelectric energy. Electricity's share of these services increased during Stage II, but it never became very large. Meanwhile, the composition of electricity use was shifting. The rate of increase in the use of electricity for materials conversion was even greater than that for machine drive. Indeed, by the end of Stage II, materials conversion was accounting for a sizable one-fourth of the electricity used in manufacturing, compared to 15 percent at the beginning of that period.

Utility and Self-Generated Electricity (Figure 12.11). During Stage II the trend favoring the use of utility-generated over manufacturer-generated electricity became pronounced, with three out of every four of the electrical Btus added coming from utilities. By the end of Stage II, utilities provided over three-fifths of the electricity used in manufacturing. The relative growth in utility supply is also shown by the fact that primary motors (i.e., motors that use utility-generated electricity) accounted for three-fourths of total manufacturing motor capacity by the end of Stage II.

Figure 12.9 Electricity and Energy Use During Stage II, 1920–1948

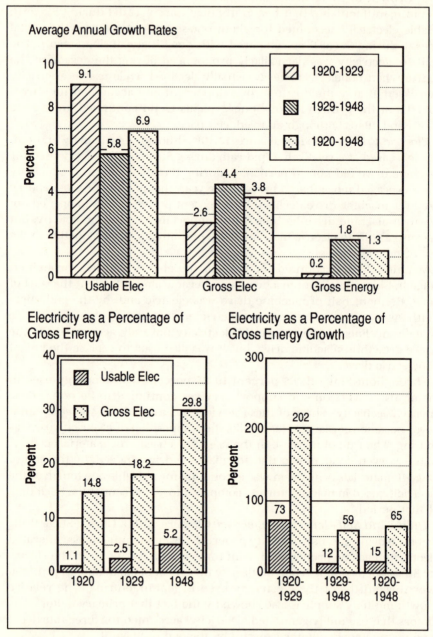

Average Annual Growth Rates

NOTE: "Usable electricity" represents the calorific value of the kilowatthours of electricity used, while "gross electricity" represents the calorific equivalent of the raw energy consumed in generating this quantity of electricity.

Figure 12.10   Energy Services During Stage II, 1920–1948

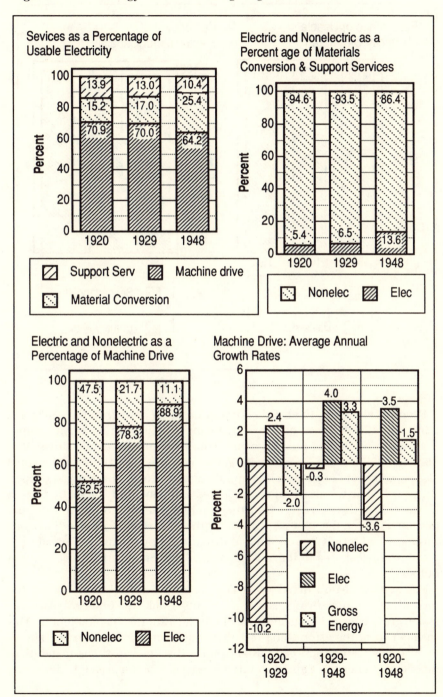

295

Figure 12.11   Sources of Electricity During Stage II, 1920–1948

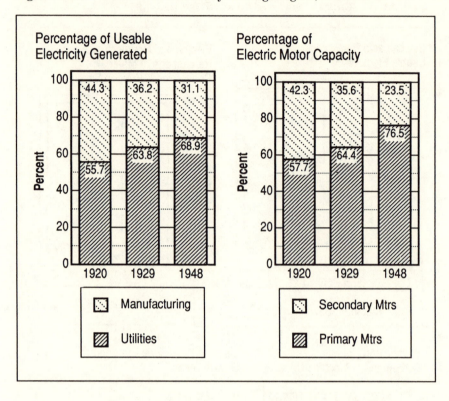

Productivity Growth and the Energy/Capital/Labor Input Mix (Figures 12.12 and 12.13). Three quarters of the increase in manufacturing output that occurred during Stage II came from productivity gains. During the 1920 to 1929 decade, productivity gains accounted for all of a substantial (4.7 percent) rate of growth in manufacturing output. There was a drop in the use of labor which was unprecedented and only partly offset by growth in capital, with the result that the combined labor and capital inputs declined. During the 1930s and 1940s, with a resurgence of growth in labor inputs, productivity gains and multifactor input growth contributed roughly equal shares to the still substantial rate of growth (3.3 percent) in output.

The spread of unit drive was accompanied by a marked change in the mixture of manufacturing production inputs. Unlike the other stages, Stage II saw gross energy inputs increase at a faster rate than capital inputs. This resulted from the astonishing fact that usable electricity increased almost eight times faster than capital (and more than eleven times faster than nonelectric energy). Capital inputs increased at a surprisingly slow rate

Figure 12.12 Output, Productivity, and Multifactor Input During
Stage II, 1920–1948

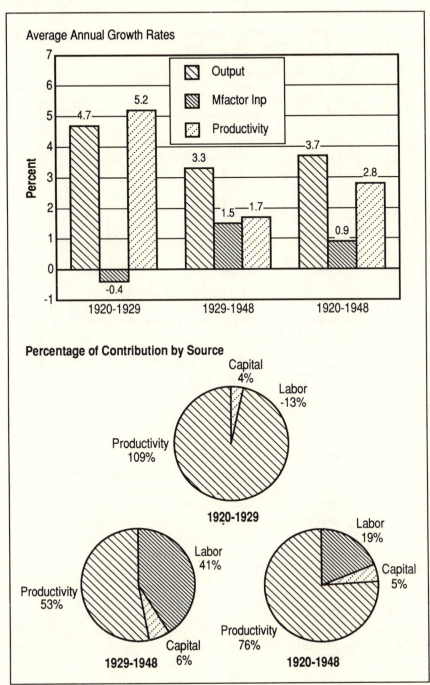

Figure 12.13 Energy and Factor Inputs During Stage II, 1920–1948, Average Annual Growth Rates

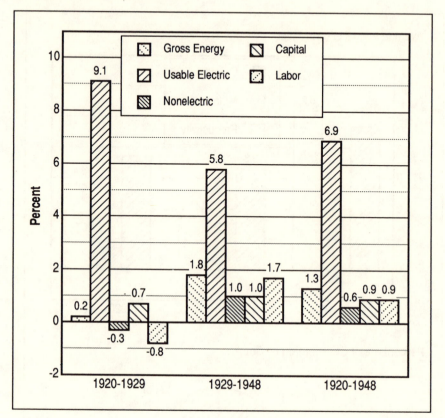

during this period of rapid productivity growth — less than 1 percent annually. The magnitude of the Stage II increase in the ratio of electricity to capital growth rates reflects the fact that the organizational/technological changes that so strongly boosted productivity required relatively small growth in capital but a very rapid increase in electrified production techniques.

Stage III (1948–1985): Broadening the Scope of Automatic Control

The Setting[22]

During Stage III, broadening the scope of automatic control in production joined speed of throughput as a major focus of managerial attention. The use of electromechanical and electronic automatic control equipment — which carry through a sequence of machine movements and production activities without human intervention — was a significant factor and, perhaps, the key element in achieving flexibility and productivity gains during these years. (Though 1985 is the last year for which data were available when this analysis was completed, the period being described is still under way.)

The evolution of automatic machine control has been described in Chapter 2.[23] Automatic control devices during Stage I were mechanical or hydraulic and generally part of the production machinery. During Stage II, electric motors were used not only to power machine functions but also to control their movements. Electromechanical devices called servomechanisms were developed that automatically measured particular machine actions, identified the difference between actual and desired actions, and responded to the difference by making adjustments to achieve the actions desired. Servomechanisms made possible tracer control, the earliest form of automatic control in which the sequence of tasks to be performed could be changed, that is, the machine reprogrammed.

The automatic controls used most widely during the early postwar years of Stage III were an extension of Stage I and II devices in that they focused on efficient control of perhaps complicated but still fixed sequences of machine movements. Gains in the speed of throughput could be achieved with these controls but mostly when they were applied to large volume production, particularly in the automotive and metal-working industries. Efficiency could be increased with these machines but often at the cost of flexibility.[24]

A significant break with the past came with the advent of numerical control; numerically controlled machines are readily reprogrammable because they utilize a mathematical model of the finished object or part. Although developed during the late 1940s and 1950s, numerical controls began to spread only during the 1960s. Besides improving work flow, these systems enabled manufacturing to depart from a past that linked efficiency gains to inflexible, large volume, and standardized output. They presented the possibility of industry becoming more productive even while increasing the production of nonstandard products in small batches.

Reprogrammable controls in the 1960s relied on the use of electromechanical or computer-controlled equipment. This equipment could automatically set, start, stop and change machine operations. Such flexibility in machine operations reduced the cost of shifting from one production task to another, eliminated the need for some production tasks, enabled general-purpose machines to perform tasks that previously required the use of special-purpose machines, and contributed to a more precise coordination of production tasks.

As described in Chapter 2, flexibility in machine operations has been greatly enhanced since the early 1970s with the spreading use of electronic, particularly microelectronic, reprogrammable control devices. Flexibility in production can be achieved because computer technology has significantly reduced the penalties traditionally associated with frequent changes in production runs. This greater flexibility has begun to tap the potential for increased efficiency in two ways: by expanding the range of tasks and the variety of products that machines can provide, and by bringing under integrated control the various phases of a business, such as linking production to marketing and research.

To a large extent, recent advances in the flexibility and scope of manufacturing operations have depended on coordinating production activities through information gathered and shared automatically during these same activities. By gathering and using information about its own performance, a production system can correct its own errors. This self-monitoring capacity has opened the way to what may become the most significant manufacturing innovation of the recent past: advances beyond the ability to preset the operations of machines to presetting the objectives of production.

Selected Quantitative Findings

The Growth of Electricity Use (Figure 12.14). During Stage III, as in the earlier stages, the use of energy in manufacturing has shifted toward electricity. Between 1948 and 1985, usable electricity more than doubled its share of total gross energy used in manufacturing from about 5 percent to 12 percent. Usable electricity also accounted for about one-quarter of the growth in the gross energy total. Measured in terms of energy inputs required for its own generation, electricity's share increased from about 25 percent to 40 percent, while accounting for 70 percent of the growth in total gross energy use.

The rate of growth of electricity used in manufacturing during Stage III (4.1 percent) slowed down substantially from its Stage II pace (6.9 percent). However, most of this slowdown occurred after 1973; in the early years of Stage III (1948–1960), electricity's average annual growth of 7 percent was actually higher than the roughly 6 percent rate during the later years (1929–1948) of Stage II.

Electric Energy Services (Figure 12.15). Even though machine drive was nine-tenths electrified at the end of Stage II, electricity's share of machine drive continued to grow and by 1985 it stood at 95 percent. Usable electricity in machine drive grew at about the same rate as its use for other purposes (4.1 percent), so that throughout Stage III machine drive accounted for about three-fifths of the electricity used in manufacturing. Just as during Stage II, three out of every five Btus of electricity added in manufacturing during Stage III went to machine drive.

Figure 12.14 Electricity and Energy Use During Stage III, 1948–1985

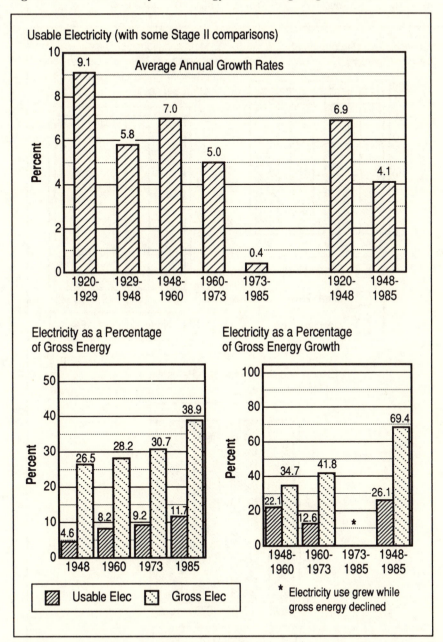

NOTE: "Usable electricity" represents the calorific value of the kilowatthours of electricity used, while "gross electricity" represents the calorific equivalent of the raw energy consumed in generating this quantity of electricity.

Figure 12.15   Energy Services During Stage III, 1948–1985

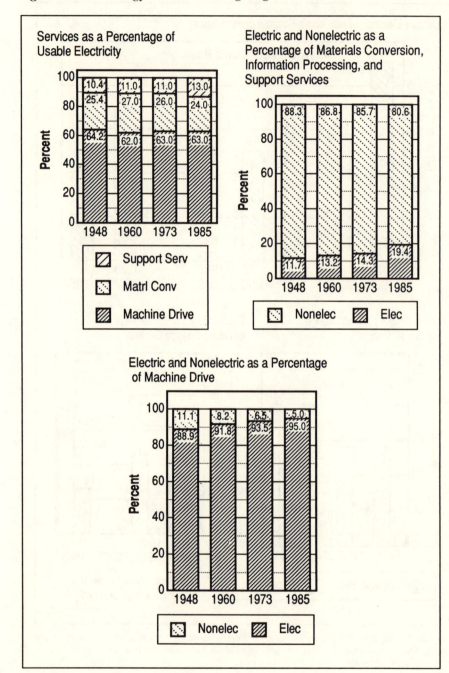

The aggregate of nonmachine drive energy services (which includes materials conversion, information processing, and support services) has continued, as in the past, to be heavily dependent on nonelectric energy. This is because of the heavy use of nonelectric energy for materials conversion, the major energy use in this heterogeneous aggregate. However, electricity's role increased substantially during Stage III. By 1985, it accounted for almost one-fifth of the energy used in providing this group of services, having started the period at about 12 percent. This indicates that in addition to its domination of machine drive and information processing, electricity is beginning to claim a substantially increasing portion of energy use in materials conversion and support services, with materials conversion still taking the lion's share.

Utility and Self-Generated Electricity (Figure 12.16). It was pointed out earlier that Stage II was a crucial period for utilities' rise to dominance in providing electricity to manufacturing. Three out of every four Btus added in manufacturing during that period came from utility sales. These trends have continued into Stage III, with almost all of the additions to manufactur-

Figure 12.16   Sources of Electricity During Stage III, 1948–1985

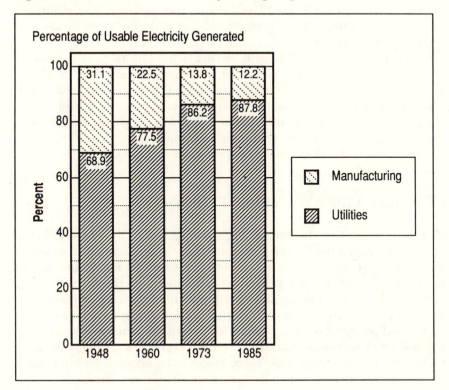

ing electricity use until recent years being provided by utilities. By the start of the 1980s, almost 90 percent of all electricity used in manufacturing was coming from utility suppliers.

Productivity Growth and the Energy/Capital/Labor Input Mix (Figures 12.17 and 12.18). Over Stage III's entire span, multifactor productivity growth accounted for almost three-fifths of the increase in manufacturing output. But the pace at which output, multifactor inputs, and productivity grew varied widely among the subperiods. From 1948 to 1960 growth rates were much the same as they had been in the 1929 to 1948 period; and, in both periods about half of the output growth could be attributed to productivity gains. During the 1960 to 1973 period the output, multifactor input, and productivity growth rates picked up speed as compared with the preceding 1948 to 1960 period. But during the 1973 to 1985 interval, the growth rates slowed considerably.

In conjunction with the spread of automated production techniques during Stage III, there has been a marked shift in the relation between capital, energy, and electricity. During Stage II gross energy increased faster than capital, but during Stage III capital increased at more than twice the gross energy rate. Furthermore, even though the Stage III rate of capital growth has been almost four times that of Stage II, usable electricity growth has been below the Stage II rate.[25] As a result, the growth rate of electricity, which was eight times faster than that of capital during Stage II, was only slightly faster during Stage III.

Earlier, we characterized Stage II (particularly 1920–1929) as a period in which organizational changes based upon electrified unit drive produced large productivity gains while requiring relatively little growth in capital. In contrast, Stage III may be characterized as a period in which organizational change based upon automatic controls and electronic information techniques has required substantial capital investment, but up to 1985 (when our data end) had yielded only moderate gains in productivity.

PROGRESS THROUGH THREE PERIODS

Our emphasis in this chapter has been on changes in the organization of production during the twentieth century tied to the evolution of technologies that are rooted in the use of electricity. Crucially important tools of production and control are dependent upon energy in this form.

In the pre-electrical organization of production, mechanical power was distributed to individual machines through a complex system of belts, shafts, and pulleys. This system imposed severe constraints on factory layout and work sequence that hampered the growth in productivity. Even though electrically driven machinery began to be adopted during Stage I, the overall production system itself was not significantly changed.

Figure 12.17 Output, Productivity, and Multifactor Input During Stage III, 1948–1985

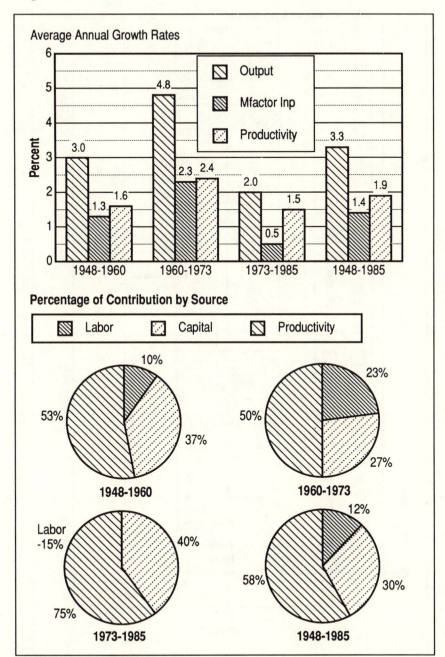

Figure 12.18 Energy and Factor Inputs During Stage III, 1948–1985

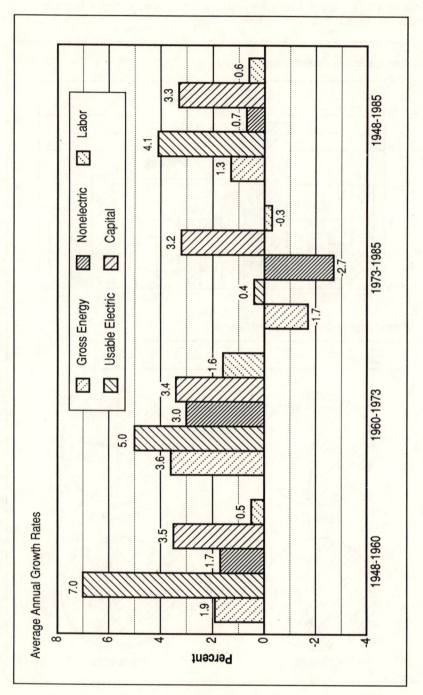
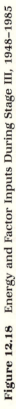

During Stage II, the use of electricity plus the electric motor ushered in a new era in the organization of production by permitting the conversion to mechanical power to take place at the point of its use — the factory machine. As a result, factory systems could be reorganized in such a way as to speed up the flow of production.

During Stage III, electricity was increasingly used not only as a source of mechanical power but also as a means of achieving automatic control of production operations.[26] No application better emphasizes the uniqueness of electricity as an energy form than its use to carry and process information. Indeed, carrying information (via telegraph) was the first important use of electricity, preceding lighting and motors by several decades. But while lighting and mechanical power can be provided by other energy forms, although with great disadvantages, instantaneous communication and large-scale information management can be achieved only with electricity.

As a result, some production steps that previously were too difficult to mechanize can now be mechanized economically. Processes formerly requiring manual control are now being monitored and controlled automatically. And operations once performed independently can now be electronically integrated into a coordinated production unit.

Indeed, electronic information management should increasingly allow manufacturers to schedule and coordinate to an unprecedented degree the use of machines and inventory, and even other company operations, to achieve greater integration of the entire production enterprise. Thus, while the energy distribution revolution brought by unit drive triggered the reorganization and integration of production processes within the factory, innovations in the use of information could facilitate the reorganization and integration of the entire manufacturing production system, a system that extends well beyond the factory floor.[27]

APPENDIX: KEY INDICATORS AND MEASUREMENT ISSUES

In this appendix we summarize and define some of the key quantitative indicators presented in Chapter 12 and discuss several important measurement issues related to these data. Table 12.A shows the average annual growth rates for selected variables during each of our three stages. Appendix II provides a complete description of the data shown in Table 12.A, including the sources and methodology used in their derivation.

Output

Output refers to the constant (1972) dollar gross product originating in manufacturing. Gross product originating is a net measure of output,

Table 12.A Indicators of Economic Growth and Energy Use in Manufacturing, Average Annual Growth Rates

	Total period	Stages			Stage 1		Stage 2		Stage 3				
	1899 to 1985	1899 to 1920	1920 to 1948	1948 to 1985	1899 to 1913	1913 to 1920	1920 to 1929	1929 to 1948	1948 to 1960	1960 to 1973	1973 to 1985	1973 to 1981	1981 to 1985
Output	3.7	4.3	3.7	3.3	4.9	3.0	4.7	3.3	3.0	4.8	2.0	1.1	4.0
Inputs													
Multifactor input	1.7	3.4	0.9	1.4	3.7	2.9	-0.4	1.5	1.3	2.3	0.5	0.8	-0.1
Real labor	1.3	3.0	0.9	0.6	3.2	2.6	-0.8	1.7	0.5	1.6	-0.3	-0.2	-0.5
Real capital	3.2	5.7	0.9	3.5	6.4	4.2	0.7	1.0	3.2	4.1	3.2	3.9	1.6
Weighted labor	1.0	2.5	0.7	0.4	2.7	2.0	-0.6	1.3	0.3	1.1	-0.3	-0.2	-0.4
Weighted capital	0.8	0.9	0.2	1.0	1.0	0.9	0.2	0.2	1.1	1.3	0.8	1.0	0.4
Capital per labor hour	1.9	2.6	0.0	2.9	3.1	1.5	1.5	-0.7	2.6	2.5	3.5	4.2	2.2
Weighted capital service per labor hour	0.4	0.4	0.0	0.8	0.5	0.3	0.4	-0.2	0.9	0.8	0.9	1.1	0.6
Productivity													
Multifactor productivity	1.9	0.8	2.8	1.9	1.2	0.0	5.2	1.7	1.6	2.4	1.5	0.2	4.0
Labor productivity	2.4	1.2	2.8	2.7	1.6	0.4	5.5	1.6	2.5	3.2	2.4	1.3	4.6
Capital productivity	0.5	-1.3	2.8	-0.2	-1.4	-1.2	3.9	2.2	-0.2	0.6	-1.1	-2.7	2.3
Energy productivity	1.5	-0.6	2.4	1.9	-0.4	-0.9	4.3	1.4	1.0	1.4	3.6	2.4	5.9
Usable electricity													
Total usable electricity	8.3	18.4	6.9	4.1	21.1	13.0	9.1	5.8	7.0	5.0	0.4	0.3	0.5
Electric machine drive	9.3	23.3	6.5	4.1	27.2	16.0	8.9	5.4	6.7	5.1	0.4	0.3	0.5
Electric materials conversion	9.4	20.8	8.8	3.9	23.9	14.6	10.4	8.1	7.5	4.7	-0.3	-0.7	0.5
Electric information processing and support services	6.4	10.4	5.8	4.7	14.5	2.8	8.3	4.6	7.5	5.0	1.8	2.4	0.5
Utility-generated electricity	9.1	19.0	7.7	4.8	19.9	17.4	10.7	6.3	8.0	5.9	0.5	1.0	-0.5
Manufacturing-generated electricity	6.6	17.7	5.5	1.5	22.3	9.0	6.7	5.0	4.1	1.2	-0.6	-5.6	10.2

Gross energy

Total gross energy	2.2	4.8	1.3	1.3	5.3	3.9	0.2	1.8	1.9	3.6	-1.7	-1.4	-2.1
Gross electric	5.3	12.5	3.8	2.6	15.0	7.6	2.6	4.4	2.5	4.9	0.3	0.3	0.3
Gross nonelectric	1.5	4.2	0.6	0.7	4.6	3.3	-0.3	1.0	1.7	3.0	-2.7	-2.3	-3.5
Gross machine drive	2.3	3.2	1.5	2.4	4.4	0.8	-2.0	3.3	1.9	4.9	0.2	0.2	0.2
Gross electric	6.3	17.2	3.5	2.6	20.7	10.4	2.4	4.0	2.2	5.0	0.3	0.3	0.3
Gross nonelectric	-1.2	-0.3	-3.6	0.2	2.2	-5.0	-10.2	-0.3	-0.7	3.1	-2.0	-1.9	-2.3
Gross materials conversion, information processing and support services	2.1	5.4	1.2	1.0	5.7	4.8	0.7	1.4	2.0	3.3	-2.2	-1.9	-2.8
Gross electric	4.6	7.9	4.6	2.7	10.7	2.5	2.9	5.4	3.0	4.7	0.3	0.3	0.3
Gross nonelectric	1.9	5.3	0.9	0.7	5.4	4.9	0.6	1.0	1.8	3.0	-2.7	-2.3	-3.5

SOURCE: Appendix II.

consistent with the gross national product concept, obtained by subtracting the value of purchased materials and services from the value of manufacturing shipments and inventory change.

Inputs

Multifactor input refers to the sum of the weighted capital plus weighted labor inputs. The input weights refer to the average cost of each input relative to the value of output.

> Varying methods of incorporating input quality changes in the data have resulted in significant measurement differences. To the extent that input quality improvements are included as an increase in the aggregate input, then the measured contribution of multifactor or total factor productivity to output growth will be reduced.

Labor input refers to the paid hours of all employees in manufacturing plus the hours of proprietors and unpaid family workers.

> Wage rate differentials tend to be associated with such worker characteristics as age, sex, education, and occupation. If these wage differentials are assumed to reflect differences in worker skills, then compositional shifts of the work force in terms of these characteristics can be taken as indicative of changes in the average quality of labor. A number of investigators have included such compositional changes in average labor quality as part of the labor input, preferring to think of labor skills as a characteristic of workers rather than as a component of productivity. The Bureau of Labor Statistics (BLS) data do not make such an assumption, so in the data which we have used changes in the average quality of labor are assumed to affect the measures of productivity but not the measures of labor input. Historically, changes in the average quality of labor have been moderately important in their effects on output and productivity. Jorgenson et al. (1987) estimate that such changes accounted for one-tenth of the postwar growth in total U.S. output while Waldorf et al. (1986) estimated that they accounted for one-tenth of the postwar growth in multifactor productivity of the nonfarm business sector.

Capital input refers to the annual flow of capital services. Since the annual flow of capital services is assumed to be a constant proportion of the annual value of the capital stock, the rate of growth in capital input can be measured by the rate of growth in the capital stock. Capital stock refers to the constant (1972) dollar value of the net (depreciated) stock of equipment, structures, land, and inventories in manufacturing.

> Capital input measures the flow of services derived from a variety of physical assets. The marginal productivity of each asset type differs so that changes in the composition of capital goods have an effect on the average quality of the capital input. Weighting each asset type by its implicit rental price (which reflects its marginal productivity) and then aggregating the results provides a capital input measure that incorporates the effects of compositional quality changes. Most measurements of capital input, including those of the BLS,

adopt this methodology so that the estimates of multifactor productivity growth that we have presented exclude the quality effects of changes in asset composition. Jorgenson et al. (1987) estimate that these compositional quality effects accounted for about one-quarter of the postwar growth in the capital input and one-tenth of the postwar growth in U.S. output.

Weighted labor input refers to the rate of growth in labor input weighted by the average share of labor costs in the value of output. Weighted capital input refers to the rate of growth in capital input weighted by the average share of capital costs in the value of output. Capital per labor hour refers to the value of capital services provided per hour of labor input. Weighted capital services per labor hour refers to the capital-per-labor-hour ratio weighted by the average share of capital costs in output.

Productivity

Multifactor productivity represents the amount by which output growth exceeds growth in multifactor input; it is derived as the ratio of output divided by multifactor input.

> The measure of output most often applied to the total economy or to the major economic sectors is a net output concept that incorporates the value added only by the factor inputs of labor and capital. However, for specific industries a gross output concept is often used that also includes the value of all purchased inputs — not only labor and capital, but also energy, materials, and business services, thus including the industry's intermediate purchases from other industries. For the total economy (or the total and nonfarm business economy), intermediate transactions tend to cancel out, so that multifactor productivity growth can be reliably calculated as the difference between net output growth and the combined labor and capital input growth. For specific industries, however, the intermediate transactions do not cancel, and total factor productivity growth is best measured as the difference between gross output growth and the combined labor, capital, energy, materials, and business services growth.
>
> The manufacturing multifactor productivity estimates that we have used are based on measures of net output, and combined labor and capital input. Although manufacturing is a major producing sector, many of its purchases come from outside manufacturing, which means that for many purposes total factor productivity measures based on the gross output concept would be more appropriate. However, the data needed to make such estimates for years prior to 1948 are not available, which has precluded their use in our evaluation of manufacturing trends throughout the twentieth century.

Labor, capital, and energy productivity refer to the ratio of output divided by labor, capital, and gross energy inputs, respectively.

Usable Electricity

Usable electricity refers to the kWh (or equivalent [3,412] Btus) of electricity used by manufacturing. It includes utility sales to manufacturing and electricity self-generated within manufacturing. Usable electricity excludes the energy losses incurred in converting primary energy sources into electricity and the energy lost in carrying the electricity from the point of generation to the point of use.

Electric machine drive refers to the kWh (or equivalent Btus) of electricity used to run machines in manufacturing.

Electric materials conversion, information processing, and support services refer to the kWh (or equivalent Btus) of electricity used by manufacturing for all purposes other than machine drive. This total category includes electric processing of materials in bulk, electronic information processing, and support services. The latter category includes electric lighting, space heating and cooling, and the operation of electric appliances, office equipment, and electric transportation equipment within the plant.

Utility-generated usable electricity refers to the kWh (or equivalent Btus) of utility sales to manufacturing.

Manufacturing self-generated electricity refers to the kWh (or equivalent Btus) of electricity self-generated by manufacturing establishments.

Gross Energy

The gross energy total refers to the Btus in (or imputed to) the primary energy sources (fuels and hydropower) used directly by manufacturing, plus the Btus in (or imputed to) the primary energy (fuels, nuclear, and hydropower) used to produce the electricity purchased from utilities.

> The energy use data presented are physical estimates in that they are aggregations of the different types of energy consumed, measured in terms of their Btu content. However, changes in the average quality of energy can occur because of shifts in the mix of solid fuels, liquid fuels, and electricity. These compositional quality changes can be captured by weighting the various forms of energy by their prices per Btu. Since the shifts have generally favored electricity, which is higher priced, over solid fuels and liquid fuels, the average quality of energy has tended to increase over time. In other words, the physical measures of energy increase at a slower rate than would the quality-adjusted measures.
>
> As a consequence, physical measures of energy result in estimates of energy productivity growth that are larger than those based on quality-adjusted measures of energy input. Alterman (1985) estimates that the 1973 to 1981 drop in the physical energy intensity for the overall economy was 30 percent faster than the quality-adjusted energy intensity decline; Berndt, et al. (1986) estimate an almost 50 percent faster drop for the 1974 to 1981 physical energy intensity in manufacturing. (See also Chapter 13, note 1.)

Gross electric energy refers to the Btus in (or imputed to) the primary energy sources (fuels and hydropower) used to produce the electricity consumed in manufacturing, both purchased and self-generated.

Gross nonelectric energy refers to the primary energy used directly by manufacturing, excluding that portion used for self-generated electricity.

Gross energy for machine drive refers to the Btus of gross energy used in the application of electric and nonelectric energy for running machines in manufacturing. Gross electric machine drive and gross nonelectric machine drive refer to the Btus of gross energy used for electric and nonelectric machine drive, respectively.

Gross energy for materials conversion, information processing, and support services refers to the Btus of gross energy used for all purposes other than machine drive.

Notes

1. The quantitative analysis is based upon historical statistics derived from standard data sources but recast so as to focus on the particular set of relationships that is central to the concerns of this study. The data and their sources are detailed in Appendix II.

Table 12.A, which appears in the appendix to this chapter, presents the average annual growth rates between selected years for key indicators of economic growth and energy use in manufacturing. All textual comparisons of growth rates are referenced to Table 12.A. The appendix also discusses some measurement issues that are relevant to an understanding of these data.

2. Empirical evaluations of the relationship among technological progress, productivity, and output growth are customarily based on an explicit or implicit production function that quantifies the maximum output that can be obtained from alternative quantities of inputs in the prevailing state of technological knowledge. Within a production function framework, multifactor productivity growth, which refers to improvements in the combined use of all inputs, is defined as growth in output less growth in the aggregate input which, in turn, is defined as a weighted sum of growth in the individual inputs (where the weights are usually each input's share in the total cost of production). Such production functions can be applied to specific industries, to broad producing sectors such as manufacturing, or to the economy as a whole.

3. Although we speak of technological progress as being incorporated in new plant and equipment there are, in fact, several different interrelated aspects involved: (1) the introduction of advanced technology that is embodied in the new equipment and facilities; (2) the evolution of learning processes associated with the use of the new technologies; and (3) the unfolding of reorganizational effects made possible by the new equipment and plant. As the data we use cannot distinguish among the three effects, our measures of multifactor productivity growth include all three.

4. Except for the latest year, 1985, each period begins and ends with a business cycle peak year in order to remove short-term economic fluctuations as a possible source of disturbance in the long-term comparisons.

5. Differences in the way that output and inputs are measured have resulted in very different perspectives on the sources of growth in the American economy. For example, increases in the use of capital and labor (particularly capital) are identified

as the "driving force" behind the economic expansion since the end of World War II in Dale W. Jorgenson, Frank Gollop, and Barbara Fraumeni, *Productivity and U.S. Economic Growth* (Cambridge: Harvard University Press, 1987). In their measurements, multifactor productivity growth accounted for about one-quarter of the 1948 to 1979 growth in output. The data that we have used, which are based largely on U.S. Bureau of Labor Statistics measurement techniques, tell a different story. In these measurements, multifactor productivity gains represent the principal source of economic expansion, accounting for about half the 1948 to 1979 growth in manufacturing and overall U.S. output, and since 1979 these gains have become even more important in their relative contribution to manufacturing output growth.

6. Our analysis describes but does not measure the connection between energy and productivity growth. However, several important aspects of this connection have recently been measured, primarily through the econometric decomposition of gross output production functions.

The impact of changes in the use of energy or electricity on productivity growth is often transmitted through changes in the quality of capital. One such change occurs as a result of sudden and substantial energy price increases. These increases are likely to lower the utilization rate of some existing capital so that the capital services provided are reduced, which is equivalent to a decline in the quality of capital. Since conventional measures do not incorporate this decline, the measure of capital is overstated vis-a-vis the services provided, and such an overstatement results in an understatement of the growth in total factor productivity. Such changes in the quality of capital have been measured, and it was found that they had little effect on manufacturing productivity growth prior to 1973 but that they had an appreciable effect after 1973, causing the 1973 to 1981 conventional measure of total factor productivity growth to be understated by 20 percent (Ernst R. Berndt and David O. Wood, "Energy Price Shocks and Productivity Growth: A Survey," in *Energy Markets and Regulation: What Have We Learned*, Festschrift in Honor of Morris A. Adelman, eds. Richard L. Gordon, Henry D. Jacoby, and Martin B. Zimmerman [Cambridge: MIT Press, 1986]).

Capital quality also changes when there is an increased use of motive power, which enhances the performance characteristics of capital goods. In effect, a new mix of energy and capital embodies capital quality improvements. The labor productivity consequences of such capital quality improvements have been calculated using the ratio of primary (electric plus nonelectric) horsepower capacity per unit of capital stock as a measure of embodied technological change through increased motive power, and the ratio of electric horsepower capacity per unit of capital stock as a measure of embodied technological change through increased electrification. It was estimated that from 1899 to 1918 embodied technological change in the form of increased motive power accounted for one-fifth of the growth in manufacturing labor productivity (Ernst R. Berndt, "Electrification, Embodied Technical Progress and Labor Productivity Growth in U.S. Manufacturing, 1899–1939," in *Electricity Use, Productive Efficiency, and Economic Growth*, proceedings of a workshop sponsored by the Electric Power Research Institute, eds. Sam H. Schurr and Sidney Sonenblum [Palo Alto, CA: Electric Power Research Institute, 1986]); but, this effect derived mostly from the increased use of nonelectric energy. In the years 1919 to 1939, embodied technological change involved the electrification of motive power, and Berndt estimates that such increased electrification of horsepower per unit of capital stock accounted for one-third of the growth in manufacturing labor productivity.

Jorgenson has found that both technological progress and substitution of energy for other inputs are important sources of growth in the use of electric and nonelectric energy (Dale W. Jorgenson, "The Role of Energy in Productivity Growth," in Schurr and Sonenblum, *Electricity Use*). Using data for the 1958 to 1979 period, he established that technological progress (which is calculated as a function of the passage of time) had an electricity-using bias in twenty-three of the thirty-five industries he studied and a nonelectric-energy-using bias in twenty-eight industries. What this means is that technological advance was accompanied by a proportionately smaller savings in the use of electricity or in the use of nonelectric energy than in the average of all other inputs combined. This indicates that increases in the amount of electricity or nonelectric energy used by new technologies have raised the associated total factor productivity gains, that increases in the real price of electricity or nonelectric energy have slowed these gains, and that technological progress encourages a more rapid growth in the use of electricity or nonelectric energy than in the combined use of all other inputs.

7. For discussions of the technological and organizational setting prior to and during Stage I, see, in particular: Robert T. Averitt, *Dual Economy: The Dynamics of American Industry Structure* (New York: Norton, 1968), 6, 8–10, 23–27; James R. Bright, *Automation and Management* (Cambridge: Harvard University Press, 1958), 13–15; Walter Buckingham, *Automation: Its Impact on Business and People* (New York: Harper and Brothers, 1961), 4, 6–7; Roger Burlingame, *Backgrounds of Power* (New York: Charles Scribner, 1949), 14–15, 60–61, 111–112, 138–147, 214–228; Alfred D. Chandler, Jr., "Technology and the Transformation of Industrial Organization," in *Technology, the Economy and Society*, eds. Joel Colton and Stuart Bruchey (New York: Columbia University Press, 1987), 57–60; Alfred D. Chandler, Jr., *The Visible Hand: The Managerial Revolution in American Business* (Cambridge: Harvard University Press, Belknap Press, 1977), 6–11, 52–58, 235–258, 345–347, 363–376, 484–490; Alfred D. Chandler, Jr., "Recent Developments in American Business Administration and Their Conceptualization," *Business History Review* 35, no. 1 (1961): 9–15; Edward Cressy, *Discoveries and Inventions of the Twentieth Century* (London: G. Routledge and Sons; New York: E. P. Dutton and Co., 1923), 134–137; Sigfried Giedion, *Mechanization Takes Command* (New York: Oxford University Press, 1948), 5, 31–32, 38–41; Stanley Hetzler, *Technological Growth and Social Change: Achieving Modernization* (Henley-on-Thames: Routledge and Kegan Paul Limited, 1969), 134–135; David A. Hounshell, *From the American System to Mass Production, 1800–1932* (Baltimore, MD: Johns Hopkins University Press, 1984), 3–4, 8–10; Joseph A. Litterer, "Systematic Management: The Search for Order and Integration," *Business History Review* 35, no. 4 (1961): 461–469; Lewis Mumford, *Technics and Civilization* (New York: Harcourt, Brace and Co., 1934), 5, 10–11, 26–28, 53–54, 80, 90–99, 102–104, 212–219; Michael J. Piore and Charles F. Sabel, *The Second Industrial Divide* (New York: Basic Books, 1984), 20–22, 26–31, 49; John Rae, "The Rationalization of Production," in *Technology in the 19th Century* I, eds. Melvin Krantzberg and Carroll Pursell, Jr. (New York: Oxford University Press, 1967): 37–42; Wolfgang P. Strassman, *Risk and Technological Innovation: American Manufacturing in the 19th Century* (Ithaca, NY: Cornell University Press, 1956), 116–127; A. P. Usher, *A History of Mechanical Innovations* (Cambridge: Harvard University Press, 1954), 116–119, 358–381; Thorstein Veblen, *The Instinct of Workmanship* (New York: Kelley, 1964. Rev. ed. reprint of 1918), 299–330; Harold F. Williamson, "Mass Production for Mass Consumption," in Krantzberg and Pursell, eds., *Technology*, I:678–686.

8. "Reliance on traditional power for much of the nineteenth century is one of the most outstanding, if one of the most neglected, characteristics of the industrializing process in Britain and in the United States" (Dolores Greenberg, "Reassessing the Power Patterns of the Industrial Revolution," *Historical Review* 87, no. 5 [December 1982]: 1261).

9. For discussions of the role of energy in manufacturing prior to and during Stage I, see, in particular: Lynwood Bryant, "The Beginnings of the Internal Combustion Engine," in Krantzberg and Pursell, eds., *Technology*, II:648–663; Burlingame, *Backgrounds of Power*, 34–54, 118–122; Chandler, *Visible Hand*, 50–51, 59–63, 76–78, 245; Cressy, *Discoveries*, 5–6, 16–30; Greenberg, "Power Patterns", 1237–1261; David Hamilton, *Technology, Man and the Environment* (New York: Charles Scribner, 1973), 124–130; Fred Henderson, *The Economic Consequences of Power Production* (London: Unwin Brothers, 1933), 9–19; Hetzler, *Technological Growth*, 95, 154–157, 182; Elting Morison, *Men, Machines and Modern Times* (Cambridge: MIT Press, 1966), 1–16, 157; Mumford, *Technics*, 112–148, 161–163, 196–199, 235–242; Bruce C. Netschert, "Developing the Energy Inheritance," in Krantzberg and Pursell, eds., *Technology*, II:237–247; Carroll W. Pursell, Jr. and Melvin Krantzberg, "Epilogue," in Krantzberg and Pursell, eds., *Technology*, I:741–742; John Rae, "Energy Conversion," in Krantzberg and Pursell, eds., *Technology*, II:330–338; Nathan Rosenberg, "The Effects of Energy Supply Characteristics on Technology and Economic Growth," in *Energy, Productivity and Economic Growth*, eds. Sam H. Schurr, Sidney Sonenblum, and David O. Wood (Cambridge, MA: Oelgeschlager, Gunn and Hain, 1983), 279–289; Sam H. Schurr, Bruce C. Netschert et al., *Energy in the American Economy, 1850–1975* (Baltimore, MD: Johns Hopkins Press, 1960), 31–192; Charles Susskind, *Understanding Technology* (Baltimore, MD: Johns Hopkins Press, 1973), 7–9, 17–22; Brinley Thomas, "Towards an Energy Interpretation of the Industrial Revolution," *Atlantic Economic Journal* (March 1980), 1–3, 11–13; Usher, *Mechanical Innovations*, 160, 187, 332–357, 382–399, 406–411.

10. "In production the first modern managers came in those industries and enterprises where technology permitted several processes of production to be carried on within a single factory or works. In those industries output soared as energy was used more intensively and machinery, plant design and administrative procedures were improved" (Chandler, *Visible Hand* [1977], 486).

11. Scaling-up the productive capacity of factories was consistent with the "American system" of production that was being widely used in manufacturing by the late 1800s. This system involved the fabrication and assembly of standardized and interchangeable components. The making of machines, and even consumer products, through the assembly of components required a consistency and exactitude in the fabrication of parts that was generally not possible with even skilled hand labor. Only with exactly machined components was it possible to draw from a stock of interchangeable parts and, thereby, to produce in large quantities at low costs.

For further discussion of the use of interchangeable and standardized components in production prior to and during Stage I, see, in particular: Burlingame, *Backgrounds of Power*, 108–109, 123–127; Giedion, *Mechanization*, 46–50; Hounshell, *Mass Production*, 3–8; Harry Jerome, *Mechanization in Industry* (New York: National Bureau of Economic Research, 1934), 408; Litterer, "Systematic Management," 465–466; Mumford, *Technics*, 90; Rae, "Rationalization," in *Technology*, I:41; Usher, *Mechanical Innovations*, 378–380; Veblen, *Instinct*, 9–19.

12. Bright, *Automation and Management*, 20.

13. For discussions of the role of electricity in manufacturing prior to and during Stage I, see, in particular: Ernst R. Berndt, "Electrification, Embodied Technical Progress and Labor Productivity Growth in U.S. Manufacturing, 1899–1939," in *Electricity Use, Productive Efficiency and Economic Growth*, eds. Sam H. Schurr and Sidney Sonenblum (Palo Alto, CA: Electric Power Research Institute, 1986), 105–111; Burlingame, *Backgrounds of Power*, 268–270; Chandler, *Visible Hand*, 210–211; Cressy, *Discoveries*, 11, 76, 102, 122–134, 161–163, 287–289; Abram John Foster, *The Coming of the Electrical Age to the United States* (New York: Arno Press, 1979),

291–302, 343–348; Mumford, *Technics*, 221–229; Rosenberg, "Energy Supply," 289–293; Harold I. Sharlin, "Applications of Electricity," in Krantzberg and Pursell, eds., *Technology*, I:563–578; Schurr, Netschert et al., *Energy*, 180–189; Strassman, *Risk*, 158–170, 197, 212–213; Susskind, *Understanding Technology*, 18–19, 23–24; Usher, *Mechanical Innovations*, 399–406.

14. Of course, electricity's share of total energy is substantially higher when measured in terms of the heat content of the energy sources used in the generation of electricity — a measure which we call "gross electricity." Figure 12.4 shows that gross electricity accounted for over 20 percent of the growth in total energy use during Stage I and by 1920 was almost 15 percent of total energy used in that year. However, when measured in such gross terms, electricity's share of the total energy put to use in manufacturing is exaggerated, particularly in these early years because of great changes in the efficiency with which raw fuel was converted into delivered electric power. For example, at the turn of the century it took over 100,000 Btus of energy to deliver 1 kWh of usable electricity; took about 40,000 Btus in 1920; and takes about 10,000 Btus today.

15. However, at the start of Stage I, primary motors had accounted for less than two-fifths of the electric motor capacity and secondary motors for over three-fifths. Why should primary motor capacity have grown at a faster rate than utility sales to manufacturing during Stage I? In part, this occurs because of definitional differences. Primary motor capacity refers only to machine drive, while electricity purchases from utilities additionally include electricity used for materials conversion and support services. Since these latter uses increased at a slower rate than machine drive during Stage I, primary motor capacity increased at a faster rate than purchases from utilities. In addition, group drive required less motor capacity than unit drive because the motors that drove groups of machines could be sized to take advantage of load diversity. Therefore, the ratio of motor capacity to electricity use for unit drive was probably higher than for group drive. Since unit drive (which came to a dominant position during Stage II) also increased at a faster rate than group drive during Stage I, the tendency would be for primary motor capacity to increase faster than electricity use.

16. Capital growth within manufacturing was slowed down by the fact that utilities, rather than manufacturers themselves, were providing one-half of the electricity used in manufacturing by the end of Stage I, a substantial increase over their one-third share at the turn of the century. Thus a growing share of the plant and equipment needed for electricity was installed by utilities rather than within manufacturing, thereby slowing down the growth in manufacturing capital vis-à-vis growth in the use of electricity. Therefore, some part of the greater growth in electricity relative to capital in manufacturing is attributable to this shift.

17. For discussion of the technological and organizational setting in manufacturing during Stage II, see, in particular: Averitt, *Dual Economy*, 25–29; Bright, *Automation and Management*, 29–30, 63; Walter Buckingham, "Principles of Automation," in *Automation, Alienation and Anomie*, ed. Simon Marcson (New York: Harper and Row, 1970), 9–11; Chandler, "Technology," in Colton and Bruchey, eds., *Technology*, 61; Chandler, *Visible Hand*, 469–479; Alfred D. Chandler, Jr., "The Structure of American Industry in the Twentieth Century: A Historical Review," *Business History Review* 43, no. 3 (1969): 274–281; Stephen S. Cohen and John Zysman, *Manufacturing Matters: The Myth of the Post Industrial Economy* (New York: Basic Books, 1987), 135–139; Giedion, *Mechanization*, 41–43; Hounshell, *Mass Production*, 10–13; Jerome, *Mechanization in Industry*, 3–8, 58–120; Mumford, *Technics*, 383–391; Harry T. Oshima, "The Growth of U.S. Factor Productivity: The Significance of New Technologies in the Early Decades of the Twentieth Century," *Journal of Economic History*

44, no. 1 (March 1984): 166–169; Piore and Sabel, *Industrial Divide*, 54–63, 76–77, 111–124; Rae, "Rationalization," in Krantzberg and Pursell, eds., *Technology*, I:42–51; Charles Walker, "The Social Effects of Mass Production," in Krantzberg and Pursell, eds., *Technology*, II:92–99.

18. "With the coming of the moving assembly line the processes of production in the metal mass production industries had become almost as continuous as those in petroleum and other refining industries. . . . Such economies came more from the ability to integrate and coordinate the flow of materials through the plant than from greater specialization and subdivision of work within the plant" (Chandler, *Visible Hand*, 280–281).

19. For discussion of continuous flow and speed of throughput in manufacturing, see, in particular: Bright, *Automation and Management*, 13–17, 100–104; Buckingham, *Automation*, 9–10, 36–38, 284–286; Chandler, *Visible Hand*, 279–281; Cressy, *Discoveries*, 137–138; D. J. Davis, "Automation in the Automotive Industry," in *Automation and Society*, eds. Howard B. Jacobson and Joseph S. Roucek (New York: Philosophical Library, 1959), 34–36; Giedion, *Mechanization*, 77–96, 115–121; Jacobson and Roucek, *Automation and Society*, 7–9; Jerome, *Mechanization in Industry*, 179–205.

20. In addition to new ways of laying out the workplace, organizing a coordinated production process required new worker skills and new inventions relating to the processing, fabricating, and assembling as well as handling and moving of materials, parts, and products. These various ingredients were available in some manufacturing industries during the second decade of the twentieth century, but they came together by the start of Stage II around the growing use of conveyors and assembly lines.

21. For discussions of the role of electricity in manufacturing during Stage II, see, in particular: Berndt, "Electrification," 105–111; Chandler, *Visible Hand*, 269–270; Hamilton, *Technology*, 131–134; Hetzler, *Technological Growth*, 206–209; Jerome, *Mechanization in Industry*, 205–255; Mumford, *Technics*, 224–229; Netschert, "Energy Inheritance," in Krantzberg and Pursell, eds., *Technology*, II:247–254; Oshima, "Factor Productivity," 165–166; Rosenberg, "Energy Supply," 295–298; Schurr, Netschert, et al., *Energy*, 180–189.

22. For discussions of the technological and organizational setting in manufacturing during Stage III, see, in particular: Bright, *Automation and Management*, 17, 18–21, 60, 83, 100–103, 132–145; James R. Bright, "The Development of Automation," in Krantzberg and Pursell, eds., *Technology*, II:635–655; Buckingham, *Automation*, 5–6, 20–21, 35–38, 66, 105–107; Cohen and Zysman, *Manufacturing Matters*, 130–135, 145–149, 150–170; Ralph J. Cordiner, "Automation in Manufacturing Industries," in Jacobson and Roucek, eds., *Automation and Society*, 19–21; Richard M. Cyert, "The Plight of Manufacturing: What Can Be Done," *Issues in Science and Technology* (Summer 1985), 91–94, 98; Robert Dubin, "Automation, The Second Industrial Revolution," in Marcson, ed., *Automation*, 152–156; *The Economist*, "Factory of the Future: A Survey" (May 30, 1987), 3–13; Karl H. Ebel, "Social and Labour Implications of Flexible Manufacturing Systems," *International Labor Review* vol. 124, no. 2 (March–April 1985): 140–143; Georges Friedmann, "Three Stages of Automation," in Marcson, ed., *Automation*, 118–119; Thomas G. Gunn, "The Mechanization of Design and Manufacturing," *Scientific American* 247, no. 3 (September 1982): 115–124; National Research Council, Manufacturing Studies Board, *Toward a New Era in U.S. Manufacturing* (Washington, D.C.: National Academy Press, 1986), 5–7, 37–42, 50–51, 102–123; Piore and Sabel, *Industrial Divide*, 233, 247, 258–265, 282–284; Edward Shils, "Automation: Technology or Concept," in Marcson, ed., *Automation*, 139–152; William Walker, "Information Technology and the Use of

Energy," *Energy Policy* (October 1985), 463, 470; William Walker, "Information Technology and the Energy Supply," University of Sussex, England (1986), mimeograph, 6–9.

23. For further discussions of control and feedback devices, see, in particular: Bright, *Automation and Management*, 100, 120, 172, 224; Buckingham, *Automation*, 8, 11–12, 55–57; Ebel, "Implications," 133–136; National Research Council, *New Era*, 33–37, 88–97; Rae, "Rationalization," in Krantzberg and Pursell, eds., *Technology*, I:512; Shils, "Automation," in Marcson, ed., *Automation*, 143–147; Walker, "Information Technology," 5–6.

24. This was particularly the case in metal-working industries that used these automatic devices to control the fixed but complex movements of transfer machines. The transfer machine is an integrated piece of machinery that mounts a number of individually powered tools on a single machine base and includes a transfer mechanism to move the work from tool to tool at the split-second timing required. Controls for the tools and the transfer mechanism are usually consolidated into one panel on the transfer machine. In describing these machines, Bright says "Great advantages also are apparent: high productivity, rapid processing, simplified handling, reduction in space requirement, reduction of labor requirement, and the reduction of in-process inventory. Nevertheless such a machine system must be uniquely designed for each job. It is, therefore, a complex, expensive and probably highly inflexible mechanism" (Bright, *Automation and Management*, 100).

For further discussions of transfer machines, see Bright, "Automation," in Krantzberg and Pursell, eds., *Technology*, II:643; Buckingham, "Automation" in Marcson, ed., *Automation*, 11; Dubin, "Automation," in Marcson, ed., *Automation*, 155; Davis, "Automation," in Jacobson and Roucek, eds., *Automation and Society*, 36–38; National Research Council, *New Era*, 38.

25. The pickup in the pace of capital growth was not matched by the Stage III labor input, which increased at only two-thirds the Stage II rate. As a result, the amount of capital used per labor hour, which remained stable during Stage II, has almost tripled so far in Stage III.

26. For further discussions of the role of electricity in manufacturing during Stage III, see, in particular: Adam Kahane and Ray Squitieri, "Electricity Use in Manufacturing," *Annual Energy Review* (Palo Alto, CA: Annual Reviews, 1987), 223–229; National Academy of Engineering, Committee on Electricity in Economic Growth, *Electricity in Economic Growth* (Washington, D.C.: National Academy Press, 1986), 9–10, 16–40, 71–78, 89–91, 115–119; Philip S. Schmidt, "The Form Value of Electricity: Some Observations and Cases," in Schurr and Sonenblum, eds., *Electricity Use*, 199–204; Philip S. Schmidt, *Electricity and Industrial Productivity: A Technical and Economic Perspective*, EPRI EM-3640 (Palo Alto, CA: Electric Power Research Institute, March 1985), 1–1 to 1–10; Shils, "Automation," in Marcson, ed., *Automation*, 142–143; Walker, "Information Technology," 19–22, 26–30; Walker, "Use of Energy," 459–460, 471–472.

27. "We are in the midst of an industrial revolution no less fundamental than the one in which modern technology was born some two hundred years ago. The present revolution involves the communication and use of information just as the first one involved the transport and use of energy supplies" (Hamilton, *Technology*, 22).

For further discussions of the relation between information and electricity use, see, in particular: Kahane and Squitieri, "Electricity Use," 244–245; Walker, "Information Technology," 1–4; Walker, "Use of Energy," 465–471; Alvin M. Weinberg, "On the Relation Between Information and Energy Systems," *Interdisciplinary Science Reviews* 7, no. 1 (1982): 47–52.

Works Cited

Jack A. Alterman, *A Historical Perspective on Changes in U.S. Energy-Output Ratios,* EPRI EA-3997 (Palo Alto, CA: Electric Power Research Institute, 1985).

Robert T. Averitt, *Dual Economy: The Dynamics of American Industry Structure* (New York: Norton, 1968).

Ernst R. Berndt, "Electrification, Embodied Technical Progress and Labor Productivity Growth in U.S. Manufacturing, 1899–1939," in *Electricity Use, Productive Efficiency and Economic Growth,* proceedings of a workshop sponsored by the Electric Power Research Institute, eds. Sam H. Schurr and Sidney Sonenblum (Palo Alto, CA: Electric Power Research Institute, 1986).

Ernst R. Berndt, Naoto Sagawa, Takamitsu Sawa, and David O. Wood, "Energy Intensity and Productivity in U.S. and Japanese Manufacturing Industries," paper presented at Eighth Annual North American Conference of the International Association of Energy Economists, Cambridge, MA, 1986.

Ernst R. Berndt and David O. Wood, "Energy Price Shocks and Productivity Growth: A Survey," in *Energy Markets and Regulation: What Have We Learned,* Festschrift in Honor of Morris A. Adelman, eds. Richard L. Gordon, Henry D. Jacoby, and Martin B. Zimmerman (Cambridge: MIT Press, 1986).

James R. Bright, "The Development of Automation," in *Technology in the 19th Century* II, eds. Melvin Krantzberg and Carroll Pursell, Jr. (New York: Oxford University Press, 1967).

James R. Bright, *Automation and Management* (Cambridge: Harvard University Press, 1958).

Lynwood Bryant, "The Beginnings of the Internal Combustion Engine," in *Technology in the 19th Century* II, eds. Melvin Krantzberg and Carroll Pursell, Jr. (New York: Oxford University Press, 1967).

Walter Buckingham, *Automation: Its Impact on Business and People* (New York: Harper and Brothers, 1961).

Walter Buckingham, "Principles of Automation," in *Automation, Alienation and Anomie,* ed. Simon Marcson (New York: Harper and Row, 1970).

Roger Burlingame, *Backgrounds of Power* (New York: Charles Scribner, 1949).

Alfred D. Chandler, Jr., "Technology and the Transformation of Industrial Organization," in *Technology, the Economy and Society,* eds. Joel Colton and Stuart Bruchey (New York: Columbia University Press, 1987).

Alfred D. Chandler, Jr., *The Visible Hand: The Managerial Revolution in American Business* (Cambridge: Harvard University Press, Belknap Press, 1977).

Alfred D. Chandler, Jr., "The Structure of American Industry in the Twentieth Century: A Historical Review," *Business History Review* 43, no. 3 (1969).

Alfred D. Chandler. Jr., "Recent Developments in American Business Administration and Their Conceptualization," *Business History Review* 35, no. 1 (1961).

Stephen S. Cohen and John Zysman, *Manufacturing Matters: The Myth of the Post Industrial Economy* (New York: Basic Books, 1987).

Joel Colton and Stuart Bruchey, eds., *Technology, The Economy and Society* (New York: Columbia University Press, 1987).

Ralph J. Cordiner, "Automation in Manufacturing Industries," in *Automation and Society*, eds. Howard B. Jacobson and Joseph S. Roucek (New York: Philosophical Library, 1959).

Edward Cressy, *Discoveries and Inventions of the Twentieth Century* (London: G. Routledge and Sons; New York: E. P. Dutton and Co., 1923).

Richard M. Cyert, "The Plight of Manufacturing: What Can Be Done," *Issues in Science and Technology* (Summer 1985).

D. J. Davis, "Automation in the Automotive Industry," in *Automation and Society*, eds. Howard B. Jacobson and Joseph S. Roucek (New York: Philosophical Library, 1959).

Robert Dubin, "Automation, the Second Industrial Revolution," in *Automation, Alienation and Anomie*, ed. Simon Marcson (New York: Harper and Row, 1970).

Karl H. Ebel, "Social and Labour Implications of Flexible Manufacturing Systems," *International Labor Review* 124, no. 2 (March–April 1985).

The Economist, "Factory of the Future: A Survey" (May 30, 1987).

Abram John Foster, *The Coming of the Electrical Age to the United States* (New York: Arno Press, 1979).

Georges Friedmann, "Three Stages of Automation," in *Automation, Alienation and Anomie*, ed. Simon Marcson (New York: Harper and Row, 1970).

Sigfried Giedion, *Mechanization Takes Command* (New York: Oxford University Press, 1948).

Dolores Greenberg, "Reassessing the Power Patterns of the Industrial Revolution," *Historical Review* 87, no. 5 (December 1982).

William Gullickson and Michael J. Harper, "Multifactor Productivity in U.S. Manufacturing," *Monthly Labor Review* (October 1987).

Thomas G. Gunn, "The Mechanization of Design and Manufacturing," *Scientific American* 247, no. 3 (September 1982).

David Hamilton, *Technology, Man and the Environment* (New York: Charles Scribner, 1973).

Fred Henderson, *The Economic Consequences of Power Production* (London: Unwin Brothers, 1933).

Stanley Hetzler, *Technological Growth and Social Change: Achieving Modernization* (Henley-on-Thames: Routledge and Kegan Paul, 1969).

David A. Hounshell, *From the American System to Mass Production, 1800–1932* (Baltimore, MD: Johns Hopkins University Press, 1984).

Howard Boone Jacobson and Joseph S. Roucek, eds., *Automation and Society* (New York: Philosophical Library, 1959).

Harry Jerome, *Mechanization in Industry* (New York: National Bureau of Economic Research, 1934).

Dale Jorgenson, "The Role of Energy in Productivity Growth," in *Electricity Use, Productive Efficiency and Economic Growth*, eds. Sam H. Schurr and Sidney Sonenblum (Palo Alto, CA: Electric Power Research Institute, 1986).

Dale W. Jorgenson, Frank Gollop, and Barbara Fraumeni, *Productivity and U.S. Economic Growth* (Cambridge: Harvard University Press, 1987).

Adam Kahane and Ray Squitieri, "Electricity Use in Manufacturing," *Annual Energy Review* (Palo Alto, CA: Annual Reviews, 1987).

Melvin Krantzberg and Carroll Pursell, Jr., eds., *Technology in Western Civilization* I, and *Technology in the 19th Century* II (New York: Oxford University Press, 1967).

Joseph A. Litterer, "Systematic Management: The Search for Order and Integration," *Business History Review* 35, no. 4 (1961).

Simon Marcson, *Automation, Alienation and Anomie* (New York: Harper and Row, 1970).

Elting Morison, *Men, Machines and Modern Times* (Cambridge: MIT Press, 1966).

Lewis Mumford, *Technics and Civilization* (New York: Harcourt, Brace and Co., 1934).

National Academy of Engineering, Committee on Electricity in Economic Growth, *Electricity in Economic Growth* (Washington, D.C.: National Academy Press, 1986).

National Research Council, Manufacturing Studies Board, *Toward a New Era in U.S. Manufacturing* (Washington, D.C.: National Academy Press, 1986).

Bruce C. Netschert, "Developing the Energy Inheritance," in *Technology in the 19th Century* II, eds. Melvin Krantzberg and Carroll Pursell, Jr. (New York: Oxford University Press, 1967).

Harry T. Oshima, "The Growth of U.S. Factor Productivity: The Significance of New Technologies in the Early Decades of the Twentieth Century," *Journal of Economic History* 44, no. 1 (March 1984).

Michael J. Piore and Charles F. Sabel, *The Second Industrial Divide* (New York: Basic Books, 1984).

Carroll Pursell, Jr. and Melvin Krantzberg. "The Promise of Technology," in *Technology in the 19th Century* II, eds. Melvin Krantzberg and Carroll Pursell, Jr. (New York: Oxford University Press, 1967).

Carroll Pursell, Jr. and Melvin Krantzberg, "Epilogue," in *Technology in Western Civilization* I, eds. Melvin Krantzberg and Carroll Pursell, Jr. (New York: Oxford University Press, 1967).

John Rae, "Energy Conversion," in *Technology in Western Civilization* I, eds. Melvin Krantzberg and Carroll Pursell, Jr. (New York: Oxford University Press, 1967).

John Rae, "The Rationalization of Production," in *Technology in the 19th Century* II, eds. Melvin Krantzberg and Carroll Pursell, Jr. (New York: Oxford University Press, 1967).

Nathan Rosenberg, "The Effects of Energy Supply Characteristics on Technology and Economic Growth," in *Energy, Productivity and Economic Growth*, eds. Sam H. Schurr, Sidney Sonenblum, and David O. Wood (Cambridge: Oelgeschlager, Gunn and Hain, 1983).

Philip S. Schmidt, "The Form Value of Electricity: Some Observations and Cases," in *Electricity Use, Productive Efficiency and Economic Growth*, eds. Sam H. Schurr and Sidney Sonenblum (Palo Alto, CA: Electric Power Research Institute, 1986).

Philip S. Schmidt, *Electricity and Industrial Productivity: A Technical and Economic Perspective*, EPRI EM-3640 (Palo Alto, CA: Electric Power Research Institute, March 1985).

Sam H. Schurr, Bruce C. Netschert, Vera Eliasberg, Joseph Lerner, and Hans H. Landsberg, *Energy in the American Economy, 1850–1975* (Baltimore, MD: Johns Hopkins Press, 1960).

Sam H. Schurr and Sidney Sonenblum, eds., *Electricity Use, Productive Efficiency and Economic Growth*, proceedings of a workshop sponsored by the Electric Power Research Institute (Palo Alto, CA: Electric Power Research Institute, 1986).

Sam H. Schurr, Sidney Sonenblum, and David O. Wood, eds., *Energy, Productivity, and Economic Growth*, proceedings of a workshop sponsored by the Electric Power Research Institute (Cambridge: Oelgeschlager, Gunn and Hain, 1983).

Harold I. Sharlin, "Applications of Electricity," in *Technology in Western Civilization* I, eds. Melvin Krantzberg and Carroll Pursell, Jr. (New York: Oxford University Press, 1967).

Edward Shils, "Automation: Technology or Concept," in *Automation, Alienation and Anomie*, ed. S. Marcson (New York: Harper and Row, 1970).

Wolfgang P. Strassman, *Risk and Technological Innovation: American Manufacturing in the 19th Century* (Ithaca, NY: Cornell University Press, 1956).

Charles Susskind, *Understanding Technology* (Baltimore, MD: Johns Hopkins Press, 1973).

Brinley Thomas, "Towards an Energy Interpretation of the Industrial Revolution," *Atlantic Economic Journal* (March 1980).

A. P. Usher, *A History of Mechanical Innovations* (Cambridge: Harvard University Press, 1954).

Thorstein Veblen, *The Instinct of Workmanship* (New York: Kelley, 1964. Rev. ed. reprint of 1918).

Thorstein Veblen, *The Theory of Business Enterprise* (New York: Charles Scribner and Sons, 1904).

William H. Waldorf, Kent Kunze, Larry S. Rosenblum, and Michael B. Tannen, "New Measures of the Contribution of Education and Experience to U.S. Productivity Growth," paper presented at annual meetings of the American Economic Association, New Orleans (December 1986).

Charles Walker, "The Social Effects of Mass Production," in *Technology in the 19th Century* II, eds. Melvin Krantzberg and Carroll Pursell, Jr. (New York: Oxford University Press, 1967).

William Walker, "Information Technology and the Use of Energy," *Energy Policy* (October 1985).

William Walker, "Information Technology and the Energy Supply," University of Sussex, England (1986), mimeograph.

Alvin M. Weinberg, "On the Relation Between Information and Energy Systems," *Interdisciplinary Science Reviews* 7, no. 1 (1982).

Harold F. Williamson, "Mass Production for Mass Consumption," in *Technology in Western Civilization* I, eds. Melvin Krantzberg and Carroll Pursell, Jr. (New York: Oxford University Press, 1967).

13 | *Electricity Use and Energy Conservation*

Sidney Sonenblum and Sam H. Schurr

QUANTITATIVE OVERVIEW

This chapter explores the strong linkages between the growing share of electricity as a component of total energy consumption and long-term improvements in the efficiency of overall energy use. That there has been a persistent historical association of these two trends, roughly since the end of World War I, can be plainly seen in Figures 13.1, 13.2, and 13.3. The relationship holds for the total U.S. economy, for the private domestic business economy, and most strikingly for the manufacturing sector. Are these historical relationships merely coincidental, or have there been identifiable causal factors at work?

Our analysis points to strong causality based on the role that electricity has played as an agent of technological progress. Nevertheless, the coexistence of declining energy intensity[1] and increasing electricity use is puzzling. More Btus are contained in the fuels used to generate electricity than in the electricity that is produced. So growing reliance on electricity over time would seem to imply to an increase in the intensity of overall energy use, not a decline.

Figure 13.1 shows a very rapid drop in energy intensity during the energy conservation years that followed the 1973 oil embargo. The figure also shows, however, that energy intensity has declined during most of the twentieth century and that in some periods the strength of the decline has rivaled the drop since 1973. Why energy intensity declined during the long "pre-energy conservation period" of U.S. economic development and why this decline was related to growth in the importance of electricity use are the topics to which we now turn.

We will focus on energy use in manufacturing to show how this apparently paradoxical outcome came about. Manufacturing is of fundamental importance to a modern industrial economy, making it a logical choice for more detailed analysis. For additional data covering the private domestic business economy and the total U.S. economy, refer to Appendix II.

Figure 13.1   Energy Intensity and Electricity's Share of Total Energy Use:
Total United States, 1899–1985

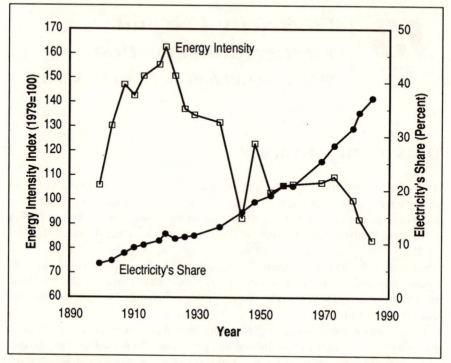

SOURCE: Appendix II.
   NOTE: Energy intensity measured as ratio between total energy input in U.S. and the
gross national product (GNP). Data points refer to business cycle peak years, except for final
year shown.

MANUFACTURING: A CASE STUDY

Figure 13.3 shows that since 1920 the intensity of energy use in manufacturing has, with few exceptions, persistently declined, while the electricity
share of energy use has increased. Why should such a long-term improvement in the efficiency of energy use be related, as it seems to be, to electricity's increasing importance in the manufacturing energy mix?

   The key, we believe, is technological progress tied to the use of electrotechnologies. Our answer, in broad outline, is based on the following set of
interrelated facts drawn from the analysis of the historical record for manufacturing in Chapter 12:

(1) During much of the twentieth century, technological progress has been
    a major factor supporting productivity and output growth;

Figure 13.2   Energy Intensity and Electricity's Share of Total Energy Use:
Private Domestic Business Economy, 1899–1985

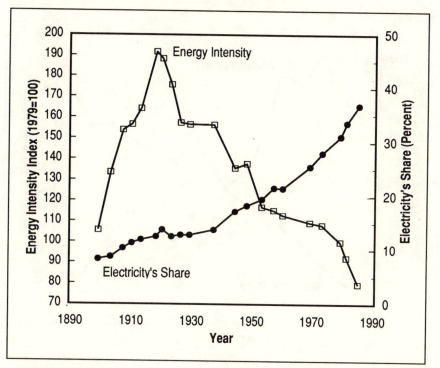

SOURCE: Appendix II.
NOTE: Energy intensity was calculated by dividing the index of energy input in the non-residential economy by the index of output for the private domestic business economy. In calculating the electricity share, data for the nonresidential economy were used for both electricity and total energy use. Data points refer to business cycle peak years, except for final year shown.

(2) The electrification of production operations, particularly since the end of World War I, has been an important contributor to technological progress through its effects on the organization of manufacturing production and the performance characteristics of capital (plant and equipment);

(3) When productivity grows, fewer production inputs are in general required to produce a given level of output;

(4) Overall energy use is one of the production inputs that has fallen relative to output, as a result of productivity growth;

(5) Taken together, these interacting relationships have produced a situation in which the growth of electricity as an input into the energy mix, and the decline of overall energy use relative to output, have gone hand in hand.

Figure 13.3 Energy Intensity and Electricity's Share of Total Energy Use:
Manufacturing, 1899–1985

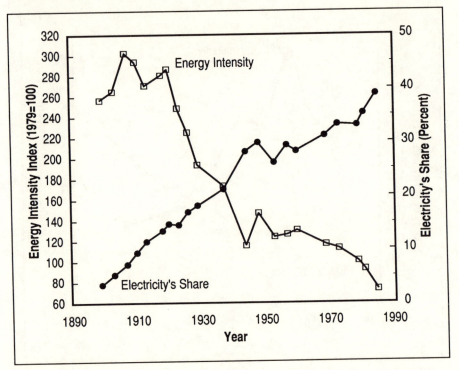

SOURCE: Appendix II.
  NOTE: Energy intensity was calculated by dividing the index of manufacturing energy
input by the index of manufacturing output. Data points refer to business cycle peak years,
except for final year shown.

   Although electrification is only one among a number of interrelated influences that affected twentieth-century technological progress and productivity growth, we believe it to be of great importance because crucially
significant advances in technology have been based upon techniques of
production that are in turn strongly dependent on the use of electricity.

Productivity Growth and Energy Intensity: The Long-Term Historical Record

The historical record can be examined in terms of the shorter time intervals
shown in Table 13.1. First we turn to the evidence showing the historical
parallelism between technological progress as reflected in multifactor productivity growth and decline in the intensity of overall energy use relative to
output. Multifactor productivity growth refers to all sources of output

growth other than increases in labor and capital inputs. It is measured as growth in output less the weighted sum of growth in labor and capital, where the weights are shares of each input in the total costs of production. It is interpreted here as improvement in the productive efficiency of the entire system.

Table 13.1 divides the record into four broad historical periods that may be characterized as follows:

1899–1920: A period of increasing energy intensity and slow productivity growth

1920–1953: A period of rapidly declining energy intensity and rapid productivity growth

1953–1973: A period of comparative stability in energy intensity and moderate productivity growth

1973–1985: A period of rapidly declining energy intensity and slow productivity growth

Data from the three periods before 1973 strongly support our general theme, showing an inverse relationship between energy intensity and productivity growth. From 1899 to 1920, the one period during which energy intensity rose most of the time, productivity grew very slowly. (As if to emphasize the point, when, indeed, there was an interval of comparatively rapid productivity growth from 1910 to 1913, there was also a comparatively high decline in energy intensity.) Between 1920 and 1953, in sharp contrast to the 1899 to 1920 period, energy intensity generally declined rapidly while productivity grew rapidly. (Again, as if to emphasize the point, 1944 to 1948, the one interval in which energy intensity grew rapidly, was also a period of considerable productivity decline.)

The 1953 to 1973 period shows comparative moderation in both variables. Energy intensity tended to decline, or to increase slowly, while productivity growth, although good compared to the years immediately following, was moderate compared to the 1920 to 1953 period. Finally, the interval from 1973 to 1985 departs sharply from the pattern for all of the earlier periods. It shows a sharp decline in energy intensity, but it is accompanied, in general, by only a very slow increase in productivity growth.

In order to elucidate the role of electricity as an explanatory factor in these trends, we will concentrate on two periods that show rapid declines in energy intensity, but sharply different records of productivity growth: 1920 to 1929 and 1973 to 1985 (see Figure 13.4). A comparison of these periods serves also to show the part played by energy price trends, as well as to highlight changes that have taken place in the role of electricity as an agent of technological progress in manufacturing during the twentieth century.

Table 13.1 Productivity Growth and Energy Intensity in Manufacturing, Average Annual Growth Rates

	Output	Energy use	Energy prices	Productivity	Energy intensity
	(1)	(2)	(3)	(4)	(5)
Increasing energy intensity and slow productivity growth:					
1899–1920	4.3	4.8	1.6	0.8	0.6
1899–1903	6.5	7.4	7.9	0.9	0.9
1903–1907	4.4	8.2	− 5.0	0.2	3.6
1907–1910	2.3	1.0	− 6.2	0.5	− 1.2
1910–1913	6.1	3.1	7.9	3.5	− 2.8
1913–1918	5.3	6.2	1.3	0.4	0.8
1918–1920	− 2.8	− 1.6	7.4	− 0.9	1.2
Rapidly declining energy intensity and rapid productivity growth:					
1920–1953	4.1	1.8	− 1.6	2.8	− 2.5
1920–1923	5.2	− 0.1	− 8.8	7.2	− 5.0
1923–1926	3.9	0.7	1.0	4.4	− 3.1
1926–1929	5.1	0.0	− 5.6	3.9	− 4.8
1929–1937	0.4	− 1.0	0.6	2.0	− 1.4
1937–1944	12.3	5.9	− 1.2	3.9	− 5.7
1944–1948	− 5.7	0.3	− 0.3	− 2.4	6.3
1948–1953	6.1	2.5	− 1.4	3.0	− 3.4
Comparatively stable energy intensity and moderate productivity growth:					
1953–1973	3.4	2.9	− 1.4	1.8	− 0.5
1953–1957	0.9	1.4	− 1.2	0.7	0.5
1957–1960	0.7	1.8	− 3.0	0.7	1.1
1960–1969	5.2	3.9	− 2.2	2.2	− 1.3
1969–1973	3.7	2.8	1.7	2.8	− 0.9
Rapidly declining energy intensity and slow productivity growth:					
1973–1985	2.0	− 1.7	6.4	1.5	− 3.6
1973–1979	1.9	0.0	11.4	0.5	− 1.9
1979–1981	− 1.5	− 5.5	19.3	− 0.6	− 3.9
1981–1985	4.0	− 2.1	− 6.3	4.0	− 5.9

SOURCE: Appendix II.

NOTE: Average annual rates of growth are between cyclical peak years (except for 1985). These time periods are defined by business cycle peak years in order to minimize the effects of short-term economic fluctuations. The data cited in this chapter are further described in Appendix II.

Figure 13.4 Productivity Growth and Energy Intensity: A Comparison of 1920–1929 and 1973–1985, Average Annual Rates of Growth

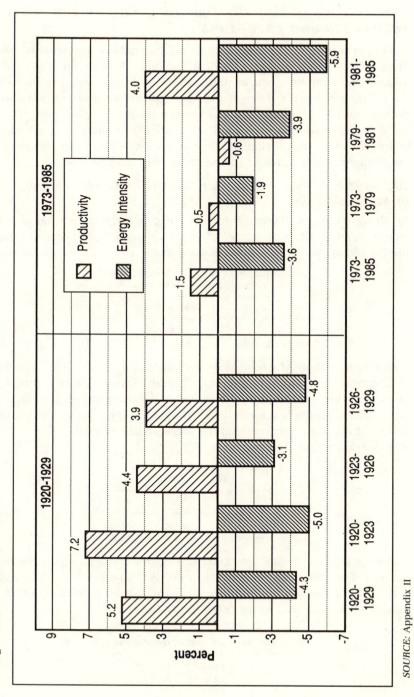

SOURCE: Appendix II

Two Periods of Rapidly Declining Energy Intensity: 1920–1929 and 1973–1985

It is important to examine the 1920 to 1929 period because what happened during those years provides an outstandingly clear example of how the interactions between electrification and technological progress led to both the rise in electricity's share of total energy use and to the decline in the overall intensity of energy use relative to manufacturing output. The 1920 to 1929 information is presented schematically in order to highlight the factors at work. That period will then be used as a backdrop for examining the changes represented by the events during 1973 to 1985.

1920–1929: A Period Characterized by the Electrification of Machine Drive

Productivity Rises, Energy Intensity Falls. In a sharp reversal of earlier trends, productivity and energy efficiency grew at very fast rates during the decade of the 1920s, as can be seen in the following comparisons:

Productivity Growth and Energy Intensity
(average annual growth rates)

	1899–1920	1920–1929
Productivity	+0.8%	+5.2%
Energy intensity	+0.6%	−4.3%

Electrification of Machine Drive Expands. During the 1920s, as detailed in Chapters 1 and 12, electrified unit drive expanded rapidly within manufacturing, thus establishing the technological foundation for a radical reorganization of production operations. This growth is partially reflected in the following statistics documenting the increase in electricity's share of total energy use in machine drive:[2]

Share of Total Energy Used in
Machine Drive

	1920	1929
Electric	53%	78%
Nonelectric	47%	22%

Energy is Saved in Machine Operations. One consequence of the electrification of machine drive was to create "direct" energy savings because the efficiency of energy use in performing the machine drive function was enhanced, as indicated in the following statistics:

Energy Inputs Required for Machine Drive
(million Btus used per unit of horsepower capacity)

1899	82
1920	53
1929	31

Electricity's Share of Energy Use Rises. A second consequence was to raise the electricity share of total energy use. Almost all of this increase resulted from the growth of electricity use and the decline of nonelectric energy use in machine drive:

Electricity's Share of Total Energy Use
in Manufacturing

1920	14.8%
1929	18.2%

Electricity Use Grows Relative to Other Production Inputs. Electricity's importance as an agent of technological progress is reflected in the fact that its use rose sharply relative to that of other inputs. Particularly important was its rise relative to capital, the production factor that embodies those advances in technology represented by changes in plant and equipment. This electrification of production equipment made a major contribution to the reorganization of factory production. Unlike electricity, nonelectric energy declined relative to capital.

Electrification of Capital
(average annual growth rates)

	1920–1929
Electricity/capital	8.3
Nonelectric energy/capital	−1.0

Use of Inputs Declines Relative to Output. So far as energy efficiency is concerned, the most important consequences of the electrification of machine drive were the "growth-related" energy savings achieved because of the productivity increases that resulted from the reorganization of factory operations. These effects show up not only in the sharply declining use of inputs of labor and capital relative to output (the standard components of the multifactor productivity measure) but also in the declines of nonelectric and overall energy use in relation to output. Note, in particular, the rapid declines in the intensity of nonelectric energy use during the 1920s compared to the two previous decades. These declines account for most of the curtailment of growth in total energy use.

Input Intensities Relative to Output
(average annual growth rates)

	1899–1920	1920–1929
Labor/output	−1.2	−5.5
Capital/output	+1.3	−3.9
Nonelectric energy/output	0.0	−4.8
Total energy/output	+0.6	−4.3

1973–1985: A Period Characterized by Strong Increases in Energy Prices

The 1973 to 1985 period is distinctive in that a rapid decline in energy intensity for the period as a whole is associated with a relatively slow growth in productivity. Thus, compared to the period from 1920 to 1929 and to the immediately preceding period of 1960 to 1973, productivity growth slowed down considerably, while the decline in energy intensity was quite strong.

Productivity Growth and Energy Intensity
(average annual growth rates)

	1920–1929	1960–1973	1973–1985
Productivity	5.2	2.4	1.5
Energy intensity	– 4.3	– 1.1	– 3.6

However, this pattern actually prevailed only from 1973 to 1981. The 1981 to 1985 period returned to a more usual pattern in which relatively high productivity growth and relatively rapid improvements in energy efficiency occurred together. In comparison with 1920 to 1929, however, there is a difference in that energy intensity fell somewhat more sharply, while the rate of productivity growth was slower.

Productivity Growth and Energy Intensity
(average annual growth rates)

	1920–1929	1973–1981	1981–1985
Productivity	5.2	0.2	4.0
Energy intensity	– 4.3	– 2.4	– 5.9

The electrification of machine drive was a fundamental influence in the 1920 to 1929 decline in energy intensity because it created direct energy savings in performing the machine drive function through the displacement of nonelectric by electric energy, and energy savings relative to output growth through its influence on productivity. The leveraging effect of productivity growth was the major source of the gain in energy efficiency during that period.

The forces at work from 1973 to 1985 have been more complex and less easy to track. Sharp and unparalleled rises in energy prices, as well as conscious efforts to conserve in response to public policy, contributed to both increased electrification and declining energy intensity during much of this period. Despite the continued electrification, reductions in energy intensity as a result of productivity growth, a distinguishing characteristic of most of the historical record, played a minor role from 1973 to 1981, but they emerged once again as a significant factor during the 1980s.

Rising Energy Prices. Higher energy prices were an exceedingly important influence in all of the years following 1973. Indeed the price changes

that began with the oil embargo ushered in a new era of energy conscious-
ness in the United States and throughout the world.

Real (inflation-adjusted) energy and electricity prices generally declined
prior to 1973. The electricity price decline was usually greater than the
decline in fuel prices, so that the price of electricity tended to fall relative to
that of fuels during this period. Real prices for energy increased particularly rapidly from 1973 to 1981, with electricity prices rising less than fuel
prices, so the relative price of electricity continued its long-term decline.
Fuel prices peaked in 1981 and then began to decline again, while electricity
prices continued to rise (although at a reduced rate) so that, contrary to
long-term historical experience, the relative price of electricity increased
during the first half of the 1980s.

Energy and Electricity Prices (constant dollars)
(average annual growth rates)

	1960–1973	1973–1985	1973–1981	1981–1985
Energy prices[a]	– 1.0	6.4	13.3	– 6.3
Electricity prices	– 1.7	3.8	5.2	1.1
Electricity relative to energy prices	– 0.7	– 2.4	– 7.2	7.9

[a] Includes both direct use of fuels and electricity.

Growth in the Electricity Share of Energy Use. Price movements produced a significant effect on long-term energy consumption trends. Just
prior to 1973 the absolute level of energy use was growing at a rapid rate.
Indeed, with the exception of the years of the Great Depression, manufacturing energy use tended to rise continuously over time. After 1973, however, the level of energy use began to decline. The declines were mostly in
nonelectric energy, where the largest price increases also occurred. The use
of electricity did not decline, but its growth from 1973 to 1985 was quite
slow by historical standards.

Energy and Electricity Use
(average annual growth rates)

	1960–1973	1973–1985	1973–1981	1981–1985
Total energy	3.6	– 1.7	– 1.4	– 2.1
Electricity	4.9	0.3	0.3	0.3
Nonelectric energy	3.0	– 2.7	– 2.3	– 3.5

Even though real energy prices began falling again after 1981, nonelectric
energy use continued to decline. The use of electricity, despite its price rise
relative to other forms of energy, continued its modest rate of increase.

As a result of these disparate trends, the electricity share of total energy
used in manufacturing increased. However, the rise in the electricity share
was not a reflection of rapid increases in one strongly emerging technology,

as it had been during the 1920 to 1929 period (i.e., electrified machine drive). Instead, it resulted from the combination of a relatively fast decline in nonelectric energy, plus sluggish growth in electricity use and continued general growth in the electrification of a variety of production processes.

**Electricity's Share of Total Energy
Use in Manufacturing**

1960	28%
1973	31%
1981	35%
1985	39%

Increasing Electrification of Production Processes. Technological progress based on the use of electrified production techniques did play a part in the declining energy intensity of the entire period. However, it was not machine drive that was undergoing further electrification because this function had been virtually totally electrified for more than twenty-five years. Instead, it was materials processing and information management. Electrification of these energy services has accelerated since 1973. Partly this is because materials processing, which still uses mostly nonelectric technologies, has become increasingly electrified; and, partly it is because the use of electrified information processing and support services (such as air conditioning and heating) has become more widespread.

**Electricity's Share of Energy Used
(average annual growth rates)**

	1920–1929	1960–1973	1973–1985	1973–1981	1981–1985
Machine drive	4.5	0.1	0.1	0.1	0.1
Materials conversion, information processing, and support services	2.1	1.4	2.6	2.3	3.2
Total manufacturing	2.3	1.3	2.0	1.8	2.5

**Electric and Nonelectric Energy Use in Materials Conversion, Information
Processing, and Support Services, (average annual growth rates)**

	1920–1929	1960–1973	1973–1985	1973–1981	1981–1985
Total energy	0.7	3.3	− 2.2	− 1.9	− 2.8
Electricity	2.9	4.7	0.3	0.3	0.3
Nonelectric energy	0.6	3.0	− 2.7	− 2.3	− 3.5

The studies presented in Part II describe a general tendency for the use of electrical processes to grow at the expense of the direct use of fuels in the materials processing industries. Some of this growth occurred in response to comparative price trends for electricity and fuels used directly, but there were, in addition, significant improvements in technology based on the use of electrified techniques. In the steel industry, for example, the use of electric furnaces in place of conventional fuel-based processes led to direct energy savings in production as well as to basic improvements in the overall productive efficiency of the industry.

It was, as we saw earlier, systemic improvements in productive efficiency leading to an overall decline in productive inputs relative to output that accounted for most of the rapid decline in manufacturing energy intensity during the 1920s. We also found that there was a sharp increase in the electrification of capital during the same time period as a result of the strong growth of electrified machine drive. Specifically the usable electricity-to-capital ratio increased at a rate of more than 8 percent per year, while nonelectric energy relative to capital declined at 1 percent per year. Both the historical literature describing the period and the aggregative quantitative evidence for manufacturing point, therefore, to strong cause and effect relationships between electric-based technological progress and the declining intensity of energy use.

No such clear thread can be found in aggregate statistical data for the most recent period. If we look at the years 1981 to 1985, when productivity resumed its upward course at a comparatively high rate of increase, we find that the electricity intensity of capital was falling, not rising. To be sure, it fell at a far slower rate than in the period from 1973 to 1981 and at a slower rate than for nonelectric energy relative to capital. But still it fell. Why?

Electrification of Capital
(average annual growth rates)

	1920–1929	1960–1973	1973–1985	1973–1981	1981–1985
Capital	0.7	4.1	3.2	3.9	1.6
Electricity/capital	8.3	0.9	− 2.7	− 3.5	− 1.1
Nonelectric energy/ capital	− 1.0	− 1.1	− 5.7	− 5.9	− 5.1

As we know from the materials presented in Chapter 2, the information and automatic control technologies that constitute the new wave of technological progress throughout manufacturing are completely dependent on energy in the form of electricity. We know also that it is the ability to subdivide electricity precisely so as to meet the needs of microcomputers that is making new systemic changes in production technologies possible, just as was the case for motorized unit drive in an earlier era.

What is quite different, however, are the quantitative demands that these new technologies place on the use of electricity. Compared to electric motors, the demand imposed by computers is microscopic. Meanwhile, the capital cost of this complex new equipment is far greater. The two trends, taken together, have led to a much lower growth in intensity of electricity use relative to capital than has been the case historically. Nevertheless, technologies that are uniquely dependent upon electricity are providing the basis for the systemic advances in production techniques that will result, we believe, in long-term gains in energy efficiency.

Lessons from History

The long historical record for the twentieth century provides a rich source of information for understanding the interrelationships among total energy use, electricity use, and productivity growth in the United States. Unfortunately, most research in recent years has tended to disregard the long-term record and to focus instead on the years since 1973. Energy conservation became a national policy at that time due to the sudden disruptions in the U.S. oil supply. Consequently, there has been a general failure to recognize that recent trends showing a decline in the intensity of energy use relative to output are far from being a brand new phenomenon in the United States. Instead, they reflect a persistent theme that has prevailed in American industry during much of the twentieth century.

The mistaken idea that declines in the intensity of energy use relative to output are a new occurrence has also led to two other mistaken beliefs: (1) that energy is used most effectively only when a conscious effort is made to conserve it; and (2) that because electricity is a secondary energy form whose production, even now, entails the loss of two out of three of the thermal units contained in the fuels used in its generation, its growing use over time must have been a factor leading to a higher intensity of energy use relative to the output produced than would otherwise have been the case.

That energy intensity has in fact declined for much of the twentieth century is evident from the historical data shown in Figures 13.1 through 13.3, as is the fact that the electricity share of total energy has risen steeply and persistently. The main purpose of this chapter has been to answer two questions concerning these trends: Why did the intensity of energy use relative to output decline? And how could this decline occur in conjunction with a rise in the electricity share of the energy total?

The answer to both questions lies in technological progress. Productivity growth, resulting largely from technological progress, explains why productive inputs (including aggregate energy use) have, generally, fallen relative to national output. The rise in the electricity share is also largely explained by technological progress because many advances in industrial technology during the twentieth century have relied on electricity-based production techniques. Combining these answers, we come to the conclusion that the electrification of production processes has been an important source, perhaps the most important source, of the long run growth in energy efficiency in manufacturing.

It is only when we reach the very last period (1973–1985) that the basis for these generalizations becomes uncertain. The unparalleled rise in energy prices, and the consequent heavy emphasis on energy conservation, resulted in an essentially different set of relationships for an interval running from about 1973 to 1981. During these years energy intensity fell while productivity growth languished. But during 1981 to 1985 the persistent historical pattern appears to have been restored, with increases in productivity and declines in energy intensity occurring in tandem.

The electricity share of the energy total continued to rise during all of the 1973 to 1985 period. A principal reason for this is that within the materials processing industries, by far the heaviest energy users in the manufacturing sector, new technologies based upon electrified production processes had been replacing technologies based upon the direct burning of fuels.

When manufacturing productivity resumed its rapid growth after 1981, some of the new technologies that were being adopted depended on smaller quantities of electricity to run capital equipment than was the case during earlier technological regimes. This is a basic characteristic of the new automatic control and information technologies that are now becoming pervasive. Although they are totally dependent on electricity as their energy source, only minute amounts are required relative to the costs of additions to capital.

What implications this has for the absolute growth of electricity use in these applications in the future will depend not just on per unit usage but also on how rapidly equipment is added as the use of these technologies multiplies and as output grows. But what will happen to overall energy intensity appears to be clear: it will fall as a result of declining per unit usage, and also as a result of rising productivity linked to the growing electrification of production operations. In brief, there will be a return to the classical pattern for American manufacturing.

Notes

1. Following customary practice, we use a physical rather than a dollar measure of energy intensity, such as is preferred by some economists. Our physical measure relates the Btu quantity of energy use to a real (constant dollar) measure of output. Dollar measures of energy intensity weight the different forms of energy by their prices. Since a shift to electricity increases overall energy quality, the use of price weights means that some energy quality changes are incorporated in the dollar energy intensity measures. As a result, measures of dollar intensity are likely to decline at a slower rate than physical energy intensity, signifying that some part of the energy efficiency gains are attributable to changes in the quality of the energy mix.

2. It is safe to infer that the growth in electricity used for *unit drive* during the 1920s was far greater than is suggested by these statistics. Electricity was first used to drive prime movers in a manner comparable to steam. It was only with the flexibility in the placement of machinery and in the layout of factories that became possible as a result of the rapid expansion of *electrified unit drive* during the 1920s that the real transformation of manufacturing operations occurred. Available data do not permit the statistical documentation of this change, but Chapter 1 tells the story as seen by contemporary observers.

Appendix I:
The Growth of
Electricity Consumption in
Historical Perspective

In its analysis of relationships between electricity use and economic growth, this book is concerned essentially with the role of electricity as a factor in production. The central focus is on increases in productive efficiency in manufacturing and other economic sectors that result from the use of technologies that are tied to electricity because of its special attributes as an energy form.

The book does not deal with the more familiar analytic question of the extent to which the rate of growth of the national economy affects the rate at which the use of electricity increases. For the sake of completeness, this appendix presents an extended discussion of the latter subject, adapted from a 1986 report prepared under the auspices of the National Academy of Engineering.[1] We have updated some of the data included in that report and have adjusted the text, where needed, to reflect the more recent statistics.*

THE LONG-TERM HISTORICAL RECORD

The Growth of Consumption of Electricity and Other Energy Forms

Electricity is such a versatile energy form that its use and the number of its applications have grown rapidly throughout the twentieth century. Electricity consumption continues to grow more rapidly than that of other energy forms. As a consequence, the proportion of the nation's primary energy supply used as electricity has expanded substantially — from near zero at the turn of the century to 38 percent in 1987. The growing importance of electricity as a component of total energy supply can be seen in Figure AI.1, which shows the growth in primary energy inputs, in British thermal units (Btu), for the production of electricity.

*Updated and edited for this book by Milton F. Searl

Figure AI.1 Historical Trends in U.S. Energy Consumption, 1902–1987

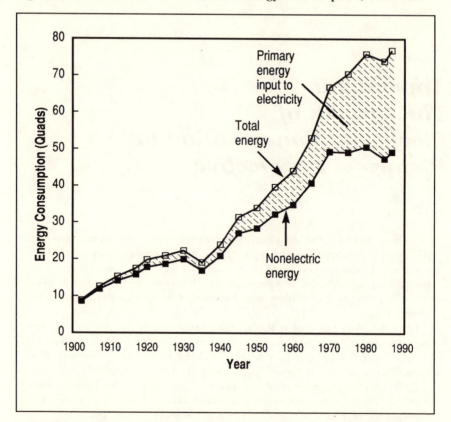

SOURCES: U.S. Bureau of Mines, as presented in *Towards Project Independence: Energy in the Coming Decade,* prepared for the Joint Committee on Atomic Energy, U.S. Congress, 94th, 1st sess. (December 1975); Edison Electric Institute, *Historical Statistics of the Electric Utility Industry,* and *Statistical Yearbook of the Electric Utility Industry* (Washington, D.C.: EEI, various editions); U.S. Department of Energy, *Annual Energy Review* (Washington, D.C.: GPO, various issues).

Another way of measuring the comparative growth of electrical and non-electrical energy is by directly comparing the energy delivered by electricity (instead of by the primary energy consumed in its generation) and the energy delivered by other forms (mostly coal and coke, oil products, and natural gas). The average annual growth rates shown in Table AI.1 are based on such data. Clearly electricity consumption has grown at a higher rate than has the consumption of other energy forms throughout the twentieth century, including the 1980s. Nevertheless, the rate of growth of electricity consumption itself fell sharply during the recent past compared with its growth rates in all earlier periods.

Table AI.1 Average Annual Growth Rates in Total Energy, Electricity, and Nonelectric Energy Consumption for Selected Periods, 1902–1987 (Percentage per Year)

Period	Total energy	Electricity	Nonelectric energy
1902–1912	6.1	15.5	5.6
1912–1920[a]	2.9	10.8	3.3
1920–1930[a]	1.2	7.3	1.3
1930–1940	0.7	4.6	0.5
1940–1950	3.5	7.9	3.0
1950–1960	2.8	8.1	2.2
1960–1973	4.1	6.7	3.4
1973–1987	0.2	2.4	−0.9

SOURCES: U.S. Bureau of Mines, as presented in *Towards Project Independence: Energy in the Coming Decade,* prepared for the Joint Committee on Atomic Energy, U.S. Congress, 94th Congress, 1st sess. (December 1975); Edison Electric Institute, *Historical Statistics of the Electric Utility Industry Through 1970,* EEI 73–34 (Washington, D.C.: EEI, 1973), and *Statistical Yearbook of the Electric Utility Industry* (Washington, D.C.: EEI, various issues); U.S. Department of Energy, *Annual Energy Review* (Washington, D.C.: GPO, various issues).

[a] Average annual growth rates for total energy and for nonelectric energy were computed from British thermal units (Btu) of consumption. The growth rate for electricity was computed from kilowatthour figures. Because of the rapidly improving efficiency of electric power generation in the early years, electricity kilowatthours grew much faster than Btu input for electricity generation. This aspect of the computation is the reason that, for these two periods, the growth rate of total energy appears to be lower than the growth rate of both of its components.

That the rate of growth of electricity use has tended to decline, particularly compared with the early periods of its introduction, is not surprising. As the base from which growth is measured becomes larger, even ever-growing absolute increments translate into smaller percentage growth rates. By the same token, the early rates of growth of a newly emerging service or industry, such as electricity in the first part of this century, will loom large compared to those of already established quantities, such as population, GNP, or the use of other energy forms.

Electricity Growth in Relation to the Growth of the Gross National Product

How should the relationship between electricity use and economic growth be expressed? Our approach is to look first at the relationship in aggregate terms, that is, in terms of total electricity made available in the United States, regardless of the source, and of the GNP, expressed in constant dollars. Later this relationship will also be examined, although only for the years following World War II, in terms of the major sectors of electricity use: residential, commercial, and industrial. The discussion addresses the nature of the aggregative relationship, changes in this relationship over time, and sectoral relationships compared to the aggregative, or national, one.

When standard statistical techniques are used to measure the relationship between annual levels of electricity use and the GNP (in constant dollars), certain regular features of the historical record appear. Perhaps the most significant characteristic of the relationship is its stability over appreciable segments of time. This stability is indicated in Figure AI.2 (with additional detail in Figure AI.3), which displays lines of regression for four periods covering most of the twentieth century to date.[2] For any point on these lines one may calculate an average electricity intensity, that is, electricity consumption per unit of the GNP. Changes along these lines relate increments in electricity use to increments in the GNP; each period is marked by a stable linear relationship, which, though showing some annual fluctuations, indicates a strong tendency toward a constant incremental intensity of electricity use within each period.

Equally significant is the fact that there are only a few changes in the slope (and level) of the regression line over the long historical record. Clearly discernible changes in slope occurred following World Wars I and II. Even the Great Depression did not result in a change in slope; the level of the regression line did shift upward, however, reflecting the fact that during this period the GNP decreased relatively more than did electricity use. The record of the post-oil-embargo years poses the question of whether we are once again witnessing a change in slope, an issue that is addressed later in this chapter.

The specific findings embodied in Figures AI.2 and AI.3 may be summarized as follows:

(1) From 1902 to 1912, the national economy tended to use an additional 0.29 kilowatthours (kWh) per additional dollar of the GNP, measured in constant (1972) dollars.

(2) A transition to a new slope occurred between 1912 and 1920; the 1917 observation shown on Figure AI.3 appears to be transitional.

(3) Between 1920 and 1929 the incremental use of electricity per unit of the GNP averaged 0.58 kWh per dollar, twice the value that prevailed between 1902 and 1912.

(4) Following 1929, the GNP dropped by almost one-third, while electricity use declined only slightly. Consequently, average electricity intensity increased. However, the slope of the line for the years 1930 through 1946 did not change significantly from that for the years 1920 through 1929. Thus, the incremental intensity of electricity use remained the same, even though average electricity intensity rose.

(5) Another transition occurred following World War II, and the new trend line has persisted ever since (with a critical question remaining about the most recent past). The new slope (through 1983) shows on the average an increment of 2.12 kWh per additional constant (1972) dollar of the GNP, about three and one-half times that characterizing the relationship observed between 1920 and 1946.

Figure AI.2 Electricity Consumption versus the GNP in the United States, with Lines of Regression by Periods, 1902–1983

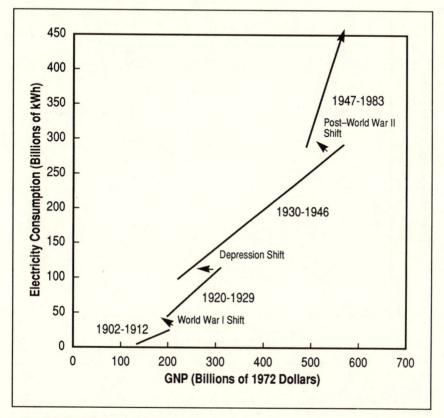

SOURCE: Compilation and figure by Energy Study Center, Electric Power Research Institute, Palo Alto, California.

NOTES: The GNP is expressed in constant (1972) dollars. Data for 1902 through 1928 have been converted from constant (1958) dollars in U.S. Bureau of the Census, *Historical Statistics of the United States, Colonial Times to 1970*, Bicentennial Ed., Part 2 (Washington, D.C.: GPO, 1975), Series F1–5, 224; for 1929 through 1980 from the Council of Economic Advisers, *Economic Report of the President* (Washington, D.C.: GPO, February 1984), 222; for 1981 through 1983 from the Council of Economic Advisers for the Joint Economic Committee, *Economic Indicators* (March 1985), 2.

Electricity consumption is expressed as "electricity made available in the United States." Conceptually this quantity includes utility generation and nonutility generation (industrial self- and co-generation), and net imports. Electricity data sources are Edison Electric Institute, *Historical Statistics of the Electric Utility Industry Through 1970*, EEI 73–34 (Washington, D.C.: EEI, 1973), and *Statistical Yearbook of the Electric Utility Industry, 1983* (Washington, D.C.: EEI, 1984).

Figure AI.3   Electricity Use and the GNP: The Transitions

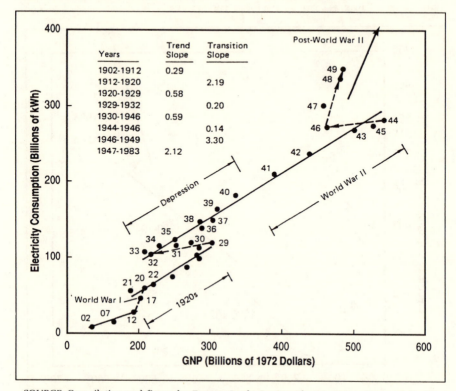

Years	Trend Slope	Transition Slope
1902-1912	0.29	
1912-1920		2.19
1920-1929	0.58	
1929-1932		0.20
1930-1946	0.59	
1944-1946		0.14
1946-1949		3.30
1947-1983	2.12	

SOURCE: Compilation and figure by Energy Study Center, Electric Power Research Institute, Palo Alto, California.

NOTES: The GNP is expressed in constant (1972) dollars. Data for 1902 through 1928 have been converted from constant (1958) dollars in U.S. Bureau of the Census, *Historical Statistics of the United States, Colonial Times to 1970,* Bicentennial Ed., Part 2 (Washington, D.C.: GPO, 1975), Series F1-5, 224; for 1929 through 1983 from the Council of Economic Advisers, *Economic Report of the President* (Washington, D.C.: GPO, February 1984), 222.

Electricity consumption is expressed as "electricity made available in the United States." This quantity includes utility generation and nonutility generation (industrial self- and co-generation), and net imports. Electricity data sources are Edison Electric Institute, *Historical Statistics of the Electric Utility Industry Through 1970,* EEI 73-34 (Washington, D.C.: EEI, 1973), 9, and *Statistical Yearbook of the Electric Utility Industry* (Washington, D.C.: EEI, 1983).

POST–WORLD WAR II TRENDS

The Growth of Electricity Use and the Gross National Product

The relationship between increases in electricity use and increases in the GNP is shown for the post–World War II period in Figure AI.4. The relationship appears to have persisted through the entire period, with the possible exception of a break since the mid-1970s. Although observations for the most recent years fall below the trend line, this fact is still consistent with a characteristic feature of the relationship, that is, a tendency for individual years to exhibit a cyclical pattern around the long-term trend line, as the figure shows.

However, to conclude that the data points after the mid-1970s are nothing more than a manifestation of a persistent cyclic pattern is only one way of interpreting the record for recent years. There has been a strong, decreasing trend in the ratio of the annual percentage growth of electricity use to that of the GNP, and this fact is frequently cited as evidence that the relationship between the two has changed.

Figure AI.5 shows that the ratio of the five-year moving averages of percentage electricity growth and percentage GNP growth has fallen from an average of 2 before 1973 to about 1 today. This tendency toward convergence is consistent with the postwar linear relationship between the two variables. For example, the electricity use-GNP line of regression for 1947 to 1983 shows an increment of 2.12 kWh of electricity for every constant (1972) dollar increment of the GNP. In the early postwar period, when average electricity intensity was comparatively low (about 0.6 to 0.8 kWh per dollar), the high incremental electricity intensity (2.12 kWh per dollar) led to much higher electricity growth rates than GNP growth rates. As average electricity intensity has increased — to 1.57 kWh per constant (1972) dollar in 1983 — the relative effect of the incremental electricity intensity (2.12 kWh per dollar) has decreased, leading toward a convergence in growth rates.[3]

A critical question before us, then, is whether the long-standing post–World War II trend has been broken by another of the historically infrequent transitions, but for the first time toward a decline in the incremental intensity of electricity use. To shed more light on this question, we next examine some of the underlying forces that determine electricity use in relation to national output. Such influences include the trends of electricity use in the major consuming sectors, the effects of changes in the composition of national output, and the effects of changes in energy prices.

Figure AI.4 **Electricity Consumption versus the GNP in the United States, 1947–1984**

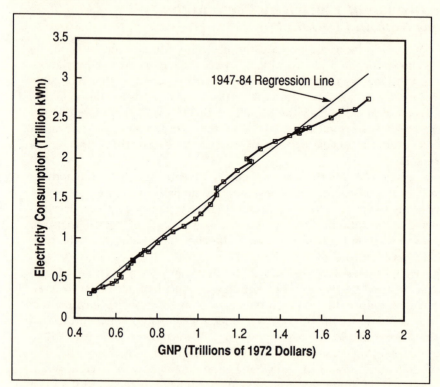

SOURCE: Compilation and figure by Energy Study Center, Electric Power Research Institute, Palo Alto, California.

NOTES: The GNP is expressed in constant (1972) dollars. Data from the Council of Economic Advisers, *Economic Report of the President* (Washington, D.C.: GPO, February 1984), 222; Council of Economic Advisers for the Joint Economic Committee, *Economic Indicators* (March 1985), 2.

Electricity consumption is expressed as "electricity made available in the United States." This quantity includes utility generation and nonutility generation (industrial self- and co-generation), and net imports. Electricity data sources are Edison Electric Institute, *Historical Statistics of the Electric Utility Industry Through 1970*, EEI 73–34 (Washington, D.C.: EEI, 1973), 9, and *Statistical Yearbook of the Electric Utility Industry, 1987* (Washington, D.C.: EEI, 1988).

Figure AI.5 (A) Growth Rates of U.S. Electricity Use and the GNP,
1950–1985; (B) Ratio of the Growth Rates, 1950–1985

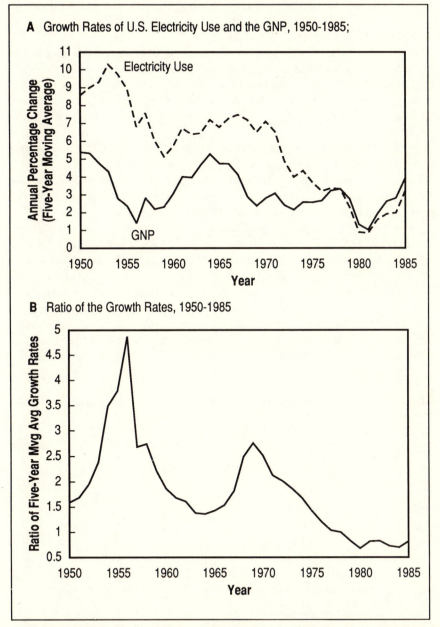

SOURCES: Based on data from Edison Electric Institute, *Statistical Yearbook of the Electric Utility Industry* (Washington, D.C.: EEI, various issues); U.S. Bureau of Economic Analysis, *The National Income and Product Accounts of the United States, Statistical Tables,* Supplement to *Survey of Current Business* (Washington, D.C.: GPO, various issues).

Electricity Use in the Major Consuming Sectors

Electricity use is ordinarily classified by three major consuming categories:
- Industrial, that is, agriculture, mining, construction, and manufacturing[4]
- Residential, that is, private households
- Commercial and all other activities

Table AI.2 shows the changing importance of each sector as reflected by its percentage of total electricity consumption over the postwar period. The residential sector sharply increased its share of electricity use from one-fifth to about one-third of the total. The commercial sector's share increased sharply during the 1960s, and in 1987 it stood at about 28 percent of the total. The industrial share, starting at 59 percent of the total, dropped dramatically after 1955, to 38 percent of the total by 1983, but it has since increased about a percentage point. (Because the statistics in Table AI.2 include industrial self-generation of electricity, the sectoral shares differ from those shown elsewhere in this book — Chapter 11 in particular — where the data refer only to electricity sales by utilities.)

These postwar trends in sectoral shares of electricity use parallel the underlying trends in the economic measures for each sector. That is, growth in disposable personal income (DPI), the residential sector surrogate for gross product originating (GPO) in the other two sectors and commercial sector growth outpaced that of industrial output over the entire period.[5] In fact, if we examine the relationships between electricity use in these sectors and their respective economic measures, as in Figure AI.6, we find the same stable linear relationship (with cyclical variation) as is seen in the aggregate economy. Also, as in analyzing the aggregate case, one gains a different

Table AI.2 U.S. Electricity Use by Sector, 1950–1987 (Percentage of Total)[a]

Year	Residential	Commercial[b]	Industrial[c]
1950	20.6	20.1	59.3
1955	22.3	17.3	60.4
1960	25.4	18.4	56.2
1965	26.6	22.6	50.7
1970	29.9	24.7	45.4
1975	32.2	26.7	41.1
1980	33.5	27.2	39.3
1983	33.9	28.3	37.8
1987	32.5	28.5	39.0

SOURCES: Edison Electric Institute, *Statistical Yearbook of the Electric Utility Industry* (Washington, D.C.: EEI, various issues).

[a] Includes industrial self-generation.

[b] Small light and power, street and highway lighting, other public authorities, railroads and railways, and interdepartmental transfers.

[c] Large light and power, and industrial self-generation.

Figure AI.6 Electricity Use–Economic Output Relationships By Economic Sector, 1947–1987

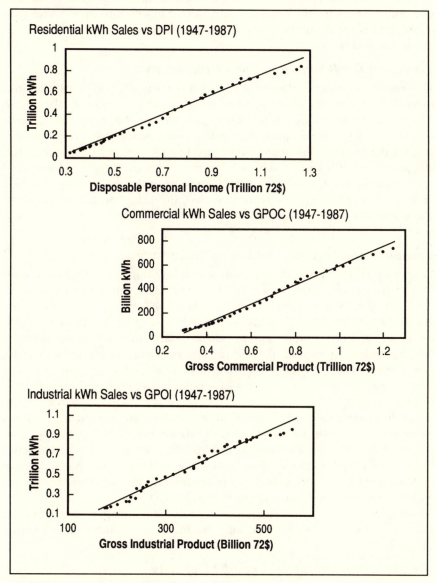

SOURCE: Compilation and figure by Energy Study Center, Electric Power Research Institute, Palo Alto, California.

NOTE: Different scales are used for the three sectors to highlight the linearity of the electricity use–economic output relationship within sectors. Based on data from Edison Electric Institute, *Statistical Yearbook of the Electric Utility Industry* (Washington, D.C.: EEI, various issues); U.S. Bureau of Economic Analysis, *The National Income and Product Accounts of the United States, Statistical Tables,* Supplement to *Survey of Current Business* (Washington, D.C.: GPO, various issues).

perspective when comparing the ratio of the percentage growth rates of electricity use and economic output measures. Figure AI.7 shows the same trend toward convergence between the sectoral percentage growth rates as was observed in the total economy. For further insight into these trends, we examine in more detail the postwar patterns of electricity use within each of the three sectors.

Trends in the Residential Use of Electricity

The trend in residential electricity consumption since World War II falls into three distinct periods, corresponding roughly to the decades of the 1950s, 1960s, and 1970s, as shown in Figure AI.7. During the 1950s, the growth rate of electricity consumption (five-year moving average) was very high but steadily decreasing from about 14 percent to 8 percent per year. During the 1960s, the growth rate slowly accelerated to about 10 percent per year toward the end of the decade. It then dropped in the post-embargo period to an average of about 5 percent per year, until the late 1970s and early 1980s when it dropped further, reaching its minimum in 1983 and rising since.

Trends in the Commercial Use of Electricity

Electricity consumption in the commercial sector has grown faster than that in the other sectors since 1960 (Figure AI.7). The large increase in commercial electricity use between 1960 and 1973 (at a 9.5 percent average annual growth rate) has been attributed to increases in the use of mechanical air-conditioning systems and to new standards of building illumination, which resulted in increased lighting requirements.[6] From 1973 to 1980, commercial electricity use continued to increase, although the (five-year moving average) growth rate decreased. Since 1980 this growth rate has been increasing.

The declining growth of commercial sector electricity use during a period when output growth remained relatively strong points to improvements in the efficiency of use. New commercial buildings are generally constructed to be more energy efficient than were older buildings.[7] Improved lighting systems and the introduction of computerized energy management systems will also increase the efficiency of energy use. On the other hand, the increasing growth rate in recent years probably reflects a trend toward greater use of electric heating and the increased automation of office services.

Trends in the Industrial Use of Electricity

Electricity use in the industrial sector grew at an average rate of 8 percent per year between 1950 and 1960. From 1960 to 1973, the growth rate of industrial electricity use averaged 4.7 percent per year.[8] By 1980, the five-year moving average growth rate had become negative, but by 1985 it was back to the level of the mid-1970s.

Figure AI.7 Growth Rates of Electricity Sales and of Sectoral Output
Indicators, 1950–1985: (a) residential, (b) commercial, (c) industrial

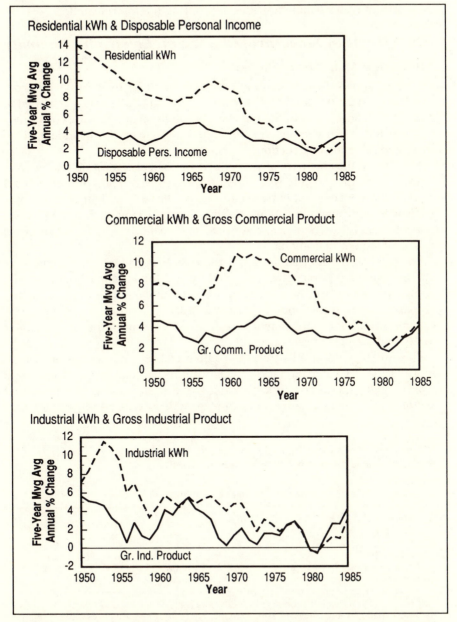

SOURCES: Based on data from Edison Electric Institute, *Statistical Yearbook of the Electric Utility Industry* (Washington, D.C.: EEI, various issues); U.S. Bureau of Economic Analysis, *The National Income and Product Accounts of the United States, Statistical Tables*, Supplement to *Survey of Current Business* (Washington, D.C.: GPO, various issues).

Electricity represents nearly 35 percent of the gross energy (13 percent of net energy) consumed in the industrial sector. Manufacturing accounts for about 85 percent of total electricity use in this sector, with agriculture, mining, and construction activities accounting for the remainder.

The Effects of Structural Change on Electricity Intensity

Measuring Structural Change

Table AI.3 shows that the share of GPO accounted for by industry fell from 38 percent in 1950 to about 32 percent in 1980, to which level it returned in 1987 after falling lower in the interim. Although manufacturing constitutes its most important component, the industrial figures also include agriculture, mining, and construction.

The GPO in the commercial sector, a very broad classification that encompasses all output originating outside the industrial sector, grew from 62 percent to 68 percent between 1950 and 1987. These figures encompass transportation; communications; electric, gas, and sanitary services; wholesale and retail trade; finance, insurance, and real estate; personal services; and government operations.

However, GPO composition is not the only way of looking at structural change. Another perspective is gained by looking at employment trends, shown in Table AI.4. As a share of total employment, industrial sector employment fell from about 45 percent in 1950 to about 27 percent in 1987. Commercial sector employment rose from 55 percent of total employment in 1950 to about 73 percent in 1987.

This information tells us that the commercial sector has been absorbing more of the growing labor force. This trend is especially true of part-time workers. On the other hand, the rapidly increasing share of commercial sector employment relative to output growth in that sector implies that

Table AI.3   Gross Product Originating (GPO) in the U.S. Economy, Selected Years, 1950–1987 (Percentage of Total)

Sector	1950	1960	1970	1980	1987
Industrial	38.0	36.0	34.2	32.0	32.1
Agriculture	5.5	4.4	3.2	2.7	2.5
Mining	2.1	1.8	1.8	1.5	3.1
Construction	5.5	6.3	5.0	3.7	4.6
Manufacturing	24.8	23.5	24.3	24.1	21.9
Commercial	62.0	64.0	65.8	68.0	67.9

SOURCES: U.S. Bureau of Economic Analysis, *The National Income and Product Accounts of the United States, Statistical Tables,* Supplement to *Survey of Current Business* (Washington, D.C.: GPO, various issues).
NOTE: All data have been rounded.

Table AI.4 Employment in the U.S. Economy, Selected Years, 1950–1987
(Percentage of Total)

Sector	1950	1960	1970	1980	1987
Industrial	44.9	38.9	34.7	28.7	26.8
Agriculture	10.9	7.0	4.0	3.3	2.8
Mining	1.6	1.1	0.8	1.0	0.7
Construction	5.8	5.3	5.3	5.1	5.7
Manufacturing	26.5	25.4	24.6	19.3	17.6
Commercial	55.1	61.1	65.3	71.3	73.2

SOURCES: U.S. Bureau of Economic Analysis, *The National Income and Product Accounts of the United States, Statistical Tables,* Supplement to *Survey of Current Business* (Washington, D.C.: GPO, various issues). Data series is "Persons Engaged in Production by Industry."
NOTE: All data have been rounded.

growth of labor productivity (that is, output per unit of labor input) has been slower in this sector than in the industrial sector.

Of these two different ways of looking at structural change, the more significant for analyzing electricity use is the GPO (value-added) measure, since it provides a measure of total productive activity — embracing both labor and capital inputs — within any particular sector. The GPO analysis shows manufacturing to have generally maintained its share of output over the postwar period. The shift in output has been from agriculture, mining, and construction toward selected commercial sector activities. The following two sections adopt the GPO as the best measure of structural change.

Electricity Intensity of the Sectors

Electricity intensity is defined as total kilowatthours (kWh) of electricity consumed divided by the aggregate economic output measure of a sector, that is, a measure of average electricity use per unit of output. Again, constant-dollar GPO is used to measure real output in the industrial and commercial sectors and constant-dollar DPI is used to measure real income for residential users. Industrial output, at 1.74 kWh per constant (1972) dollar, is now (1987) almost three times as electricity intensive as commercial output, at 0.60 kWh per constant (1972) dollar. The residential sector consumed 0.67 kWh per constant (1972) dollar of DPI in 1987.

For the postwar period as a whole, the intensity of electricity use in all three sectors has increased, as shown in Figure AI.8. Between 1950 and 1983, industrial electricity intensity increased 80 percent, commercial sector intensity increased more than 180 percent, and residential intensity increased about 260 percent. Electricity use was growing faster than output or income in every sector.

Although there were large increases in average electricity intensity over this total period, almost all of the growth occurred prior to 1973. Industrial

Figure AI.8   Electricity Intensities in the U.S. Economy, Total and by Sectors, 1947–1987

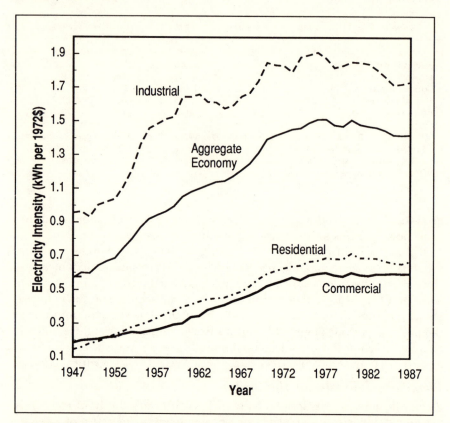

SOURCES: Based on data from Edison Electric Institute, *Statistical Yearbook of the Electric Utility Industry* (Washington, D.C.: EEI, various issues); U.S. Bureau of Economic Analysis, *The National Income and Product Accounts of the United States, Statistical Tables,* Supplement to *Survey of Current Business* (Washington, D.C.: GPO, various issues).

and commercial sector electricity intensities increased slightly in some years after 1973, but by 1983 they had fallen back to their 1973 levels. In the residential sector, electricity intensity was about 8 percent higher in 1983 than in 1973, but it has not shown an appreciable increase since 1977.

Changes in Energy Prices

Price Movements

The trend in energy prices for the forty-year period before 1973, as illustrated in Figure AI.9, was one of generally stable or decreasing prices for most fuels. Electricity prices, in particular, declined throughout the entire

Figure AI.9   Trends in Real Energy Prices to U.S. Personal Consumers, 1935–1987

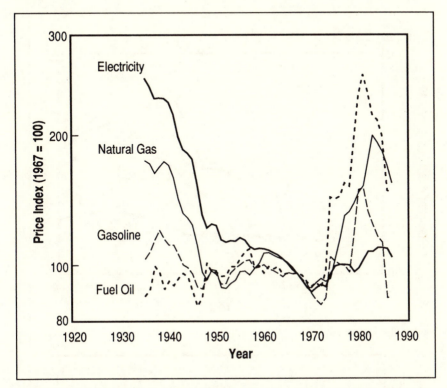

SOURCES: U.S. Bureau of the Census, *Historical Statistics of the United States, Colonial Times to 1970*, Bicentennial Ed. (Washington, D.C.: GPO, 1975); U.S. Bureau of Labor Statistics, *Handbook of Labor Statistics*, Bulletin 2000 (Washington, D.C.: GPO, 1979), and *Monthly Labor Review* (various issues).

period. The rapid price decline for electricity has been attributed to the increasing economies of scale in electricity generation and distribution over this period and to improvements in the efficiency (heat rate) of generation. Electricity prices were also favorably affected by the stability of primary energy input costs over the period.

Since 1973 a number of forces have combined to reverse the historical trend of declining electricity prices. First, there was the great increase in oil prices that accompanied the Arab oil embargo of 1973. This event drove petroleum product prices up immediately and also adversely affected the price of electricity in those regions of the country that depended on oil for a significant portion of their generation requirements. Figure AI.10 compares energy price trends over this period for personal consumers and industrial users. The 1973 change was followed during the rest of the 1970s by a

Figure AI.10 Trends in Real Energy Prices, 1967–1987

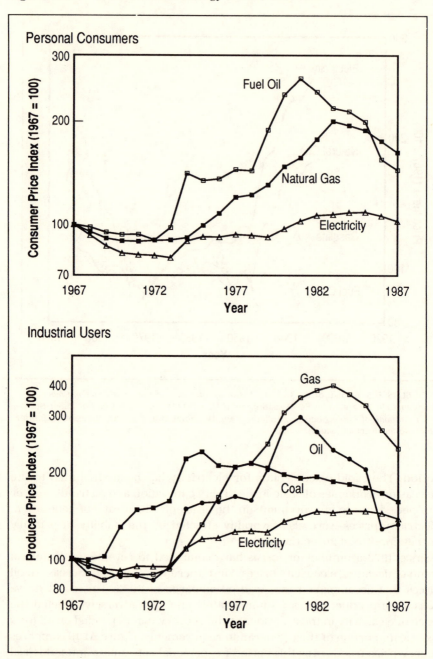

SOURCES: U.S. Bureau of Labor Statistics, *Handbook of Labor Statistics*, Bulletin 2000 and Bulletin 2175 (Washington, D.C.: GPO, 1979 and 1983), and *Monthly Labor Review* (various issues).

sustained rise in the price of natural gas and the second oil price shock of 1979 to 1980. The electric utility industry undertook programs during this period (with federal prodding) to cut back on both oil and natural gas as generation sources.

A second fundamental change occurred during this period: the apparent exhaustion of the economies of scale and improvements in heat rate that had led to lower per unit costs of generation over the longer period. Power plant construction projects were also being increasingly affected by inflation and delays so that the average cost of electricity generation in many utility systems has risen as many of these new plants have come on line. An additional factor was the sharp increase in environmental regulations for power plants during the 1970s and early 1980s.

Throughout this troubled period, however, electricity price increases, on average, were only moderate. In real terms, the consumer price index (CPI) for electricity rose only 18 percent between 1973 and 1983, while the indexes for both fuel oil and natural gas more than doubled. The same trends are evident for producers. The producer price index (PPI) for electricity rose 44 percent in real terms from 1973 to 1983, while the index for petroleum rose 136 percent and the index for natural gas more than tripled.

Price increases for electricity undoubtedly have led to increases in the efficiency of electricity utilization.[9] However, there has also been a trend toward greater electricity use resulting from the substitution of electricity for oil and gas, which have been increasing in price much faster than electricity.

Figure AI.11 shows that the ratios of electricity price to the prices of oil and natural gas continued their historical decreasing trend during much of the post-embargo period. Studies have shown that electric resistance heating becomes more cost effective in residences than fuel heating at existing furnace efficiencies when the ratio of electricity price to competing fuel prices reaches three or below.[10] In the aggregate, the ratio of electricity price to heating oil price has been below three between 1978 and 1985, while the ratio of electricity to natural gas price approached three in 1983 but has since turned upward as a result of higher electricity and lower natural gas prices. Of course, using electric heat pumps, electricity can be cost effective at electricity price ratios above three. The same price trends are apparent in the industrial sector. Although energy-using technologies in this sector are quite diverse, studies have shown that the same cost-effective price ratio thresholds for electricity and oil and electricity and natural gas were achieved in several materials processing industries in the early 1980s.[11]

Figure AI.11   Electricity Price Ratios in the United States, 1960–1987

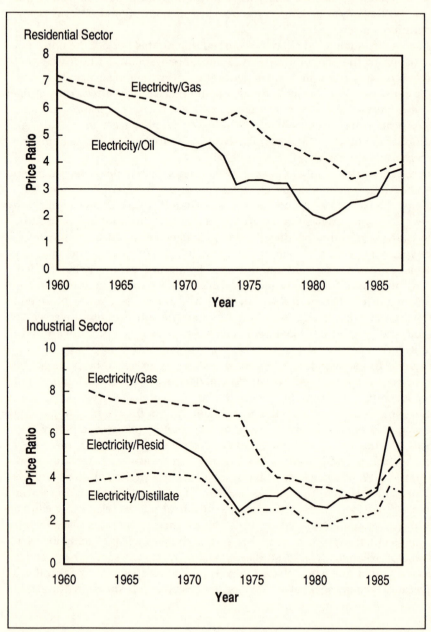

SOURCES: Based on data from Edison Electric Institute, *Statistical Yearbook of the Electric Utility Industry* (Washington, D.C.: EEI, various issues); American Gas Association, *Gas Facts* (Washington, D.C.: AGA, various issues); U.S. Bureau of the Census, *Annual Survey of Manufactures: Fuels and Electric Energy Consumed* (Washington, D.C.: GPO, various years); U.S. Energy Information Administration, *Monthly Energy Review* (various issues).

NOTE: Price ratio calculated as ratio of actual prices in dollars per million Btu.

Price Elasticities

The basically favorable relative price ratios of electricity to other fuels must be weighed against the fact that electricity price increases themselves also tend to discourage electricity use. It is not simply a matter of dividing a fixed energy market among electricity and other energy forms; the total market for energy may be diminished as a result of rises in all energy prices.

Econometricians have tried for years to disentangle the complex of own-price elasticity, cross-price elasticity (with other energy forms), delays in price responses, and nonprice factors such as income and changes in end-use technology. Powerful statistical tools are employed, but the results leave much to be desired. The available data base does not yet cover enough experience with high energy prices, and, in addition, theory provides very little guidance in this complex task.

Bohi[12] provides a critical review of the methods, data, and results of the leading econometric analyses of the demand for energy forms. He finds that even for electricity, which has been subject to extensive study, "there is wide disagreement about the responsiveness of demand to changes in prices and incomes, and surprisingly broad gaps in the understanding of the nature of this process."[13]

Each sector is considered separately by Bohi. In twenty-five studies of the residential sector, the spread of long-range, own-price elasticities was found to be −0.45 to −1.89, with a consensus value of −1.0 (that is to say, a 10 percent price increase would produce a 10 percent consumption decrease). However, after considering the methods and data employed in the studies, Bohi concludes that the best estimate for long-range residential price elasticity is −0.70[14] (long-range effects are usually considered as achieved in up to ten years).

Five studies on the commercial sector are reported. Some used more than one approach, so that there are nine different sets of results. Price elasticities ranged from −1.0 to −1.60. Bohi declines to choose the most likely value, saying simply that "commercial demand appears to be price elastic in the long run. . . ."[15]

Review of a broad range of industrial elasticity demand studies, which used a variety of approaches, yielded a range from −0.51 to −1.82 with a consensus estimate "around −1.30." But Bohi notes that "one has difficulty in placing much confidence in the consensus estimate."[16] His own judgment, after examining the various studies, is that the price elasticity of industrial demand is between −0.5 and −1.0.[17]

Sweeney[18] concludes that "the long-run delivered price elasticity of demand for electricity probably exceeds [that is, is more negative than] unity but may be as low as −0.7."[19]

Bohi discusses cross-elasticity estimates but does not present numerical values. He notes that problems in the data tend to make estimates of cross elasticities unreliable.

The most that can be safely concluded, therefore, is that own- and cross-price elasticities exist that are nonnegligible, but they are hard to establish precisely. As a result there are counteracting price influences on electricity demand — in particular, electricity's own price and electricity's price movements compared with those of other energy forms. In addition, of course, there are the sizeable effects on the growth in electricity demand produced by the overall growth in the national output of goods and services. The net aggregate effects of all of these forces are assessed in the next section comparing pre- and post-embargo trends in electricity consumption.

CONTINUITY AND CHANGE: PRE- AND POST-EMBARGO TRENDS

The foregoing discussion shows that growth rates of electricity use have slowed in recent years from the high growth rates of earlier periods. GNP growth, averaged over recent years, has also slowed from the higher rates achieved over most of the postwar period.

In light of the strong association that has long been observed between electricity and the GNP, viewing electricity growth rates only with respect to time can give misleading impressions. Nevertheless, the ratio of electricity growth rates to GNP growth rates has been gradually declining (see Table AI.5), a point to which many analysts have drawn attention. Although this trend is consistent in principle with a linear relationship between electricity use and the GNP, the question remains whether the degree and rate of convergence are consistent with the trend that has characterized the entire postwar period.

Electricity price changes are frequently cited as the reason for a shift in the relationship. The econometric studies summarized above show that when the price of electricity increases it tends to slow the growth of electricity demand. However, the more recent historical period over which these statistical analyses were performed also contained the counteracting influences of rising competing fuel prices, which tend to counterbalance to some degree the effect of the electricity price increases. The extent of the competing fuel price influences on the historical relationship is not well established.

Our examination of the data leads us to believe that by far the most important contributor to the slower growth rates in electricity demand over the last decade has been lower economic growth. Others have come to a similar conclusion. The econometric analysis of Hogan[20] shows that the

Table AI.5 Average Annual Growth Rates of Electricity and the Gross National Product (GNP) and Their Ratios Over Selected Postwar Periods

Period	Electricity growth rate	GNP growth rate	Ratio of electricity and the GNP growth rates
1947–1960	8.07	3.52	2.29
1960–1973	6.70	4.17	1.61
1973–1983	1.99	2.04	0.98
1983–1987	3.31	4.07	0.81
1973–1987	2.36	2.44	0.97

SOURCES: Based on data from Edison Electric Institute, *Statistical Yearbook of the Electric Utility Industry* (Washington, D.C.: EEI, various issues); U. S. Bureau of Economic Analysis, *The National Income and Product Accounts of the United States, Statistical Tables*, Supplement to *Survey of Current Business* (Washington, D.C.: GPO, various issues).

primary reason for the lower growth rates in electricity demand during the 1970s was slower economic growth. He attributes only about 30 percent of the drop in electricity demand since 1972 to electricity price increases. The Edison Electric Institute[21] reached similar conclusions regarding the magnitude of price effects at the aggregate level. However, Hogan notes that his results capture "only part of the eventual adjustment we can expect in the gradual replacement of energy-using equipment."[22] Thus, it can be expected that the energy price changes already experienced will continue to affect demand growth in the future.

The central question is, of course, what the net effect of all factors — price, income, structural change, technological advance, and so forth — has been on electricity demand in recent years and what these influences portend for the future. In our judgment, at the present time there is no clear answer to this question. Figure AI.12.b identifies three different possible interpretations (depicted by lines A, B, and C) of the recent trend in electricity use as a function of the GNP.

The first interpretation (line A) in Figure AI.12.b is that no shift has occurred in the underlying long-term relationship but that the data for recent years represent a "down phase" of the cycle that has persistently characterized the long-term relationship. In fact, over the postwar period it could have been inferred several times that shifts had occurred in the slope of the relationship given its cyclic movements (Figure AI.12.a). Nevertheless, these inferences would have been incorrect, as shown by data for subsequent years. In 1987 and 1988, the growth rate of electricity exceeded that of the GNP, and if the cycle continues as before, there will be a long period when the consumption of electricity will exceed trend values. However, if the growth rate of electricity use in the next few years does not continue to

Figure AI.12 Electricity versus the GNP: (a) The 1947–1983 Record, (b) Some Possible Future Relationships

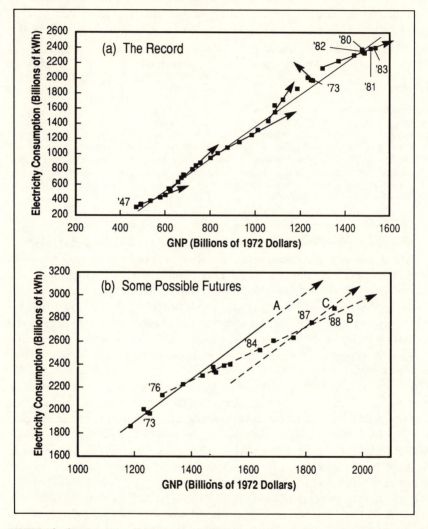

NOTES: The long-term trend line in figure (a) is the same as in Figure AI.4. The lines with the arrowheads in this figure indicate how the trend seemed to be changing at various times in the past based on short-term movements of the data. However, these movements turned out to be aberrations, and there was always a reversion to the underlying long-term trend.

Figure (b) depicts three possible interpretations of the recent past, none of which can be proved or disproved at the present time. Line A is a continuation of the basic long-term trend shown in figure (a); line B is a continuation of the short-term trend starting in 1976: line C is a new trend line. Line C assumes that the decline in average electricity intensity that occurred through 1984 represents a lasting change but that incremental intensities will revert to the basic trend (see text for further discussion and for a historical parallel). Cyclic and random variations (e.g., because of weather) around any future trend line will still occur, no matter what the trend line turns out to be.

Figure AI.13 Gross Energy Use by Economic Sector, 1960–1984

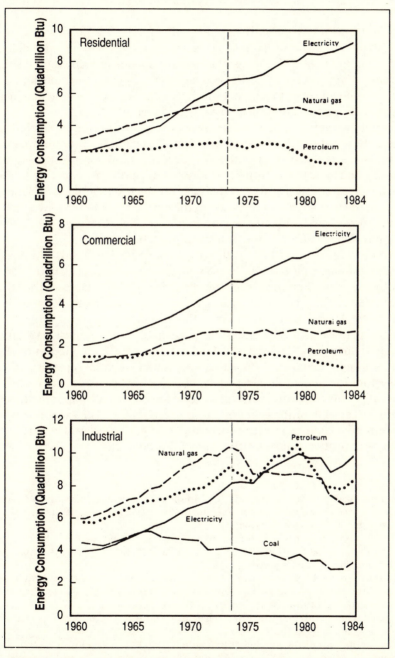

SOURCES: U.S. Energy Information Administration, *State Energy Data Report*, DOE/EIA-0214(83) (Washington, D.C.: U.S. GPO, 1983), and *Monthly Energy Review* (various issues).

exceed the growth rate of the GNP (that is, if the "down phase" is not succeeded by an "up phase"), this interpretation must be considered incorrect.

A second interpretation is that a permanent shift has occurred in the relationship, one toward a diminished increase in electricity use per dollar increment in the GNP. This interpretation corresponds to a downward shift in the slope of the electricity-GNP trend line, and in fact the relationship can be read in such a way as to support this belief (Figure AI.12.b, line B, which is an extension of the uppermost arrow on Figure AI.12.a).

Still another interpretation is that the increase in the rate of structural change between the industrial and commercial sectors (and within manufacturing) in recent years will neither be corrected nor proceed at the same rate in the future. If in the future the structural shift were to revert to the slower historical postwar rate and the sectoral electricity intensity relationships continue to hold, then the effect on the electricity-GNP trend would be a parallel downward shift in the postwar trend line (an intercept shift), leaving the slope coefficient intact (Figure AI.12.b, line C).[23]

It will be several years before these questions are resolved. The post-embargo years are still too few to provide definitive answers about trend shifts. In the meantime, however, the historical record suggests that the electrification of the economy will continue. Indeed, electricity use has continued to increase in all sectors over the post-embargo period while fuel consumption more generally either has been stable or, as in most cases, has fallen, as is shown in Figure AI.13. Furthermore, our examination of the major consuming sectors indicates that substantial potential remains within these sectors for the continued penetration of electricity in many uses.

Generally, the rates of growth in electricity use will depend on the strength and growth of the economy. That much is clear. The exact quantification of this relationship for the current period and its relevance to future trends are important questions that remain to be settled.

Notes

1. National Academy of Engineering, Commission on Engineering and Technical Systems, *Electricity in Economic Growth* (Washington, D.C.: National Academy Press, 1986), Chapter 2, "Historical Perspectives."

2. A regression is the best relationship of a given functional type between two (or more) correlated variables as judged by a particular statistical criterion, such as the criterion of ordinary least squares.

3. In other words, because of the nonzero intercept in the relationship (which appears as the offset of the regression line from the origin in Figure AI.4), the percentage growths of electricity and the GNP along the regression line are more nearly equal where both quantities are large, as in the later years, than where both quantities are small, as in earlier years.

4. These definitions differ somewhat from the Edison Electric Institute (EEI) sector definitions. However, for our purposes, the differences are not great enough to

warrant concern, and so we have used the EEI statistics, with no change, to fit our categories.

5. Gross product originating (GPO), the statistic used for both commercial and industrial outputs, is a measure of value added derived from the national income and product accounts; it emphasizes the sectoral origin of gross national product (GNP).

6. Solar Energy Research Institute, *A New Prosperity: Building a Sustainable Energy Future* (Andover, MA: Brick House Publishing, 1981).

7. U.S. Energy Information Administration, *Energy Conservation Indicators, 1983 Annual Report*, DOE/EIA-0441(83) (Washington, D.C.: U.S. GPO, October 1983).

8. The electricity consumption of the federal government's gaseous diffusion plants for uranium enrichment represented about 20 percent of industrial sector demand growth in the 1950s but has declined since then, except for a brief increase in the early and mid-1970s. Uranium enrichment operations are not expected to have an important effect on industrial electricity consumption in the near future. Excluding electricity consumption for uranium enrichment, the average annual growth rates of industrial sector electricity demand are 6.8 percent per year from 1950 to 1960, 5.5 percent per year from 1960 to 1973, and 0.7 percent per year from 1973 to 1983.

9. Edison Electric Institute, "What Happened to the 40 Percent?," *Electric Perspectives* (Spring 1984), 28–31.

10. C. C. Burwell, et al. *Electric Home Heating: Substitution for Oil and Gas*, ORAU/IEA-82-3(M) (Oak Ridge, TN: Institute for Energy Analysis, Oak Ridge Associated Universities, March 1982).

11. C. C. Burwell, *Industrial Electrification: Current Trends*, ORAU/IEA-83-4(M) (Oak Ridge, TN: Institute for Energy Analysis, Oak Ridge Associated Universities, February 1983).

12. D. R. Bohi, *Analyzing Demand Behavior: A Study of Energy Elasticities* (Baltimore, MD: Johns Hopkins University Press for Resources for the Future, 1981).

13. Ibid., 55.

14. Ibid., Table 7.1, 159.

15. Ibid., 79.

16. Ibid., 90

17. Ibid., Table 7.1

18. J. S. Sweeney, "The Response of Energy Demand to Higher Prices: What Have We Learned?," *AEA Papers and Proceedings* 74, no. 2:31–37.

19. Ibid., 36.

20. W. W. Hogan, *Patterns of Energy Use*, E-84-04 (Cambridge: Energy and Environmental Policy Center, Harvard University, May 1984).

21. Edison Electric Institute, "What Happened to the 40 Percent?" 28–31.

22. Hogan, *Patterns of Energy Use*, 27.

23. If this result occurs, it would be a mirror image of the transition from the relationship of the 1920s to that of the 1930s. Incremental electricity intensity did not change between the two periods, as can be seen from the parallel lines of regression for the two periods (Figure AI.3). Because the percentage decreases in the GNP in the years immediately following 1929 were larger than the percentage decreases in electricity use, average electricity intensity actually increased. However, as the years passed the trend of the 1930s turned out to be parallel to that of the 1920s. The net result of the change in average intensity was simply a shift in the line parallel to itself (mathematically, a shift of the intercept). At the time that the shift occurred, there was of course no way of knowing that this outcome would result. Likewise, only time will tell whether the line of the future will fall below and

parallel to the line representing 1947 through 1973. Such behavior would reflect a decrease in average intensity as a result of price effects, conservation, and the permanent decline of some energy-intensive industries, with any lingering effects of past price rises being offset by the tendencies toward increased electrification already discussed.

Appendix II:
Basic Statistical Data:
Long-Term Quantitative Trends
Sidney Sonenblum

Introduction

The long-term quantitative analyses in Chapter 12 and 13 are based on the statistics presented in this appendix. In Tables AII.1 through AII.7 data are shown for several variants of the U.S. economy:

- The **total U.S. economy** is defined to include all private business sectors in the economy plus the government and household sectors.
- The **nonresidential economy** is defined to include the private business and government sectors, but excludes households.
- The **private domestic business economy** is defined to include only privately owned, profit-oriented enterprises.
- **Manufacturing** is defined to include only the manufacturing industries (i.e., *Standard Industrial Classification* industries 20–39).

Data are shown only for business cycle peak years because in delineating long-term trends we wanted to remove the influence of short-term fluctuations resulting from cyclical factors. Peak years are dated in accordance with the National Bureau of Economic Research (NBER) reference system, as reported in *Business Conditions Digest* (U.S. Department of Commerce, February 1982). A peak year is defined as the year in which the NBER peak month occurs, except that 1979 is associated with the January 1980 peak month. Also, the final year covered in our data series — 1985 — does not include an NBER-reported peak month. In some cases (such as 1920) a peak month occurs in a year which, on an overall basis, may not have been at a cyclical peak.

The data shown were drawn primarily from the sources listed at the end of this appendix. Numerous gaps in the data were encountered, particularly in the very early years. Also, some of the basic statistical sources, such as the periodic censuses, are available only for certain years. Problems also resulted from changes in the definitions and measurement approaches for various classes of data over time, along with shifts in responsibility among the reporting agencies.

As a result, in developing our long-term statistical series, it was necessary to make estimates in order to fill in gaps and to deal with inconsistencies among the data sources. The notes to these tables report on most of our estimating procedures. A detailed explanation of the steps followed in developing the statistics shown is contained in the underlying data files for this study at EPRI.

ENERGY CONSUMPTION AND ENERGY SERVICES

Tables AII.1, AII.2, and AII.3

Tables AII.1 and AII.2 cover data for the consumption of total energy, non-electric energy, and electricity in the total U.S. economy (Tables AII.1.1 and AII.2.1), in the nonresidential economy (Tables AII.1.2 and AII.2.2) and in manufacturing (Tables AII.1.3 and AII.2.3). Tables AII.3 shows data for the manufacturing sector only, including energy services (Table AII.3.1), electricity services (Table AII.3.2) and mechanical power capacity (Table AII.3.3).

In each of these tables, panel A presents the data in Btus for 1899 to 1985 business cycle peak years; panel B shows indexes for the same years; and panel C presents the intercyclical average annual growth rates. The energy and electricity time series in panel A (Tables AII.1A, AII.2A, AII.3A) contain discontinuities in the years 1948 and 1973, which we have shown by entering two estimates for 1948 (i.e., 1948E and 1948U) and for 1973 (i.e., 1973U and 1973R). We deal with these discontinuities by using ratios of the two entries in each overlapping year for adjusting the index numbers in panel B (Tables AII.1B, AII.2B, AII.3B) so as to link all the years into a continuous time series.[1] The intercyclical average annual growth rates shown in panel C (Tables AII.1C, AII.2C, AII.3C) are derived from these indexes.

The 1899 through 1948E estimates were mostly based on Bureau of Mines (BoM) data while the 1948U through 1985 estimates were based on a more detailed set of data provided by the Energy Information Administration (EIA). The methodology used to develop the 1899 through 1948E estimates differ in a number of ways from the methodology used to develop the 1948U to 1973U estimates. These methodological differences are the major reason for the 1948 discontinuity.[2] The 1973 discontinuity occurs because of EIA methodological revisions that were carried forward from 1973R to 1985 but were not incorporated in the 1948U to 1973U data.[3]

Tables AII.1
Energy Use in the Total U.S.
Economy, the Nonresidential U.S.
Economy, and Manufacturing

Table AII.1.1   Energy Use in the Total U.S. Economy

Panel A: Trillion Btu

| Year | Gross energy | Gross nonelec. | Gross elec. | Gross electricity from | |
				Utilities	Industry
	(1)	(2)	(3)	(4)	(5)
1899	6913	6483	430	182	248
1903	10245	9548	697	292	405
1907	13831	12700	1131	470	662
1910	14800	13444	1356	605	751
1913	16719	15095	1624	798	826
1918	20436	18288	2148	1341	807
1920	19768	17442	2326	1620	705
1923	21667	19305	2362	1694	667
1926	22471	19944	2527	1860	667
1929	23720	20994	2726	2153	574
1937	22713	19742	2971	2412	559
1944	31724	26689	5035	4110	925
1948E	33862	27838	6024	5056	968
1948U	33862	27848	6014	5056	958
1953	36780	29779	7001	6010	991
1957	40810	32299	8511	7490	1021
1960	44080	34890	9190	8230	960
1969	64530	48110	16420	15250	1170
1973U	74610	53503	21107	20010	1097
1973R	74282	53364	20918	19853	1065
1979	78897	54030	24867	24128	739
1981	73991	48557	25434	24760	674
1985	73964	46501	27463	26482	981

NOTE: Col. 1 = col. 2 + col. 3. Col. 3 = col. 4 + col. 5. The data have been rounded.

Panel B: Index: 1979 = 100

1899	8.7	12.0	1.7	0.7	32.3
1903	12.9	17.6	2.8	1.2	52.7
1907	17.5	23.5	4.5	1.9	86.1
1910	18.7	24.8	5.4	2.5	97.7
1913	21.1	27.9	6.5	3.3	107.5
1918	25.8	33.8	8.5	5.5	105.0
1920	24.9	32.2	9.3	6.7	91.8
1923	27.3	35.7	9.4	7.0	86.8
1926	28.4	36.8	10.1	7.6	86.8
1929	29.9	38.8	10.8	8.9	74.6
1937	28.7	36.5	11.8	9.9	72.7
1944	40.0	49.3	20.0	16.9	120.3
1948E	42.7	51.4	24.0	20.8	125.9
1948U	42.7	51.4	24.0	20.8	125.9
1953	46.4	55.0	27.9	24.7	130.3
1957	51.5	59.6	33.9	30.8	134.1
1960	55.6	64.4	36.6	33.8	126.1
1969	81.4	88.8	65.4	62.7	153.7
1973U	94.2	98.8	84.1	82.3	144.1
1973R	94.2	98.8	84.1	82.3	144.1
1979	100.0	100.0	100.0	100.0	100.0
1981	93.8	89.9	102.3	102.6	91.2
1985	93.7	86.1	110.4	109.8	132.8

Table AII.1.1 (Continued)

Panel C: Annual percentage growth rates

Period	*Gross energy*	*Gross nonelec.*	*Gross elec.*	Gross electricity from	
				Utilities	*Industry*
	(1)	*(2)*	*(3)*	*(4)*	*(5)*
1899–1985	2.8	2.3	5.0	6.0	1.7
1899–1920	5.1	4.8	8.4	11.0	5.1
1899–1913	6.5	6.2	10.0	11.1	9.0
1899–1903	10.3	10.2	12.9	12.6	13.0
1903–1907	7.8	7.4	12.9	12.6	13.0
1907–1910	2.3	1.9	6.2	8.8	4.3
1910–1913	4.1	3.9	6.2	9.6	3.2
1913–1920	2.4	2.1	5.3	10.7	−2.2
1913–1918	4.1	3.9	5.8	11.0	−0.5
1918–1920	−1.6	−2.3	4.1	9.9	−6.5
1920–1948	1.9	1.7	3.5	4.1	1.1
1920–1929	2.0	2.1	1.8	3.2	−2.3
1920–1923	3.1	3.4	0.5	1.5	−1.8
1923–1926	1.2	1.1	2.3	3.2	0.0
1926–1929	1.8	1.7	2.6	5.0	−4.9
1929–1948	1.9	1.5	4.3	4.6	2.8
1929–1937	−0.5	−0.8	1.1	1.4	−0.3
1937–1944	4.9	4.4	7.8	7.9	7.5
1944–1948	1.6	1.1	4.6	5.3	1.1
1948–1985	2.1	1.4	4.2	4.6	0.1
1948–1960	2.2	1.9	3.6	4.1	0.0
1948–1953	1.7	1.3	3.1	3.5	0.7
1953–1957	2.6	2.1	5.0	5.7	0.7
1957–1960	2.6	2.6	2.6	3.2	−2.0
1960–1973	4.1	3.3	6.6	7.1	1.0
1960–1969	4.3	3.6	6.7	7.1	2.2
1969–1973	3.7	2.7	6.5	7.0	−1.6
1973–1985	0.0	−1.1	2.3	2.4	−0.7
1973–1981	0.0	−1.2	2.5	2.8	−5.6
1973–1979	1.0	0.2	2.9	3.3	−5.9
1979–1981	−3.2	−5.2	1.1	1.3	−4.5
1981–1985	0.0	−1.1	1.9	1.7	9.9

Table AII.1.2   Energy Use in the Nonresidential U.S. Economy

Panel A: Trillion Btu

Year	Gross energy	Gross nonelec.	Gross elec.	Gross electricity from	
				Utilities	Industry
	(1)	(2)	(3)	(4)	(5)
1899	5028	4610	418	170	248
1903	7721	7047	674	269	405
1907	10577	9491	1086	424	662
1910	11472	10172	1300	549	751
1913	13079	11524	1555	729	826
1918	16225	14189	2036	1229	807
1920	15752	13580	2172	1467	705
1923	17523	15339	2184	1517	667
1926	18057	15744	2313	1646	667
1929	19199	16742	2457	1883	574
1937	18520	15969	2551	1992	559
1944	25068	20764	4304	3379	925
1948E	26177	21397	4780	3812	968
1948U	26177	21397	4780	3822	958
1953	27486	22128	5358	4367	991
1957	29638	23246	6392	5371	1021
1960	31145	24446	6699	5739	960
1969	44194	32921	11273	10103	1170
1973U	50249	36190	14059	12962	1097
1973R	49921	36051	13870	12805	1065
1979	53516	36886	16630	15891	739
1981	49735	33015	16720	16046	674
1985	48844	30875	17969	16988	981

NOTE: Col. 1 = col. 2 + col. 3. Col. 3 = col. 4 + col. 5. All data have been rounded.

Panel B: Index: 1979 = 100

1899	9.3	12.5	2.5	1.1	32.3
1903	14.3	19.0	4.0	1.7	52.7
1907	19.7	25.4	6.4	2.6	86.1
1910	21.3	27.5	7.7	3.4	97.7
1913	24.3	31.1	9.3	4.5	107.5
1918	30.1	38.4	12.0	7.6	105.0
1920	29.2	36.7	12.9	9.1	91.8
1923	32.5	41.4	12.9	9.4	86.8
1926	33.5	42.5	13.7	10.2	86.8
1929	35.7	45.2	14.6	11.7	74.6
1937	34.4	43.1	15.1	12.4	72.7
1944	46.5	56.1	25.6	21.0	120.3
1948E	48.6	57.8	28.3	23.7	125.9
1948U	48.6	57.8	28.3	23.8	125.9
1953	51.1	59.8	31.8	27.2	130.3
1957	55.1	62.7	37.9	33.4	134.1
1960	57.9	66.0	39.8	35.7	126.1
1969	82.1	88.9	66.9	62.8	153.7
1973U	93.3	97.7	83.4	80.6	144.1
1973R	93.3	97.7	83.4	80.6	144.1
1979	100.0	100.0	100.0	100.0	100.0
1981	93.0	89.5	100.5	101.0	91.2
1985	91.3	83.7	108.1	106.9	132.8

Table AII.1.2 (Continued)

Panel C: Annual percentage growth rates

Period	Gross energy	Gross nonelec.	Gross elec.	Gross electricity from	
				Utilities	Industry
	(1)	(2)	(3)	(4)	(5)
1899–1985	2.7	2.2	4.5	5.5	1.7
1899–1920	5.6	5.3	8.2	10.8	5.1
1899–1913	7.1	6.8	9.8	11.0	9.0
1899–1903	11.3	11.2	12.7	12.2	13.0
1903–1907	8.2	7.7	12.7	12.0	13.0
1907–1910	2.7	2.3	6.2	9.0	4.3
1910–1913	4.5	4.2	6.2	9.9	3.2
1913–1920	2.7	2.4	4.9	10.5	− 2.2
1913–1918	4.4	4.2	5.5	11.0	− 0.5
1918–1920	− 1.5	− 2.2	3.3	9.3	− 6.5
1920–1948	1.8	1.6	2.9	3.5	1.1
1920–1929	2.2	2.4	1.4	2.8	− 2.3
1920–1923	3.6	4.1	0.2	1.1	− 1.8
1923–1926	1.0	0.9	1.9	2.8	0.0
1926–1929	2.1	2.1	2.0	4.6	− 4.9
1929–1948	1.6	1.3	3.6	3.8	2.8
1929–1937	− 0.4	− 0.6	0.5	0.7	− 0.3
1937–1944	4.4	3.8	7.8	7.8	7.5
1944–1948	1.1	0.8	2.7	3.1	1.1
1948–1985	1.7	1.0	3.6	4.1	0.1
1948–1960	1.5	1.1	2.9	3.5	0.0
1948–1953	1.0	0.7	2.3	2.7	0.7
1953–1957	1.9	1.2	4.5	5.3	0.7
1957–1960	1.7	1.7	1.6	2.2	− 2.0
1960–1973	3.7	3.1	5.9	6.5	1.0
1960–1969	4.0	3.4	6.0	6.5	2.2
1969–1973	3.3	2.4	5.7	6.4	− 1.6
1973–1985	− 0.2	− 1.3	2.2	2.4	− 0.7
1973–1981	0.0	− 1.1	2.4	2.9	− 5.6
1973–1979	1.2	0.4	3.1	3.7	− 5.9
1979–1981	− 3.5	− 5.4	0.3	0.5	− 4.5
1981–1985	− 0.5	− 1.7	1.8	1.4	9.9

Table AII.1.3   Energy Use in Manufacturing

Panel A: Trillion Btu

Year	Gross energy	Gross nonelec.	Gross elec.	Gross electricity from		Primary energy
				Utilities	Manufacturing	
	(1)	(2)	(3)	(4)	(5)	(6)
1899	2890	2793	97	51	46	2839
1903	3850	3645	205	90	114	3760
1907	5271	4893	378	160	218	5111
1910	5438	4927	511	222	289	5216
1913	5961	5273	688	314	374	5648
1918	8041	6958	1083	597	487	7445
1920	7789	6639	1150	673	476	7115
1923	7775	6636	1139	654	484	7121
1926	7932	6582	1350	837	513	7095
1929	7934	6491	1443	960	483	6974
1937	7339	5784	1555	1024	530	6314
1944	10938	7861	3077	2180	897	8758
1948E	11082	7783	3299	2364	934	8717
1948U	12418	9129	3289	2364	925	10054
1953	14036	10373	3664	2713	951	11323
1957	14837	10487	4351	3373	977	11464
1960	15642	11235	4407	3487	919	12154
1969	22124	15236	6888	5768	1120	16356
1973U	24730	16523	8206	7156	1050	17574
1973R	26431	18311	8120	7100	1020	19331
1979	26369	17677	8691	7984	708	18385
1981	23571	15248	8323	7678	645	15893
1985	21637	13217	8419	7479	940	14158

NOTE: Col. 1 = col. 2 + col. 3; col. 6 + col. 4. Col. 3 = col. 4 + col. 5. Col. 6 = col. 2 + col. 5.
All data have been rounded.

Panel B: Index: 1979 = 100

Year	(1)	(2)	(3)	(4)	(5)	(6)
1899	13.1	20.5	1.1	0.6	6.3	19.6
1903	17.5	26.8	2.3	1.1	15.6	25.9
1907	23.9	36.0	4.3	2.0	29.6	35.3
1910	24.7	36.2	5.8	2.8	39.3	36.0
1913	27.1	38.8	7.8	3.9	50.8	39.0
1918	36.5	51.2	12.3	7.4	66.1	51.4
1920	35.4	48.8	13.1	8.4	64.7	49.1
1923	35.3	48.8	12.9	8.1	65.8	49.1
1926	36.0	48.4	15.3	10.4	69.7	49.0
1929	36.0	47.7	16.4	11.9	65.6	48.1
1937	33.3	42.5	17.6	12.7	72.0	43.6
1944	49.7	57.8	34.9	27.1	121.9	60.4
1948E	50.3	57.2	37.4	29.4	126.9	60.2
1948U	50.3	57.2	37.4	29.4	126.9	60.2
1953	56.9	65.0	41.7	33.7	130.5	67.7
1957	60.1	65.7	49.5	41.9	134.1	68.6
1960	63.4	70.4	50.2	43.3	126.1	72.7
1969	89.7	95.5	78.4	71.7	153.7	97.9
1973U	100.2	103.6	93.4	88.9	144.1	105.1
1973R	100.2	103.6	93.4	88.9	144.1	105.1
1979	100.0	100.0	100.0	100.0	100.0	100.0
1981	89.4	86.3	95.8	96.2	91.2	86.4
1985	82.1	74.8	96.9	93.7	132.8	77.0

Table AII.1.3 (Continued)

Panel C: Annual percentage growth rates

| Period | Gross energy | Gross nonelec. | Gross elec. | Gross electricity from | | Primary energy |
				Utilities	Manufacturing	
	(1)	*(2)*	*(3)*	*(4)*	*(5)*	*(6)*
1899–1985	2.2	1.5	5.3	6.0	3.6	1.6
1899–1920	4.8	4.2	12.5	13.1	11.8	4.5
1899–1913	5.3	4.6	15.0	13.9	16.1	5.0
1899–1903	7.4	6.9	20.5	15.4	25.6	7.3
1903–1907	8.2	7.6	16.6	15.4	17.5	8.0
1907–1910	1.0	0.2	10.6	11.5	9.9	0.7
1910–1913	3.1	2.3	10.4	12.3	9.0	2.7
1913–1920	3.9	3.3	7.6	11.5	3.5	3.4
1913–1918	6.2	5.7	9.5	13.7	5.4	5.7
1918–1920	− 1.6	− 2.3	3.0	6.3	− 1.1	− 2.2
1920–1948	1.3	0.6	3.8	4.6	2.4	0.7
1920–1929	0.2	− 0.3	2.6	4.0	0.2	− 0.2
1920–1923	− 0.1	0.0	− 0.3	− 1.0	0.6	0.0
1923–1926	0.7	− 0.3	5.8	8.6	1.9	− 0.1
1926–1929	0.0	− 0.5	2.3	4.7	− 2.0	− 0.6
1929–1948	1.8	1.0	4.4	4.9	3.5	1.2
1929–1937	− 1.0	− 1.4	0.9	0.8	1.2	− 1.2
1937–1944	5.9	4.5	10.2	11.4	7.8	4.8
1944–1948	0.3	− 0.2	1.7	2.0	1.0	− 0.1
1948–1985	1.3	0.7	2.6	3.2	0.1	0.7
1948–1960	1.9	1.7	2.5	3.3	0.0	1.6
1948–1953	2.5	2.6	2.2	2.8	0.6	2.4
1953–1957	1.4	0.3	4.4	5.6	0.7	0.3
1957–1960	1.8	2.3	0.4	1.1	− 2.0	2.0
1960–1973	3.6	3.0	4.9	5.7	1.0	2.9
1960–1969	3.9	3.4	5.1	5.8	2.2	3.4
1969–1973	2.8	2.0	4.5	5.5	− 1.6	1.8
1973–1985	− 1.7	− 2.7	0.3	0.4	− 0.7	− 2.6
1973–1981	− 1.4	− 2.3	0.3	1.0	− 5.6	− 2.4
1973–1979	0.0	− 0.6	1.1	2.0	− 5.9	− 0.8
1979–1981	− 5.5	− 7.1	− 2.1	− 1.9	− 4.5	− 7.0
1981–1985	− 2.1	− 3.5	0.3	− 0.7	9.9	− 2.8

Notes to Tables AII.1.1, AII.1.2, and AII.1.3

col. 1: *Gross energy* refers to the calorific (Btu) content attributed to all of the primary energy sources including those converted to electricity. Excluded are direct mechanical applications of wind and water power as well as wood and waste (except as used by electric utilities).

col. 2: *Gross nonelectric energy* refers to Btus contained in the primary energy sources, excluding those used to generate electricity.

cols. 3–5: *Gross electricity* refers to the energy inputs into electricity generation. It, therefore, includes not only the Btu content of the kilowatthours of electricity used, but also the energy content that is "lost" (see Table AII.2, col. 4) in the process of generating and transmitting electricity. It includes the energy input in the *electricity generated by utilities* (Table AII.1.1, col. 4) plus the energy input in the *electricity generated by industrial establishments* (Table AII.1.1, col. 5).

col. 6: *Primary energy* and gross energy are equivalent for the total U.S. economy and can be defined as the sum of domestic energy production plus the net of energy imports less exports and the net change in energy inventories. However, for a particular energy-using sector gross energy and primary energy are not equivalent because of the sector's electricity purchases from utilities. Therefore, primary energy in manufacturing includes the gross input of energy used directly (col. 2) as well as the energy content of fuels used to generate its own electricity (col. 5).

Tables AII.2
***Electricity Use in the Total U.S.
Economy, the Nonresidential U.S.
Economy, and Manufacturing***

Table AII.2.1 Electricity Use in the Total U.S. Economy

Panel A: Trillion Btu

Year	Usable electricity			Energy losses		
	Total	From utilities	From industry	Total	From utilities	From industry
	(1)	(2)	(3)	(4)	(5)	(6)
1899	12	5	7	418	177	241
1903	23	9	14	674	283	391
1907	46	18	28	1085	452	633
1910	64	27	37	1292	579	714
1913	89	41	48	1535	756	778
1918	150	89	60	1998	1252	746
1920	178	120	59	2148	1501	647
1923	225	156	69	2137	1538	598
1926	296	211	85	2231	1649	582
1929	362	279	84	2364	1874	490
1937	455	361	94	2516	2051	465
1944	865	690	175	4170	3420	749
1948E	1028	844	185	4996	4212	783
1948U	1028	844	185	4986	4212	773
1953	1595	1351	244	5406	4659	747
1957	2255	1965	290	6256	5525	731
1960	2652	2351	301	6538	5879	659
1969	4861	4483	377	11559	10767	792
1973U	6195	5845	350	14912	14165	747
1973R	6194	5844	350	14724	14009	715
1979	7310	7067	244	17557	17061	495
1981	7546	7326	220	17888	17434	454
1985	8204	7880	324	19259	18602	658

NOTE: Col. 1 = col. 2 + col. 3. Col. 4 = Table AII.1.1, col. 3 − col. 1; col. 4. = col. 5 + col. 6. All data have been rounded.

Panel B: Index: 1979 = 100

1899	0.2	0.1	2.9	2.3	1.0	46.6
1903	0.3	0.1	5.8	3.8	1.6	74.6
1907	0.6	0.3	11.6	6.1	2.6	120.9
1910	0.9	0.4	15.3	7.3	3.4	136.2
1913	1.2	0.6	19.7	8.6	4.4	148.5
1918	2.0	1.3	24.8	11.2	7.3	142.5
1920	2.4	1.7	24.0	12.1	8.7	123.5
1923	3.1	2.2	28.3	12.0	8.9	114.2
1926	4.0	3.0	34.8	12.5	9.6	111.1
1929	5.0	3.9	34.4	13.3	10.9	93.5
1937	6.2	5.1	38.6	14.1	11.9	88.8
1944	11.8	9.8	71.9	23.4	19.8	143.0
1948E	14.1	11.9	75.8	28.0	24.4	149.5
1948U	14.1	11.9	75.8	28.0	24.4	149.5
1953	21.8	19.1	100.2	30.4	27.0	144.5
1957	30.8	27.8	118.9	35.2	32.0	141.3
1960	36.3	33.3	123.7	36.8	34.1	127.3
1969	66.5	63.4	154.9	65.0	62.4	153.1
1973U	84.7	82.7	143.6	83.9	82.1	144.4
1973R	84.7	82.7	143.6	83.9	82.1	144.4
1979	100.0	100.0	100.0	100.0	100.0	100.0
1981	103.2	103.7	90.3	101.9	102.2	91.6
1985	112.2	111.5	133.0	109.7	109.0	132.7

Table AII.2.1   (Continued)

Panel C: Annual percentage growth rates

Period	Usable electricity			Energy losses		
	Total	From utilities	From industry	Total	From utilities	From industry
	(1)	*(2)*	*(3)*	*(4)*	*(5)*	*(6)*
1899–1985	7.9	9.1	4.6	4.6	5.6	1.2
1899–1920	13.9	16.8	10.6	8.1	10.7	4.8
1899–1913	15.7	17.0	14.7	9.7	10.9	8.7
1899–1903	18.8	18.5	19.0	12.7	12.4	12.9
1903–1907	18.8	18.5	19.0	12.6	12.4	12.8
1907–1910	11.7	14.6	9.8	6.0	8.6	4.1
1910–1913	11.6	15.4	8.7	5.9	9.3	2.9
1913–1920	10.4	16.5	2.9	4.9	10.3	− 2.6
1913–1918	11.0	16.8	4.8	5.4	10.6	− 0.8
1918–1920	9.1	15.7	− 1.6	3.7	9.5	− 6.9
1920–1948	6.5	7.2	4.2	3.1	3.8	0.7
1920–1929	8.2	9.8	4.1	1.1	2.5	− 3.0
1920–1923	8.1	9.3	5.5	− 0.2	0.8	− 2.6
1923–1926	9.5	10.5	7.2	1.5	2.3	− 0.9
1926–1929	7.0	9.8	− 0.4	1.9	4.4	− 5.6
1929–1948	5.6	6.0	4.2	4.0	4.4	2.5
1929–1937	2.9	3.3	1.4	0.8	1.1	− 0.6
1937–1944	9.6	9.7	9.3	7.5	7.6	7.1
1944–1948	4.4	5.1	1.3	4.6	5.3	1.1
1948–1985	5.8	6.2	1.5	3.8	4.1	− 0.3
1948–1960	8.2	8.9	4.2	2.3	2.8	− 1.3
1948–1953	9.2	9.9	5.7	1.6	2.0	− 0.7
1953–1957	9.0	9.8	4.4	3.7	4.4	− 0.5
1957–1960	5.6	6.2	1.3	1.5	2.1	− 3.4
1960–1973	6.7	7.3	1.2	6.5	7.0	1.0
1960–1969	7.0	7.4	2.5	6.5	7.0	2.1
1969–1973	6.3	6.9	− 1.9	6.6	7.1	− 1.5
1973–1985	2.4	2.5	− 0.6	2.3	2.4	− 0.7
1973–1981	2.5	2.9	− 5.6	2.5	2.8	− 5.5
1973–1979	2.8	3.2	− 5.9	3.0	3.3	− 5.9
1979–1981	1.6	1.8	− 5.0	0.9	1.1	− 4.3
1981–1985	2.1	1.8	10.2	1.9	1.6	9.7

Table AII.2.2 Electricity Use in the Nonresidential U.S. Economy

Panel A: Trillion Btu

Year	Total	Usable electricity From utilities	From industry	Energy losses
	(1)	(2)	(3)	(4)
1899	12	5	7	406
1903	22	8	14	652
1907	44	16	28	1042
1910	61	24	37	1239
1913	85	37	48	1470
1918	143	83	60	1893
1920	167	108	59	2005
1923	209	140	69	1975
1926	272	187	85	2041
1929	327	243	84	2130
1937	392	298	94	2159
1944	742	567	175	3562
1948E	820	635	185	3960
1848U	820	635	185	3960
1953	1226	982	244	4132
1957	1699	1409	290	4693
1960	1938	1637	301	4761
1969	3347	2970	377	7926
1973U	4140	3790	350	9919
1973R	4140	3790	350	9730
1979	4887	4643	244	11743
1981	4983	4763	220	11737
1985	5388	5064	324	12581

NOTE: Col. 1 = col. 2 + col. 3. Col. 4 = Table AII.1.2, col. 3 − col. 1. All data have been rounded.

Panel B: Index: 1979 = 100

Year	Total	From utilities	From industry	Energy losses
1899	0.3	0.1	2.9	3.5
1903	0.5	0.2	5.7	5.6
1907	0.9	0.3	11.5	8.9
1910	1.3	0.5	15.2	10.6
1913	1.7	1.0	19.7	12.5
1918	2.9	1.8	24.6	16.1
1920	3.4	2.3	24.2	17.1
1923	4.3	3.0	28.3	16.8
1926	5.6	4.0	34.8	17.4
1929	6.7	5.2	34.4	18.1
1937	8.0	6.4	38.5	18.4
1944	15.2	12.2	71.7	30.3
1948E	16.8	13.7	75.8	33.7
1948U	16.8	13.7	75.8	33.7
1953	25.1	21.2	100.0	35.2
1957	34.8	30.3	118.8	40.0
1960	39.7	35.3	123.4	40.5
1969	68.5	64.0	154.5	67.5
1973U	84.7	81.6	143.4	84.5
1973R	84.7	81.6	143.4	82.9
1979	100.0	100.0	100.0	100.0
1981	102.0	102.6	90.2	99.9
1985	110.3	109.1	132.8	107.1

Table AII.2.2 (Continued)

Panel C: Annual percentage growth rate

Period	Usable electricity			Energy losses
	Total	From utilities	From industry	
	(1)	(2)	(3)	(4)
1899–1985	7.4	8.4	4.6	4.1
1899–1920	13.4	15.8	10.7	7.9
1899–1913	15.0	15.4	14.7	9.6
1899–1903	16.4	12.5	18.9	12.6
1903–1907	18.9	18.9	18.9	12.4
1907–1910	11.5	14.5	9.7	5.9
1910–1913	11.7	15.5	9.1	5.8
1913–1920	10.1	16.5	3.0	4.5
1913–1918	11.0	17.5	4.6	5.2
1918–1920	8.1	14.1	−0.8	2.9
1920–1948	5.8	6.5	4.2	2.5
1920–1929	7.8	9.4	4.0	0.7
1920–1923	7.8	9.0	5.4	−0.5
1923–1926	9.2	10.1	7.2	1.1
1926–1929	6.3	9.1	−0.4	1.4
1929–1948	5.0	5.2	4.2	3.3
1929–1937	2.3	2.6	1.4	0.2
1937–1944	9.5	9.6	9.3	2.5
1944–1948	2.5	2.9	1.4	2.7
1948–1985	5.2	5.8	1.5	3.2
1948–1960	7.4	8.2	4.1	1.5
1948–1953	8.4	9.1	5.7	0.9
1953–1957	8.5	9.4	4.4	3.2
1957–1960	4.5	5.1	1.2	0.5
1960–1973	6.0	6.7	1.2	5.8
1960–1969	6.3	6.8	2.5	5.8
1969–1973	5.5	6.3	−1.8	5.8
1973–1985	2.2	2.4	−0.6	1.6
1973–1981	2.3	2.9	−5.6	2.4
1973–1979	2.8	3.4	−5.8	3.2
1979–1981	1.0	1.3	−5.0	0.0
1981–1985	2.0	1.5	10.2	1.8

Table AII.2.3 Electricity Use in Manufacturing

Panel A: Trillion Btu

	Usable electricity			Energy losses			Heat rates (10³ Btu/kWh)	
Year	Total	From utilities	From manufac.	Total	From utilities	From manufac.	Sales based	Production based
	(1)	(2)	(3)	(4)	(5)	(6)	(7)	(8)
1899	2.6	1.3	1.3	94	50	45	135.7	120.7
1903	7	3	4	198	87	111	110.5	98.3
1907	15	6	9	363	154	209	90.0	80.1
1910	24	10	14	487	212	275	77.2	68.7
1913	38	16	22	650	298	352	66.2	58.9
1918	76	40	36	1007	557	450	51.2	45.6
1920	89	50	40	1061	624	437	46.2	41.1
1923	110	60	50	1028	594	434	37.0	33.1
1926	160	95	65	1190	742	448	30.1	26.8
1929	195	124	71	1248	836	413	26.4	23.4
1937	243	153	89	1312	871	441	22.8	20.3
1944	536	366	170	2541	1814	727	20.3	18.0
1948E	573	395	178	2726	1970	756	20.4	17.9
1948U	573	395	178	2716	1970	747	20.4	17.7
1953	844	610	234	2820	2103	717	15.2	13.9
1957	1162	885	277	3188	2488	700	13.0	12.0
1960	1285	996	288	3122	2491	631	11.9	10.9
1969	2057	1696	361	4831	4072	759	11.6	10.6
1973U	2425	2090	335	5781	5066	715	11.7	10.7
1973R	2425	2090	335	5695	5010	685	11.6	10.4
1979	2571	2338	233	6120	5645	474	11.6	10.4
1981	2482	2272	211	5841	5406	435	11.5	10.5
1985	2536	2226	310	5884	5254	630	11.5	10.3

NOTE: Col. 1 = col. 2 + col. 3. Col. 4 = Table AII.1.3, col. 3 − col. 1; col. 4. = col. 5 + col. 6. All data have been rounded.

Panel B: Index: 1979 = 100

1899	0.1	0.1	0.6	1.5	0.9	8.9	1155.5	1121.1
1903	0.3	0.1	1.7	3.2	1.5	22.0	941.2	913.1
1907	0.6	0.3	4.0	5.8	2.7	41.6	766.6	743.8
1910	0.9	0.4	6.2	7.8	3.7	54.8	657.2	637.7
1913	1.5	0.7	9.3	10.4	5.2	70.3	563.5	546.7
1918	3.0	1.7	15.6	16.2	9.8	89.7	436.0	423.1
1920	3.5	2.1	16.9	17.0	10.9	87.0	393.6	381.8
1923	4.3	2.6	21.4	16.5	10.4	86.6	315.3	307.1
1926	6.2	4.1	28.0	19.1	13.0	89.2	256.5	249.0
1929	7.6	5.3	30.3	20.0	14.6	82.2	224.5	216.8
1937	9.4	6.6	38.2	21.0	15.3	87.9	194.1	188.3
1944	20.8	15.7	72.9	40.8	31.8	145.0	173.0	167.2
1948E	22.3	16.9	76.4	43.7	34.5	150.7	174.2	166.0
1948U	22.3	16.9	76.4	43.7	34.5	150.7	174.2	166.0
1953	32.8	26.1	100.3	45.4	36.8	144.6	129.3	130.0
1957	45.2	37.9	118.9	51.3	43.6	141.3	110.7	112.8
1960	50.0	42.6	123.7	50.3	43.6	127.3	101.7	102.0
1969	80.0	72.5	154.9	77.8	71.3	153.1	98.8	99.2
1973U	94.3	89.4	143.6	93.1	88.7	144.4	99.5	100.3
1973R	94.3	89.4	143.6	93.1	88.7	144.4	99.5	100.3
1979	100.0	100.0	100.0	100.0	100.0	100.0	100.0	100.0
1981	96.5	97.2	90.3	95.4	95.8	91.6	99.0	101.0
1985	98.6	95.2	133.0	96.1	93.1	132.7	98.4	99.9

Table AII.2.3 (Continued)

Panel C: Annual percentage growth rates

Period	Usable electricity			Energy losses			Heat rates	
	Total	From utilities	From manufac.	Total	From utilities	From manufac.	Sales based	Production based
	(1)	(2)	(3)	(4)	(5)	(6)	(7)	(8)
1899–1985	8.3	9.1	6.6	4.9	5.6	3.2	−2.8	−2.8
1899–1920	18.4	19.0	17.7	12.2	12.8	11.5	−5.0	−5.0
1899–1913	21.1	19.9	22.3	14.8	13.7	15.9	−5.0	−5.0
1899–1903	27.2	21.4	32.2	20.3	15.2	25.4	−5.0	−5.0
1903–1907	22.8	21.4	23.6	16.3	15.2	17.2	−5.0	−5.0
1907–1910	16.4	17.4	15.7	10.3	11.3	9.6	−5.0	−5.0
1910–1913	16.2	18.2	14.7	10.1	12.0	8.6	−5.0	−5.0
1913–1920	13.0	17.4	9.0	7.2	11.1	3.1	−5.0	−5.0
1913–1918	15.0	19.7	10.9	9.1	13.3	5.0	−5.0	−5.0
1918–1920	8.2	11.8	4.1	2.6	5.8	−1.5	−5.0	−5.0
1920–1948	6.9	7.7	5.5	3.4	4.2	2.0	−2.9	−2.9
1920–1929	9.1	10.7	6.7	1.8	3.3	−0.6	−6.0	−6.1
1920–1923	7.3	6.6	8.1	−1.0	−1.6	−0.2	−7.1	−7.0
1923–1926	13.2	16.3	9.3	5.0	7.7	1.0	−6.6	−6.8
1926–1929	6.8	9.4	2.7	1.6	4.0	−2.7	−4.3	−4.5
1929–1948	5.8	6.3	5.0	4.2	4.6	3.2	−1.3	−1.4
1929–1937	2.8	2.7	3.0	0.6	0.5	0.8	−1.8	−1.7
1937–1944	12.0	13.2	9.7	9.9	11.0	7.4	−1.6	−1.7
1944–1948	1.7	1.9	1.2	1.8	2.1	1.0	0.2	−0.2
1948–1985	4.1	4.8	1.5	2.2	2.7	−0.3	−1.5	−1.4
1948–1960	7.0	8.0	4.1	1.2	2.0	−1.4	−4.4	−4.0
1948–1953	8.1	9.1	5.6	0.8	1.3	−0.8	−5.8	−4.8
1953–1957	8.3	9.8	4.3	3.1	4.3	−0.6	−3.8	−3.5
1957–1960	3.4	4.0	1.3	−0.7	0.0	−3.4	−2.8	−3.3
1960–1973	5.0	5.9	1.2	4.9	5.6	1.0	−0.2	−0.1
1960–1969	5.4	6.1	2.5	5.0	5.6	2.1	−0.3	−0.3
1969–1973	4.2	5.4	−1.9	4.6	5.6	−1.5	0.2	0.3
1973–1985	0.4	0.5	−0.6	0.3	0.4	−0.7	−0.1	0.0
1973–1981	0.3	1.0	−5.6	0.3	1.0	−5.5	−0.1	0.1
1973–1979	1.0	1.9	−5.9	1.2	2.0	−5.9	0.1	−0.1
1979–1981	−1.8	−1.4	−5.0	−2.3	−2.1	−4.3	−0.5	0.5
1981–1985	0.5	−0.5	10.2	0.2	−0.7	9.7	−0.1	−0.3

Notes to Tables AII.2.1, AII.2.2, and AII.2.3

cols. 1–3: *Usable electricity* refers to the kilowatthours of electricity actually used (3,412 Btu per kWh). *Usable electricity for the total U.S. economy* (Table AII.2.1, col. 1) includes the energy content of *utility generated electricity* (Table AII.2.1, col. 2) and of *industrially generated electricity* (Table AII.2.1, col. 3). *Usable electricity for the nonresidential economy* (Table AII.2.2, col. 1) includes *utility sales to the nonresidential sector* (Table AII.2.2, col. 2) and *industrially generated electricity* (Table AII.2.2, col. 3). *Usable electricity for manufacturing* (Table AII.2.3, col. 1) includes *utility sales to manufacturing* (Table AII.2.3, col. 2) and *manufacturing generated electricity* (Table AII.2.3, col. 3).

cols. 4–6: *Losses* occur when converting fuels to electricity since some of the energy is dissipated in the form of heat (an arbitrary imputation of such losses is also made for hydropower). In addition, some energy is used up in electricity transmission and distribution (i.e., line losses from the generating site to the point of use) and in direct use by utilities (including electricity whose disposition is not otherwise accounted for). Similar losses also occur in the industrial establishments that generate electricity. *Total U.S. economy losses* (Table AII.2.1, col. 4) include the *losses created in utilities* (Table AII.2.1, col. 5) and the *losses in industrial establishments* (Table AII.2.1, col. 6). *Total nonresidential economy losses* (Table AII.2.2, col. 4) equals the nonresidential gross electricity use (Table AII.1.2, col. 3) less usable electricity (Table AII.2.2, col. 1). *Manufacturing losses* (Table AII.2.3, col. 4) include the *losses in the utility sales to manufacturing*[4] (Table AII.2.3, col. 5) plus the *losses created in manufacturing establishments* (Table AII.2.3, col. 6).

cols. 7 and 8 (Table AII.2.3): *Heat rate* refers to the average gross energy input per kilowatthour of electricity. Declines in the heat rate track improvements in the thermal efficiency of electricity generation. The *sales based heat rate* (col. 7) refers to the ratio of the utilities' gross energy input divided by the utilities' kilowatthours of sales. The *production based heat rate* (col. 8) refers to the ratio of gross energy input at the typical fossil fuel steam generating plant divided by the kilowatthours of electricity generated at that plant.[5]

Tables AII.3
Energy and Electricity Services
and Electric Power Capacity
in Manufacturing

Table AII.3.1   Energy Services in Manufacturing

Panel A: Trillion Btu gross energy				*Materials conversion, information processing, and support services*		
	Machine drive					
Year	*Total*	*Electric*	*Nonelectric*	*Total*	*Electric*	*Nonelectric*
	(1)	*(2)*	*(3)*	*(4)*	*(5)*	*(6)*
1899	808	29	779	2082	68	2014
1903	1085	75	1009	2765	129	2636
1907	1196	161	1035	4075	217	3858
1910	1218	251	966	4221	260	3961
1913	1467	407	1060	4495	281	4214
1918	1657	754	904	6384	329	6055
1920	1553	815	738	6235	334	5901
1923	1341	804	537	6434	334	6099
1926	1386	949	436	6547	400	6146
1929	1290	1010	280	6644	433	6211
1937	1252	1053	199	6087	502	5585
1944	2307	2016	291	8632	1062	7570
1948E	2383	2117	265	8699	1181	7517
1948U	2376	2111	265	10043	1178	8865
1953	2519	2286	234	11517	1378	10139
1957	2949	2697	252	11888	1653	10235
1960	2975	2732	243	12667	1675	10992
1969	4668	4340	328	17456	2549	14907
1973U	5531	5170	361	19199	3036	16162
1973R	5473	5116	357	20958	3004	17954
1979	5811	5476	335	20558	3216	17342
1981	5549	5243	306	18022	3079	14942
1985	5583	5304	279	16054	3115	12938

NOTE: Col. 1 = col. 2 + col. 3. Col. 4 = Table AII.1.3, col. 1 − col. 1. All data have been rounded.

Panel B: Index: 1979 = 100

1899	13.7	0.5	229.3	12.8	2.1	15.2
1903	18.4	1.4	297.2	17.0	4.0	19.9
1907	20.3	2.9	304.6	25.0	6.7	29.1
1910	20.7	4.5	284.5	25.9	8.0	29.9
1913	24.9	7.3	312.0	27.6	8.6	31.8
1918	28.1	13.6	266.0	39.1	10.1	45.7
1920	26.4	14.7	217.3	38.2	10.3	44.6
1923	22.8	14.5	158.1	39.4	10.3	46.1
1926	23.5	17.1	128.4	40.1	12.3	46.4
1929	21.9	18.2	82.4	40.7	13.3	46.9
1937	21.3	19.0	58.7	37.3	15.4	42.2
1944	39.2	36.3	85.6	52.9	32.6	57.2
1948E	40.5	38.2	78.2	53.3	36.2	56.8
1948U	40.5	38.2	78.2	53.3	36.2	56.8
1953	42.9	41.3	69.0	61.2	42.4	64.9
1957	50.2	48.7	74.3	63.1	50.9	65.6
1960	50.7	49.4	71.7	67.3	51.5	70.4
1969	79.5	78.4	96.9	92.7	78.4	95.5
1973U	94.2	93.4	106.6	101.9	93.4	103.5
1973R	94.2	93.4	106.6	101.9	93.4	103.5
1979	100.0	100.0	100.0	100.0	100.0	100.0
1981	95.5	95.8	91.3	87.7	95.8	86.2
1985	96.1	96.9	83.3	78.1	96.9	74.6

Table AII.3.1 (Continued)

Panel C: Annual percentage growth rates	Machine drive			Materials conversion, info. processing, & support srvs.		
Period	Total	Electric	Nonelectric	Total	Electric	Nonelectric
	(1)	*(2)*	*(3)*	*(4)*	*(5)*	*(6)*
1899–1985	2.3	6.3	− 1.2	2.1	4.6	1.9
1899–1920	3.2	17.2	− 0.3	5.4	7.9	5.3
1899–1913	4.4	20.7	2.2	5.7	10.7	5.4
1899–1903	7.6	26.9	6.7	7.4	17.5	7.0
1903–1907	2.5	20.8	0.6	10.2	13.8	10.0
1907–1910	0.6	16.0	− 2.3	1.2	6.2	0.9
1910–1913	6.4	17.5	3.1	2.1	2.7	2.1
1913–1920	0.8	10.4	− 5.0	4.8	2.5	4.9
1913–1918	2.5	13.1	− 3.1	7.3	3.2	7.5
1918–1920	− 3.2	4.0	− 9.6	− 1.2	0.7	− 1.3
1920–1948	1.5	3.5	− 3.6	1.2	4.6	0.9
1920–1929	− 2.0	2.4	− 10.2	0.7	2.9	0.6
1920–1923	− 4.8	− 0.5	− 10.1	1.0	0.0	1.1
1923–1926	1.1	5.7	− 6.7	0.6	6.2	0.3
1926–1929	− 2.3	2.1	− 13.7	0.5	2.6	0.3
1929–1948	3.3	4.0	− 0.3	1.4	5.4	1.0
1929–1937	− 0.4	0.5	− 4.2	− 1.1	1.9	− 1.3
1937–1944	9.1	9.7	5.5	5.1	11.3	4.4
1944–1948	0.8	1.2	− 2.2	0.2	2.7	− 0.2
1948–1985	2.4	2.6	0.2	1.0	2.7	0.7
1948–1960	1.9	2.2	− 0.7	2.0	3.0	1.8
1948–1953	1.2	1.6	− 2.5	2.8	3.2	2.7
1953–1957	4.0	4.2	1.9	0.8	4.7	0.2
1957–1960	0.3	0.4	− 1.2	2.1	0.4	2.4
1960–1973	4.9	5.0	3.1	3.3	4.7	3.0
1960–1969	5.1	5.3	3.4	3.6	4.8	3.4
1969–1973	4.3	4.5	2.4	2.4	4.5	2.0
1973–1985	0.2	0.3	− 2.0	− 2.2	0.3	− 2.7
1973–1981	0.2	0.3	− 1.9	− 1.9	0.3	− 2.3
1973–1979	1.0	1.1	− 1.1	− 0.3	1.1	− 0.6
1979–1981	− 2.3	− 2.1	− 4.5	− 6.4	− 2.1	− 7.2
1981–1985	0.2	0.3	− 2.3	− 2.8	0.3	− 3.5

Table AII.3.2   Electric Services in Manufacturing

Panel A: Trillion Btu

Year	Usable electricity			Gross electricity		
	Machine drive	*Materials conversion*	*Info. processing & support srvs.*	*Machine drive*	*Materials conversion*	*Info. processing & support srvs.*
	(1)	*(2)*	*(3)*	*(4)*	*(5)*	*(6)*
1899	0.8	0.3	1.5	29	10	58
1903	2	1	4	75	23	107
1907	7	2	7	161	47	170
1910	12	3	9	251	67	192
1913	22	5	10	407	95	186
1918	53	11	12	754	160	169
1920	63	14	12	815	175	160
1923	78	17	15	804	180	155
1926	113	26	21	949	221	179
1929	136	33	25	1010	245	188
1937	164	51	28	1053	325	177
1944	351	128	57	2016	734	328
1948E	368	146	60	2117	839	343
1948U	368	146	60	2111	836	342
1953	526	233	85	2286	1009	369
1957	721	325	116	2697	1218	435
1960	796	347	141	2732	1190	485
1969	1296	535	226	4340	1791	758
1973U	1528	631	267	5170	2134	903
1973R	1528	631	267	5116	2111	893
1979	1620	643	309	5476	2173	1043
1981	1564	596	323	5243	1997	1082
1985	1598	609	330	5304	2021	1095

NOTE: Col. 1 + col. 2 + col. 3 = Table AII.2.3, col. 1. Col. 4 + col. 5 + col. 6 = Table AII.1.3, col. 3. Col. 4 = Table AII.3.1, col. 2. Col. 5 + col. 6 = Table AII.3.1, col. 5. All data have been rounded.

Panel B: Index: 1979 = 100

Year	(1)	(2)	(3)	(4)	(5)	(6)
1899	0.0	0.0	0.5	0.5	0.4	5.5
1903	0.2	0.1	1.1	1.4	1.0	10.1
1907	0.4	0.3	2.2	2.9	2.1	16.1
1910	0.7	0.5	2.9	4.5	3.1	18.2
1913	1.4	0.8	3.3	7.3	4.3	17.6
1918	3.3	1.8	3.9	13.6	7.3	16.0
1920	3.9	2.1	4.0	14.7	7.9	15.1
1923	4.8	2.7	4.9	14.5	8.2	14.6
1926	7.0	4.1	6.9	17.1	10.0	17.0
1929	8.4	5.2	8.2	18.2	11.1	17.8
1937	10.1	7.9	8.9	19.0	14.8	16.7
1944	21.7	19.9	18.5	36.3	33.3	31.0
1948E	22.7	22.7	19.3	38.2	38.1	32.4
1948U	22.7	22.7	19.3	38.2	38.1	32.4
1953	32.5	36.2	27.5	41.3	46.0	35.0
1957	44.5	50.6	37.7	48.7	55.5	41.3
1960	49.2	53.9	45.8	49.4	54.2	46.0
1969	80.0	83.2	73.3	78.4	81.6	71.9
1973U	94.3	98.1	86.5	93.4	97.2	85.6
1973R	94.3	98.1	86.5	93.4	97.2	85.6
1979	100.0	100.0	100.0	100.0	100.0	100.0
1981	96.5	92.7	104.6	95.8	91.9	103.7
1985	98.6	94.7	106.8	96.9	93.0	104.9

Table AII.3.2 (Continued)

Panel C: Annual percentage growth rates

	Usable electricity			Gross electricity		
Period	Machine drive	Matrls. conv.	Info. proc. & suprt. srvs.	Machine drive	Matrls. conv.	Info. proc. & suprt. srvs.
	(1)	(2)	(3)	(4)	(5)	(6)
1899–1985	9.3	9.4	6.4	6.3	6.4	3.5
1899–1920	23.3	20.8	10.4	17.2	14.8	4.9
1899–1913	27.2	23.9	14.5	20.7	17.7	8.7
1899–1903	33.9	30.6	22.7	26.9	23.7	16.3
1903–1907	27.2	26.0	18.4	20.8	19.7	12.5
1907–1910	22.1	19.0	9.5	16.0	13.1	4.1
1910–1913	23.5	17.8	4.1	17.5	12.0	−1.1
1913–1920	16.0	14.6	2.8	10.4	9.1	−2.2
1913–1918	18.8	16.7	3.0	13.1	11.1	−1.9
1918–1920	9.2	9.7	2.1	4.0	4.4	−2.8
1920–1948	6.5	8.8	5.8	3.5	5.8	2.8
1920–1929	8.9	10.4	8.3	2.4	3.9	1.8
1920–1923	7.2	8.7	6.5	−0.5	0.9	−1.1
1923–1926	13.1	14.7	12.4	5.7	7.2	5.1
1926–1929	6.6	8.1	6.0	2.1	3.5	1.5
1929–1948	5.4	8.1	4.6	4.0	6.7	3.2
1929–1937	2.3	5.5	1.1	0.5	3.6	−0.7
1937–1944	11.5	14.1	11.0	9.7	12.3	9.2
1944–1948	1.2	3.3	1.0	1.2	3.4	1.1
1948–1985	4.1	3.9	4.7	2.6	2.4	3.2
1948–1960	6.7	7.5	7.5	2.2	3.0	3.0
1948–1953	7.4	9.8	7.4	1.6	3.8	1.5
1953–1957	8.2	8.8	8.2	4.2	4.8	4.2
1957–1960	3.4	2.1	6.7	0.4	−0.8	3.7
1960–1973	5.1	4.7	5.0	5.0	4.6	4.9
1960–1969	5.6	4.9	5.4	5.3	4.6	5.1
1969–1973	4.2	4.2	4.2	4.5	4.5	4.5
1973–1985	0.4	−0.3	1.8	0.3	−0.4	1.7
1973–1981	0.3	−0.7	2.4	0.3	−0.7	2.4
1973–1979	1.0	0.3	2.5	1.1	0.5	2.6
1979–1981	−1.8	−3.7	2.3	−2.1	−4.1	1.9
1981–1985	0.5	0.5	0.5	0.3	0.3	0.3

Table AII.3.3 Mechanical Power Capacity in Manufacturing

Panel A: Rated capacity (million horsepower)

	Available power capacity			Installed generating capacity		Electric motors		
Year	Total	Nonelectric direct drive	Electric	Total prime movers	Elec. drive generators	Total elec. motors	Primary motors	Secondary motors
	(1)	(2)	(3)	(4)	(5)	(6)	(7)	(8)
1899	9.8	9.4	0.4	9.6	0.2	0.5	0.2	0.3
1903	12.3	11.2	1.1	11.9	0.7	1.3	0.4	0.9
1907	15.8	13.0	2.8	14.8	1.8	3.2	1.0	2.2
1910	18.6	14.0	4.6	16.7	2.7	5.1	1.9	3.2
1913	20.7	14.0	6.7	17.6	3.6	7.4	3.1	4.3
1918	26.7	13.8	12.9	19.1	5.3	13.8	7.6	6.2
1920	29.2	13.3	15.9	19.4	6.1	17.0	9.8	7.2
1923	32.6	12.0	20.6	19.4	7.4	22.0	13.2	8.8
1926	36.9	9.4	27.5	19.3	9.9	29.4	17.6	11.8
1929	41.1	9.3	31.8	19.3	10.0	33.8	21.8	12.0
1937	49.3	9.7	39.6	20.8	11.1	41.4	28.5	12.9
1944	84.7	9.2	75.5	25.9	16.7	76.8	58.8	18.0
1948	86.2	13.5	72.7	30.4	16.9	73.0	55.8	17.2
1953	109.0	15.0	94.0	35.4	20.4	92.9	73.6	19.3
1957	136.8	16.8	120.0	38.8	22.0	118.6	98.0	20.6
1960	146.5	18.9	127.6	42.0	23.1	126.2	104.5	21.7
1969	*	28.3	*	53.0	24.7	*	*	*
1973	*	33.6	*	58.0	24.4	*	*	*
1979	*	41.2	*	63.0	21.8	*	*	*
1981	*	42.2	*	64.0	21.8	*	*	*

NOTE: Col. 1 = col. 2 + col. 3 = col. 4 + col. 7. Col. 2 = col. 4 – col. 5. Col. 3 = col. 5 + col. 7. Col. 6 = col. 7 + col. 8. All data have been rounded.

* The Census of Manufactures is the primary source of electric motor capacity data in manufacturing, and 1962 is the latest year for which such data are reported. As a result, we have not been able to estimate for the years after 1960 (col. 1, col. 3, cols. 6–8).

Panel B: Index: 1979 = 100

Year	Available power capacity			Installed generating capacity		Electric motors		
	Total	Nonelectric direct drive	Electric	Total prime movers	Elec. drive generators	Total elec. motors	Primary motors	Secondary motors
	(1)	(2)	(3)	(4)	(5)	(6)	(7)	(8)
1899	3.7	22.8	0.2	15.3	1.2	0.2	0.1	1.4
1903	4.7	27.1	0.5	19.0	3.5	0.6	0.2	4.4
1907	6.0	31.4	1.3	23.4	8.3	1.4	0.5	10.6
1910	7.1	34.0	2.1	26.5	12.3	2.4	1.0	15.7
1913	7.9	33.8	3.0	27.9	16.6	3.4	1.6	21.0
1918	10.2	33.5	5.8	30.3	24.3	6.3	3.8	30.3
1920	11.1	32.4	7.2	30.8	27.9	7.7	4.9	34.9
1923	12.4	29.2	9.3	30.8	33.9	10.0	6.6	42.7
1926	14.1	22.9	12.4	30.6	45.1	13.4	8.8	57.3
1929	15.7	22.6	14.4	30.7	45.9	15.4	10.9	58.6
1937	18.8	23.6	17.9	33.1	51.0	18.8	14.3	63.0
1944	32.3	22.3	34.2	41.1	76.8	34.9	29.5	87.6
1948	32.9	32.8	32.9	48.3	77.4	33.2	28.0	83.4
1953	41.5	36.5	42.5	56.2	93.4	42.2	36.9	93.7
1957	52.1	40.8	54.2	61.5	100.6	53.9	49.2	99.9
1960	55.8	46.0	57.7	66.7	105.7	57.4	52.4	105.4
1969	*	68.7	*	84.1	113.3	*	*	*
1973	*	81.6	*	92.1	111.8	*	*	*
1979	*	100.0	*	100.0	100.0	*	*	*
1981	*	102.4	*	101.6	100.0	*	*	*

Table AII.3.3 (Continued)

Panel C: Annual percentage growth rates

Year	Available power capacity			Installed generating capacity		Electric motors		
	Total	Nonelectric direct drive	Electric	Total prime movers	Elec. drive generators	Total elec. motors	Primary motors	Secondary motors
	(1)	(2)	(3)	(4)	(5)	(6)	(7)	(8)
1899–1981	4.0	1.9	7.8	2.3	5.6	7.7	8.9	5.3
1899–1920	5.3	1.7	18.7	3.4	16.3	18.6	21.0	16.4
1899–1913	5.5	2.9	21.7	4.4	20.9	21.7	22.7	21.1
1899–1903	5.8	4.5	27.0	5.5	31.7	27.9	19.2	32.4
1903–1907	6.4	3.7	25.8	5.4	24.0	25.8	29.5	24.3
1907–1910	5.7	2.7	18.0	4.2	14.1	17.5	24.4	14.1
1910–1913	3.5	-0.2	13.4	1.7	10.5	12.8	17.1	10.1
1913–1920	5.1	-0.6	13.0	1.4	7.7	12.5	17.8	7.5
1913–1918	5.2	-0.2	13.8	1.7	7.9	13.2	19.4	7.6
1918–1920	4.7	-1.7	11.1	0.8	7.2	10.9	13.8	7.2
1920–1948	3.9	0.0	5.6	1.6	3.7	5.3	6.4	3.2
1920–1929	3.9	-3.9	8.0	-0.1	5.7	8.0	9.3	5.9
1920–1923	3.8	-3.4	9.1	0.0	6.7	9.1	10.5	7.0
1923–1926	4.1	-7.8	10.0	-0.2	10.0	10.1	10.0	10.3
1926–1929	3.7	-0.5	5.0	0.1	0.6	4.8	7.4	0.7
1929–1948	4.0	2.0	4.4	2.4	2.8	4.1	5.1	1.9
1929–1937	2.3	0.5	2.8	0.9	1.3	2.6	3.4	0.9
1937–1944	8.0	-0.8	9.7	3.2	6.0	9.2	10.9	4.8
1944–1948	0.4	10.2	-1.0	4.1	0.2	-1.3	-1.3	-1.2
1948–1981	3.3	3.5	3.3	2.3	0.8	3.3	3.8	0.5
1948–1960	4.5	2.9	4.8	2.7	2.6	4.7	5.4	2.0
1948–1953	4.8	2.2	5.3	3.1	3.8	4.9	5.7	2.4
1953–1957	5.8	2.8	6.3	2.3	1.9	6.3	7.4	1.6
1957–1960	2.3	4.1	2.1	2.7	1.7	2.1	2.2	1.8

1960–1973	*	4.5	*	2.5	0.4	*	*	*
1960–1969	*	4.6	*	2.6	0.8	*	*	*
1969–1973	*	4.4	*	2.3	− 0.3	*	*	*
1973–1981	*	2.9	*	1.2	− 1.4	*	*	*
1973–1979	*	3.4	*	1.4	− 1.8	*	*	*
1979–1981	*	1.2	*	0.8	0.0	*	*	*

Notes to Tables AII.3.1, AII.3.2, and AII.3.3

Table AII.3.1 Energy Services in Manufacturing

cols. 1–6: Energy services in manufacturing refer to the tasks performed by the application of power and heat in manufacturing establishments. These include driving machines; processing materials in bulk; processing information; and such supporting tasks as lighting, heating, and cooling space, transporting materials etc., and running appliances and office equipment. The sum of the gross energy input applied to these services (cols. 1 and 4) equals the gross energy input in manufacturing (as shown in Table AII.1.3, col. 1).

Total machine drive (col. 1) refers to the Btus of gross energy input required in the application of electric and nonelectric mechanical power for the purpose of running machines. This total is separated into *electric machine drive* (col. 2) and *nonelectric machine drive* (col. 3).

The *total of materials conversion, information processing, and support services* (col. 4) refers to the Btus of gross energy used for all energy services other than machine drive. This total is separated into its gross electricity (col. 5) and gross nonelectric energy (col. 6) components.

Table AII.3.2 Electric Services in Manufacturing

cols. 1–6: Electric services refer to the tasks performed by electricity in manufacturing establishments. The categories included are the *same* as those shown *for energy services* (Table AII.3.1).

Table AII.3.3 Mechanical Power Capacity in Manufacturing

Rated power capacity refers to the potential amount of work output per unit of time capable of being provided by prime movers and electric motors in manufacturing. Capacity is ordinarily measured in terms of horsepower for prime movers, and kilowatts for electric motors (we have translated kilowatts into horsepower at 1 hp equals 0.7457 kW). The potential work output in installed capacity is normally greater than the actual work output since prime movers and electric motors do not usually operate at full capacity.

cols. 1–3: *Available power capacity* refers to the electric and nonelectric horsepower potentially available in the prime movers that provide mechanical power in manufacturing, plus the
capacity of primary motors that use electricity provided by utilities.

Total available power (col. 1) includes nonelectric direct drive and electric power capacity.
Nonelectric direct drive (col. 2), a component of prime mover capacity, includes the nonelectric
power that is used in direct drive for running machines. Available *electric power capacity* (col. 3)
includes that portion of prime mover capacity that is used for generating electricity (i.e.,
generator capacity for electric drive) plus the primary motor capacity to use utility generated
electricity.

cols. 4–5: *Prime mover capacity* (col. 4) refers to the total of installed generating capacity. It
includes the nonelectric drive capacity (col. 2) plus the generator capacity for electric drive.[6]
Electric drive generator capacity (col. 5) refers to that portion of prime mover capacity devoted to
the production of electricity used in running machines.[7]

cols. 6–8: *Total electric motor capacity* (col. 6) includes the capacity of primary and secondary
motors. *Primary motors* (col. 7) include the capacity of those motors driven by electricity
obtained from utilities; *secondary motors* (col. 8) include the capacity of those motors driven by
the electricity generated within manufacturing.[8]

ECONOMIC DATA AND ENERGY INTENSITY DATA

Tables AII.4, AII.5, and AII.6

Tables AII.4 shows output, productivity, and factor input (labor and capital) indexes for the total U.S. economy (Table AII.4.1), the private domestic business economy (Table AII.4.2), and manufacturing (Table AII.4.3). Table AII.5 combines the total energy, nonelectric energy, and electricity indexes of Tables AII.1 and AII.2 with the output indexes of Table AII.4 to provide energy intensity indexes for the total United States (Table AII.5.1), the private domestic business economy (Table AII.5.2), and manufacturing (Table AII.5.3). Table AII.6 presents energy and electricity price indexes. Tables AII.4, AII.5, and AII.6 show the cyclical peak year indexes in panel B and intercyclical average annual growth rates in panel C. (The data in Tables AII.4, AII.5, and AII.6 were initially derived as index numbers. To keep a consistent format for the panels in Tables AII.1 through AII.6, the index numbers are presented in panel B and growth rates in panel C. There is no panel A.)

There were discontinuities for the output, factor input, and productivity data in the year 1948. However, the index numbers presented (Table AII.4, panel B) already incorporate the necessary adjustments.[9]

Table AII.4.1 Output in the Total U.S. Economy

Panel B: Output Index: 1979 = 100

Year	Gross national product (1a)	Gross domestic product (1b)
1899	8.2	8.7
1903	9.9	10.5
1907	11.9	12.6
1910	13.1	13.4
1913	14.0	14.6
1918	16.6	17.4
1920	15.3	15.3
1923	18.1	18.1
1926	20.7	20.8
1929	22.2	22.5
1937	21.8	22.1
1944	43.3	43.9
1948	34.7	35.2
1953	45.0	45.5
1957	48.6	49.1
1960	52.2	52.7
1969	75.9	78.5
1973	86.0	86.5
1979	100.0	100.0
1981	101.8	101.8
1985	112.3	113.1

Panel C: Annual percentage growth rates

Period	Gross national product	Gross domestic product
1899–1985	3.1	3.0
1899–1920	3.0	2.7
1899–1913	3.9	3.8
1899–1903	4.8	4.8
1903–1907	4.7	4.7
1907–1910	3.3	2.1
1910–1913	2.2	2.9
1913–1920	1.3	0.7
1913–1918	3.5	3.6
1918–1920	-1.6	-2.5
1920–1948	3.0	3.0
1920–1929	4.2	4.4
1920–1923	5.8	5.8
1923–1926	4.6	4.7
1926–1929	2.4	2.7
1929–1948	2.4	2.4
1929–1937	-0.2	-0.2
1937–1944	10.3	10.3
1944–1948	-5.4	-5.4
1948–1985	3.2	3.2
1948–1960	3.5	3.4
1948–1953	5.3	5.3
1953–1957	1.9	1.9
1957–1960	2.4	2.4
1960–1973	3.9	3.9
1960–1969	4.2	4.5
1969–1973	3.2	2.5
1973–1985	2.2	2.3
1973–1981	2.1	2.1
1973–1979	2.5	2.5
1979–1981	0.9	0.9
1981–1985	2.5	2.7

Table AII.4.2 Output, Productivity, and Factor Inputs in the Private Domestic Business Economy

Panel B: Index: 1979 = 100

Year	Output	Labor	Capital	Multifactor input	Multifactor productivity	Labor productivity	Capital productivity	Capital per labor hour
	(1)	(2)	(3)	(4)	(5)	(6)	(7)	(8)
1899	8.8	38.3	14.3	27.5	32.0	23.0	61.5	37.3
1903	10.7	44.9	16.4	32.2	33.2	23.8	65.2	36.5
1907	12.8	50.1	18.8	36.1	35.5	25.6	68.1	37.5
1910	13.6	52.3	20.4	38.1	35.7	26.0	66.7	39.0
1913	14.8	56.1	22.2	41.0	36.1	26.4	66.7	39.6
1918	15.7	60.7	25.1	44.9	35.0	25.9	62.6	41.4
1920	15.5	59.3	26.3	44.7	34.7	26.1	58.9	44.4
1923	18.5	62.7	27.4	47.2	39.2	29.5	67.5	43.7
1926	21.3	65.8	29.7	49.9	42.7	32.4	71.7	45.1
1929	22.8	67.4	32.0	51.9	43.9	33.8	71.3	47.5
1937	22.0	59.0	30.1	46.2	47.6	37.3	73.1	51.0
1944	34.3	76.0	32.3	57.2	60.0	45.1	94.5	42.5
1948	35.3	75.5	36.3	58.3	60.6	46.7	97.1	48.1
1953	43.7	76.5	42.0	61.9	70.7	57.2	104.0	55.0
1957	47.8	76.7	47.1	64.5	74.2	62.4	101.4	61.5
1960	51.3	75.8	50.2	65.4	78.3	67.6	102.2	66.1
1969	75.1	85.2	70.0	79.6	94.3	88.1	107.2	82.1
1973	86.2	89.4	81.8	86.7	99.4	96.4	105.4	91.4
1979	100.0	100.0	100.0	100.0	100.0	100.0	100.0	100.0
1981	100.9	99.8	109.1	103.1	98.0	101.1	92.5	109.1
1985	114.9	106.3	123.7	112.1	102.5	108.1	92.9	116.3

402

Panel C: Annual percentage growth rates

Period	Output	Labor	Capital	Multifactor input	Multifactor productivity	Labor productivity	Capital productivity	Capital per labor hour	Capital intensity	Weighted Labor	Weighted Capital
	(1)	(2)	(3)	(4)	(5)	(6)	(7)	(8)	(9)	(10)	(11)
1899–1985	3.0	1.2	2.5	1.7	1.4	1.8	0.5	1.3	0.4	0.9	0.8
1899–1920	2.7	2.1	3.0	2.4	0.4	0.6	-0.2	0.8	0.2	1.5	0.8
1899–1913	3.8	2.8	3.2	2.9	0.9	1.0	0.5	0.5	0.1	2.0	0.9
1899–1903	4.9	4.1	3.7	4.0	0.9	0.8	1.2	-0.4	-0.1	3.1	0.9
1903–1907	4.8	2.8	3.4	2.9	1.8	2.0	1.4	0.6	0.2	2.0	0.9
1907–1910	2.0	1.4	2.8	1.8	0.2	0.6	-0.8	1.4	0.4	1.0	0.8
1910–1913	2.8	2.4	2.9	2.5	0.3	0.4	-0.1	0.5	0.1	1.8	0.8
1913–1920	0.7	0.8	2.4	1.3	-0.6	-0.1	-1.7	1.6	0.5	0.6	0.7
1913–1918	1.2	1.6	2.5	1.8	-0.6	-0.4	-1.3	0.9	0.2	1.1	0.7
1918–1920	-0.6	-1.2	2.3	-0.2	-0.5	0.5	-2.8	3.5	1.0	-0.8	0.7
1920–1948	3.0	0.9	1.2	0.9	2.0	2.1	1.8	0.3	0.1	0.6	0.3
1920–1929	4.4	1.4	2.2	1.7	2.7	2.9	2.1	0.8	0.2	1.0	0.6
1920–1923	6.2	1.9	1.4	1.8	4.3	4.2	4.7	-0.5	-0.1	1.5	0.3
1923–1926	4.7	1.6	2.7	1.9	2.8	3.1	2.0	1.1	0.3	1.2	0.7
1926–1929	2.3	0.8	2.6	1.3	0.9	1.4	-0.3	1.7	0.5	0.6	0.7
1929–1948	2.3	0.6	0.7	0.6	1.7	1.7	1.7	0.1	0.0	0.4	0.2
1929–1937	-0.4	-1.7	-0.8	-1.5	1.0	1.3	0.4	0.9	0.2	-1.3	-0.2
1937–1944	6.6	3.7	1.0	3.1	3.4	2.8	5.5	-2.6	-0.6	2.9	0.2
1944–1948	0.7	-0.2	2.9	0.5	0.2	0.9	-2.2	3.1	0.7	-0.1	0.6
1948–1985	3.2	0.9	3.4	1.8	1.4	2.3	-0.1	2.4	0.9	0.6	1.3
1948–1960	3.2	0.0	2.7	1.0	2.2	3.1	0.4	2.7	0.9	0.0	0.9
1948–1953	4.4	0.3	3.0	1.2	3.1	4.1	1.4	2.7	1.0	0.2	1.1
1953–1957	2.3	0.1	2.9	1.0	1.2	2.2	-0.6	2.8	1.0	0.0	1.0
1957–1960	2.4	-0.4	2.1	0.5	1.8	2.7	0.3	2.4	0.9	-0.2	0.8
1960–1973	4.1	1.3	3.8	2.2	1.9	2.8	0.2	2.5	0.9	0.8	1.4
1960–1969	4.3	1.3	3.8	2.2	2.1	3.0	0.5	2.4	0.9	0.8	1.4
1969–1973	3.5	1.2	4.0	2.2	1.3	2.3	-0.4	2.7	1.0	0.7	1.5
1973–1985	2.4	1.5	3.5	2.2	0.3	1.0	-1.1	2.0	0.7	1.0	1.2
1973–1981	2.0	1.4	3.7	2.2	-0.2	0.6	-1.6	1.5	0.8	0.9	1.3
1973–1979	2.5	1.9	3.4	2.4	0.1	0.6	-0.9	1.5	0.5	1.3	1.1
1979–1981	0.5	0.1	4.4	1.5	-1.0	0.5	-3.8	4.5	1.5	-0.1	1.5
1981–1985	3.3	1.6	3.2	2.1	1.1	1.7	0.1	1.6	0.6	1.0	1.2

Table AII.4.3 Output, Productivity, and Factor Inputs in Manufacturing

Panel B: Index: 1979 = 100

Year	Output	Labor	Capital	Multifactor input	Multifactor productivity	Labor productivity	Capital productivity	Capital per labor hour
	(1)	(2)	(3)	(4)	(5)	(6)	(7)	(8)
1899	5.1	31.4	7.8	22.3	22.9	16.2	65.4	24.8
1903	6.6	38.9	10.5	27.8	23.7	17.0	62.8	27.0
1907	7.9	45.2	13.4	32.7	24.2	17.5	59.0	29.7
1920	8.4	46.6	15.7	34.6	24.3	18.0	53.5	33.7
1913	10.0	48.8	18.8	37.2	26.9	20.5	53.2	38.5
1918	13.0	62.2	23.9	47.3	27.5	20.9	54.4	38.4
1920	12.3	58.4	25.1	45.5	27.0	21.1	49.0	43.0
1923	14.2	54.1	25.5	43.1	33.0	26.3	55.7	47.1
1926	16.0	52.5	26.1	42.3	37.8	30.5	61.3	49.7
1929	18.6	54.5	26.8	43.8	42.5	34.1	69.4	49.2
1937	19.2	48.7	23.0	38.8	49.5	39.4	83.5	47.2
1944	43.2	93.0	26.9	66.7	64.8	46.5	160.6	28.9
1948	34.2	74.6	32.5	58.2	58.7	45.8	105.2	43.6
1953	45.9	84.3	39.7	67.6	67.9	54.5	115.6	47.1
1957	47.6	81.2	45.1	68.5	69.7	58.6	105.5	55.5
1960	48.6	79.3	47.2	68.2	71.2	61.3	103.0	59.5
1969	77.0	96.7	70.8	88.9	86.6	79.6	108.8	73.2
1973	89.1	96.8	79.7	92.0	96.9	92.1	111.8	82.3
1979	100.0	100.0	100.0	100.0	100.0	100.0	100.0	100.0
1981	96.9	94.9	108.4	98.1	98.8	102.1	89.4	114.2
1985	113.3	92.7	115.6	97.9	115.8	122.2	98.0	124.7

Panel C: Annual percentage growth rates

Period	Output (1)	Labor (2)	Capital (3)	Multifactor input (4)	Multifactor productivity (5)	Labor productivity (6)	Capital productivity (7)	Capital per labor hour (8)	Capital intensity (9)	Weighted Labor (10)	Weighted Capital (11)
1899–1985	3.7	1.3	3.2	1.7	1.9	2.4	0.5	1.9	0.4	1.0	0.8
1899–1920	4.3	3.0	5.7	3.4	0.8	1.2	-1.3	2.6	0.4	2.5	0.9
1899–1913	4.9	3.2	6.4	3.7	1.2	1.6	-1.4	3.1	0.5	2.7	1.0
1899–1903	6.5	5.5	7.6	5.5	0.9	1.0	-1.0	2.0	0.1	5.3	0.2
1903–1907	4.4	3.8	6.1	4.2	0.2	0.6	-1.6	2.2	0.4	3.2	1.0
1907–1910	2.3	1.1	5.4	1.8	0.5	1.2	-3.0	4.3	0.7	0.9	0.9
1910–1913	6.1	1.6	6.4	2.5	3.5	4.4	-0.3	4.7	0.9	1.2	1.3
1913–1920	3.0	2.6	4.2	2.9	0.0	0.4	-1.2	1.5	0.3	2.0	0.9
1913–1918	5.3	4.9	4.8	4.9	0.4	0.4	0.5	-0.1	0.0	3.9	1.0
1918–1920	-2.8	-3.1	2.5	-1.9	-0.9	0.3	-5.2	5.8	1.2	-2.4	0.5
1920–1948	3.7	0.9	0.9	0.9	2.8	2.8	2.8	0.0	0.0	0.7	0.2
1920–1929	4.7	-0.8	0.7	-0.4	5.2	5.5	3.9	1.5	0.4	-0.6	0.2
1920–1923	5.2	-2.5	0.4	-1.9	7.2	8.0	4.8	3.1	0.8	-2.0	0.1
1923–1926	3.9	-1.0	0.9	-0.5	4.4	4.9	2.9	1.9	0.5	-0.8	0.2
1926–1929	5.1	1.2	0.9	1.1	3.9	3.8	4.1	-0.3	-0.1	0.9	0.2
1929–1948	3.3	1.7	1.0	1.5	1.7	1.6	2.2	-0.7	-0.2	1.3	0.2
1929–1937	0.4	-1.4	-1.9	-1.5	2.0	1.8	2.4	-0.6	-0.1	-1.1	-0.4
1937–1944	12.3	9.7	2.3	8.1	3.9	2.4	9.7	-6.7	-1.5	7.5	0.5
1944–1948	-5.7	-5.4	4.8	-3.4	-2.4	-0.3	-9.9	10.7	2.1	-4.3	0.9
1948–1985	3.3	0.6	3.5	1.4	1.9	2.7	-0.2	2.9	0.8	0.4	1.0
1948–1960	3.0	0.5	3.2	1.3	1.6	2.5	-0.2	2.6	0.9	0.3	1.1
1948–1953	6.1	2.5	4.1	3.0	3.0	3.5	1.9	1.6	0.5	1.7	1.3
1953–1957	0.9	-0.9	3.2	0.3	0.7	1.8	-2.3	4.2	0.9	-0.7	0.7
1957–1960	0.7	-0.8	1.5	-0.1	0.7	1.5	-0.8	2.3	0.8	-0.6	0.5
1960–1973	4.8	1.6	4.1	2.3	2.4	3.2	0.6	2.5	0.8	1.1	1.3
1960–1969	5.2	2.2	4.6	3.0	2.2	3.0	0.6	2.3	0.8	1.5	1.6
1969–1973	3.7	0.0	3.0	0.9	2.8	3.7	0.7	3.0	0.9	0.0	0.9
1973–1985	2.0	-0.3	3.2	0.5	1.5	2.4	-1.1	3.5	0.9	-0.3	0.8
1973–1981	1.1	-0.2	3.9	0.8	0.2	1.3	-2.7	4.2	1.1	-0.2	1.0
1973–1979	1.9	0.5	3.9	1.4	0.5	1.4	-1.8	3.3	0.9	0.4	1.0
1979–1981	-1.5	-2.6	4.1	-0.9	-0.6	1.1	-5.4	6.9	1.7	-1.9	1.0
1981–1985	4.0	-0.5	1.6	-0.1	4.0	4.6	2.3	2.2	0.6	-0.4	0.4

Notes to Tables AII.4.1, AII.4.2, and AII.4.3

Index numbers with 1979 selected as the base year are presented in panel B and growth rates in panel C. (There is no panel A as in Tables AII.4.1 through AII.4.3 because use of these data begins with index numbers.) The index estimates for years since 1948 were mostly derived from a comprehensive set of data published by the Bureau of Labor Statistics (BLS). Estimates for years prior to 1948 were derived from other sources, primarily National Bureau of Economic Research data prepared by John Kendrick. These data were linked in the overlapping year 1948 to approximate a continuous series for the years 1899 through 1985.

The *Gross National Product* (GNP) and *Gross Domestic Product* (GDP) are measures of U.S. economic output (Table AII.4.1). The GNP (col. 1a) refers to the constant dollar value of all final goods and services consumed in the United States. This includes the sum of personal consumption expenditures, gross private domestic investment expenditures (including the net change in business inventories), government expenditures, and net exports of goods and services; it excludes, however, the value of all intermediate transactions among industries. The GDP (col. 1b) refers to the constant dollar value of the nation's production, which is equivalent to the GNP less the value of net transactions with the rest of the world.[10]

Private domestic business output (Table AII.4.2, col. 1) refers to the constant dollar value of sales by nongovernmental, profit-oriented enterprises to households, government, and business investment.[11] *Manufacturing output* (Table AII.4.3, col. 1) refers to the constant dollar gross product originating in manufacturing, which is equivalent to the sum of factor incomes (i.e., labor, net interest, corporate profits, and noncorporate income) in manufacturing plus capital consumption allowances and indirect business taxes.[12]

The *labor input* (col. 2) for both manufacturing (Table AII.4.3) and for the private domestic business economy (Table AII.4.2) refers to the hours of labor input for all workers including production and nonproduction workers, self-employed, and unpaid family workers. The BLS measures, which refer to the years since 1948, refer to hours paid for rather than hours worked. Also, adjustments for shifts in the composition of the work force are not made so that changes in the quality of labor are not included. This means that shifts from less to more skilled labor are not reflected as an increase in the labor input but instead as a contribution to productivity growth.[13]

Capital input (col. 3) for both manufacturing (Table AII.4.3) and for the private domestic business economy (Table AII.4.2) refers to the annual flow of capital services obtained from the stock of physical assets. Growth in capital input is assumed to be proportional to growth in net stock (i.e., exclusive of depreciation) of capital, which includes equipment, nonresidential structures, land, and inventories.[14]

Multifactor inputs (col. 4) for both manufacturing (Table AII.4.3) and the private domestic business economy (Table AII.4.2) refers to the weighted sum of the labor and capital inputs; weights used to aggregate the inputs are approximations to each input's share in total factor income where the shares sum to unity in each period.[15]

Multifactor productivity (col. 5) includes both capital and labor as contributors to productivity growth. Multifactor productivity for both manufacturing (Table AII.4.3) and the private domestic business economy (Table AII.4.2) refers to the (value-added) output per unit of combined labor and capital input.

Labor productivity (col. 6) and *capital productivity* (col. 7) are single factor productivity indexes which refer, respectively, to the output per unit of labor input and output per unit of capital input.

Capital per labor hour (col. 8) refers to the capital services provided per unit of labor. *Capital intensity* (col. 9) refers to the average annual growth rate in capital per labor hour weighted by the share of capital income in output. The *weighted labor input* (col. 10) refers to the labor input growth rate weighted by the labor income share of output. The *weighted capital input* (col. 11) refers to the capital input growth rate weighted by the capital share of output.

The rate of growth in output is equal to the sum of the multifactor productivity, weighted labor input, and weighted capital input growth rates. The "percentage contribution" of each of these components to output growth (as shown in Chapter 12) is defined to equal its share of the output growth rate.[16]

Tables AII.5
Energy Intensity in the Total U.S.
Economy, Private Domestic Business
Economy, and Manufacturing

Table AII.5.1   Energy Intensity in the Total U.S. Economy

Panel B: Index: 1979 = 100

Year	Intensities with respect to GNP				Intensities with respect to GDP			
	Gross energy	Gross nonelec.	Gross elec.	Usable elec.	Gross energy	Gross nonelec.	Gross elec.	Usable elec.
	(1)	(2)	(3)	(4)	(1a)	(2a)	(3a)	(4a)
1899	106.1	146.3	20.7	2.4	100.0	137.9	19.5	2.3
1903	130.3	177.8	28.3	3.0	122.9	167.6	26.7	2.9
1907	147.1	197.5	37.8	5.0	138.9	186.5	35.7	4.8
1910	142.7	189.3	41.2	6.9	139.6	185.1	40.3	6.7
1913	150.7	199.3	46.4	8.6	144.5	191.1	44.5	8.2
1918	155.4	203.6	51.2	12.0	148.3	194.3	48.9	11.5
1920	162.7	210.5	60.8	15.7	162.7	210.5	60.8	15.7
1923	150.8	197.2	51.9	17.1	150.8	197.2	51.9	17.1
1926	137.2	177.8	48.8	19.3	136.5	176.9	48.6	19.2
1929	134.7	174.8	48.6	22.5	132.9	172.4	48.0	22.2
1937	131.7	167.4	54.1	28.4	129.9	165.2	53.3	28.1
1944	92.4	113.9	46.2	27.3	91.1	112.3	45.6	26.9
1948	123.1	148.1	69.2	40.6	121.3	146.0	59.1	40.1
1953	103.1	122.2	62.0	48.4	102.0	120.9	61.3	47.9
1957	106.0	122.6	69.8	63.4	104.9	121.4	69.0	62.7
1960	106.5	123.4	70.1	69.5	105.5	122.2	69.5	68.9
1969	107.2	117.0	86.2	87.6	103.7	113.1	83.3	84.7
1973	109.5	114.9	97.8	98.5	108.9	114.2	97.2	97.9
1979	100.0	100.0	100.0	100.0	100.0	100.0	100.0	100.0
1981	92.1	88.3	100.5	101.4	92.1	88.3	100.5	101.4
1985	83.4	76.7	98.3	99.9	82.8	76.1	97.6	99.2

Table AII.5.1 (Continued)

Panel C: Annual percentage growth rates

Period	Intensities with respect to GNP				Intensities with respect to GDP			
	Gross energy	Gross nonelec.	Gross elec.	Usable elec.	Gross energy	Gross nonelec.	Gross elec.	Usable elec.
	(1)	*(2)*	*(3)*	*(4)*	*(1a)*	*(2a)*	*(3a)*	*(4a)*
1899–1985	− 0.3	− 0.7	1.8	4.4	− 0.2	− 0.7	1.9	4.5
1899–1920	2.1	1.7	5.3	9.4	2.3	2.0	5.6	9.6
1899–1913	2.5	2.2	5.9	9.5	2.7	2.4	6.1	9.5
1899–1903	5.3	5.0	8.1	5.7	5.3	5.0	8.2	6.0
1903–1907	3.1	2.7	7.5	13.6	3.1	2.7	7.5	13.4
1907–1910	− 1.0	− 1.4	2.9	11.3	0.2	− 0.3	4.1	11.8
1910–1913	1.8	1.7	4.0	7.6	1.2	1.1	3.4	7.0
1913–1920	1.1	0.8	3.9	9.0	1.7	1.4	4.6	9.7
1913–1918	0.6	0.4	2.0	6.9	0.5	0.3	1.9	7.0
1918–1920	0.9	0.7	3.5	5.5	1.9	1.6	4.5	6.4
1920–1948	− 1.0	− 1.2	0.5	3.5	− 1.0	− 1.3	− 0.1	3.4
1920–1929	− 2.1	− 2.0	− 2.5	4.1	− 2.2	− 2.2	− 2.6	3.9
1920–1923	− 2.5	− 2.2	− 5.1	2.9	− 2.5	− 2.2	− 5.1	2.9
1923–1926	− 3.1	− 3.4	− 2.0	4.1	− 3.3	− 3.6	− 2.2	3.9
1926–1929	− 0.6	− 0.6	− 0.1	5.2	− 0.9	− 0.9	− 0.4	5.0
1929–1948	− 0.5	− 0.9	1.9	3.2	− 0.5	− 0.9	1.1	3.2
1929–1937	− 0.3	− 0.5	1.3	3.0	− 0.3	− 0.5	1.3	2.9
1937–1944	− 4.9	− 5.4	− 2.2	− 0.6	− 4.9	− 5.4	− 2.2	− 0.6
1944–1948	7.4	6.8	10.6	10.4	7.4	6.8	6.7	10.5
1948–1985	− 1.0	− 1.8	1.0	2.5	− 1.0	− 1.7	1.4	2.5
1948–1960	− 1.2	− 1.5	0.1	4.6	− 1.2	− 1.5	1.4	4.6
1948–1953	− 3.5	− 3.8	− 2.2	3.6	− 3.4	− 3.7	0.7	3.6
1953–1957	0.7	0.1	3.0	7.0	0.7	0.1	3.0	7.0
1957–1960	0.2	0.2	0.1	3.1	0.2	0.2	0.2	3.2
1960–1973	0.2	− 0.5	2.6	2.7	0.2	− 0.5	2.6	2.7
1960–1969	0.1	− 0.6	2.3	2.6	− 0.2	− 0.9	2.0	2.3
1969–1973	0.5	− 0.5	3.2	3.0	1.2	0.2	3.9	3.7
1973–1985	− 2.2	− 3.3	0.0	0.1	− 2.3	− 3.3	0.0	0.1
1973–1981	− 2.1	− 3.2	0.3	0.4	− 2.1	− 3.2	0.4	0.4
1973–1979	− 1.5	− 2.3	0.4	0.3	− 1.4	− 2.2	0.5	0.4
1979–1981	− 4.0	− 6.0	0.2	0.7	− 4.0	− 6.0	0.2	0.7
1981–1985	− 2.5	− 3.5	− 0.6	− 0.4	− 2.6	− 3.6	− 0.7	− 0.5

Table AII.5.2 Energy Intensity in the Private Domestic Business Economy

Panel B: Index: 1979 = 100

Year	Intensities with respect to output				Intensities with respect to capital				Intensities with respect to labor				Intensities with respect to multifactor input			
	Gross energy	Gross nonelec.	Gross elec.	Usable elec.	Gross energy	Gross nonelec.	Gross elec.	Usable elec.	Gross energy	Gross nonelec.	Gross elec.	Usable elec.	Gross energy	Gross nonelec.	Gross elec.	Usable elec.
	(1)	(2)	(3)	(4)	(5)	(6)	(7)	(8)	(9)	(10)	(11)	(12)	(13)	(14)	(15)	(16)
1899	105.7	142.0	28.4	3.4	65.0	87.4	17.5	2.1	24.3	32.6	6.5	0.8	33.8	45.5	9.1	1.0
1903	133.6	177.6	37.4	4.7	87.2	115.9	24.4	3.0	31.8	42.3	8.9	1.1	44.4	59.0	12.4	1.6
1907	153.9	198.1	50.0	7.0	104.8	135.1	34.0	4.8	39.3	50.7	12.8	1.8	54.6	70.4	17.7	2.5
1910	156.6	202.2	56.6	9.6	104.4	134.8	37.7	6.4	40.7	52.6	14.7	2.5	55.9	72.2	20.2	3.4
1913	164.2	210.1	62.8	11.5	109.5	140.1	41.9	7.7	43.3	55.4	16.6	3.0	59.3	75.9	22.7	4.2
1918	191.7	244.6	76.4	18.5	119.9	153.0	47.8	11.6	49.6	63.3	19.8	4.8	67.0	85.5	26.7	6.5
1920	188.4	236.8	83.2	21.9	111.0	139.5	49.0	12.9	49.2	61.9	21.8	5.7	65.3	82.1	28.9	7.6
1923	175.7	223.8	69.7	23.2	118.6	151.1	47.1	15.7	51.8	66.0	20.6	6.9	68.9	87.7	27.3	9.1
1926	157.3	199.5	64.3	26.3	112.8	143.1	46.1	18.9	50.9	64.6	20.8	8.5	67.1	85.2	27.5	11.2
1929	156.6	198.2	64.0	29.4	111.6	141.3	45.6	20.9	53.0	67.1	21.7	9.9	68.8	87.1	28.1	12.9
1937	156.4	195.9	68.6	36.4	114.3	143.2	50.2	26.6	58.3	73.1	25.6	13.6	74.5	93.3	32.7	17.3
1944	135.6	163.6	74.6	44.3	144.0	173.7	79.3	47.1	61.2	73.8	33.7	20.0	81.3	98.1	44.8	26.6
1948	137.7	163.7	80.2	47.6	133.9	159.2	78.0	46.3	64.4	76.6	37.5	22.3	83.4	99.1	48.5	28.8
1953	116.9	136.8	72.8	57.4	121.7	142.4	75.7	59.8	66.8	78.2	41.6	32.8	82.6	96.6	51.4	40.5
1957	115.3	131.2	68.8	72.8	117.0	133.1	69.9	73.9	71.8	81.7	42.9	45.4	85.4	97.2	51.0	54.0
1960	112.9	128.7	77.6	77.4	115.3	131.5	79.3	79.1	76.4	87.1	52.5	52.4	88.5	100.9	60.9	60.7
1969	109.3	118.4	89.1	91.2	117.3	127.0	95.6	97.9	96.4	104.3	78.5	80.4	103.1	111.7	84.0	86.1
1973	108.2	113.3	96.8	98.3	114.1	119.4	102.0	103.5	104.4	109.3	93.3	94.7	107.6	112.7	96.2	97.7
1979	100.0	100.0	100.0	100.0	100.0	100.0	100.0	100.0	100.0	100.0	100.0	100.0	100.0	100.0	100.0	100.0
1981	92.3	88.7	99.6	101.0	85.3	82.0	92.1	93.5	93.3	89.7	100.7	102.2	90.3	86.8	97.5	98.9
1985	79.5	72.8	94.1	96.0	73.8	67.7	87.4	89.2	85.9	78.7	101.7	103.8	81.4	74.7	96.4	98.4

Table AII.5.2 (Continued)

Panel C: Annual percentage growth rates

Period	Intensities with respect to output				Intensities with respect to capital				Intensities with respect to labor				Intensities with respect to multifactor input			
	Gross energy	Gross nonelec.	Gross elec.	Usable elec.	Gross energy	Gross nonelec.	Gross elec.	Usable elec.	Gross energy	Gross nonelec.	Gross elec.	Usable elec.	Gross energy	Gross nonelec.	Gross elec.	Usable elec.
	(1)	(2)	(3)	(4)	(5)	(6)	(7)	(8)	(9)	(10)	(11)	(12)	(13)	(14)	(15)	(16)
1899–1985	-0.3	-0.8	1.4	4.0	0.2	-0.3	1.9	4.5	1.5	1.0	3.3	5.8	1.0	0.6	2.8	5.4
1899–1920	2.9	2.5	5.3	9.3	2.6	2.3	5.0	9.0	3.4	3.1	5.9	9.8	3.2	2.9	5.7	9.6
1899–1913	3.2	2.8	5.8	9.1	3.8	3.4	6.4	9.7	4.2	3.9	6.9	9.9	4.1	3.7	6.7	10.0
1899–1903	6.0	5.8	7.1	8.4	7.6	7.3	8.7	9.3	7.0	6.7	8.2	8.3	7.1	6.7	8.0	9.8
1903–1907	3.6	2.8	7.5	10.5	4.7	3.9	8.7	12.5	5.4	4.6	9.5	13.1	5.3	4.5	9.3	11.8
1907–1910	0.6	0.6	4.2	11.1	-0.1	-0.1	3.5	10.1	1.2	1.2	4.7	11.6	0.8	0.9	4.5	10.8
1910–1913	1.6	1.3	3.5	6.2	1.6	1.3	3.6	6.4	2.1	1.7	4.1	6.3	2.0	1.7	4.0	7.3
1913–1920	2.0	1.7	4.1	9.6	0.2	-0.1	2.3	7.7	1.8	1.6	4.0	9.6	1.4	1.1	3.5	8.8
1913–1918	3.2	3.1	4.0	10.0	1.8	1.8	2.7	8.5	2.8	2.7	3.6	9.9	2.5	2.4	3.3	9.1
1918–1920	-0.9	-1.6	4.4	8.8	-3.8	-4.5	1.3	5.5	-0.4	-1.1	4.9	9.0	-1.3	-2.0	4.0	8.1
1920–1948	-1.1	-1.3	-0.1	2.8	0.7	0.5	1.7	4.7	1.0	0.8	2.0	5.0	0.9	0.7	1.9	4.9
1920–1929	-2.0	-2.0	-2.9	3.3	0.1	0.1	-0.8	5.5	0.8	0.9	-0.1	6.3	0.6	0.7	-3.1	6.1
1920–1923	-2.3	-1.9	-5.7	1.9	2.2	2.7	-1.3	6.8	1.7	2.2	-1.9	6.6	1.8	2.2	-1.9	6.2
1923–1926	-3.6	-3.8	-2.7	4.3	-1.7	-1.8	-0.7	6.4	-0.6	-0.7	0.3	7.2	-0.9	-1.0	0.2	7.2
1926–1929	-0.2	-0.2	-0.2	3.8	-0.4	-0.4	-0.4	3.4	1.4	1.3	1.4	5.2	0.8	0.7	0.7	4.8
1929–1948	-0.7	-1.0	1.2	2.6	1.0	0.6	2.9	4.3	1.0	0.7	2.9	4.4	1.0	0.7	2.9	4.3
1929–1947	0.0	-0.1	0.9	2.7	0.3	0.2	1.2	3.1	1.2	1.1	2.1	4.1	1.0	0.9	1.9	3.7
1937–1944	-2.0	-2.5	1.2	2.9	3.4	2.8	6.7	8.5	0.7	0.1	4.0	5.7	1.3	0.7	4.6	6.3
1944–1948	0.4	0.0	1.8	1.8	-1.8	-2.2	-0.4	-0.4	1.3	0.9	2.7	2.8	0.6	0.3	2.0	2.0
1948–1985	-1.5	-2.2	0.4	1.9	-1.6	-2.3	0.3	1.8	0.8	0.1	2.7	4.2	-0.1	-0.8	1.9	3.4
1948–1960	-1.6	-2.0	-0.3	4.1	-1.2	-1.6	0.1	4.6	1.4	1.1	2.8	7.4	0.5	0.2	1.9	6.4
1948–1953	-3.2	-3.5	-1.9	3.8	-1.9	-2.2	-0.6	5.3	0.7	0.4	2.1	8.0	-0.2	-0.5	1.2	7.1
1953–1957	-0.3	-1.0	-1.4	6.1	-1.0	-1.7	-2.0	5.4	1.8	1.1	0.8	8.5	0.8	0.2	-0.2	7.5
1957–1960	-0.7	-0.6	4.1	2.1	-0.5	-0.4	4.3	2.3	2.1	2.2	7.0	4.9	1.2	1.3	6.1	4.0

1960–1973	−0.3	−1.0	1.7	1.9	−0.1	−0.7	2.0	2.1	2.4	1.8	4.5	4.7	1.5	0.9	3.6	3.7
1960–1969	−0.4	−0.9	1.5	1.8	0.2	−0.4	2.1	2.4	2.6	2.0	4.6	4.9	1.7	1.1	3.6	4.0
1969–1973	−0.3	−1.1	2.1	1.9	−0.7	−1.5	1.6	1.4	2.0	1.2	4.4	4.2	1.1	0.2	3.4	3.2
1973–1985	−2.5	−3.6	−0.2	−0.2	−3.6	−4.6	−1.3	−1.2	−1.6	−2.7	0.7	0.8	−2.3	−3.4	0.0	0.1
1973–1981	−2.0	−3.0	0.4	0.4	−3.6	−4.6	−1.3	−1.3	−1.3	−2.4	1.0	1.0	−2.2	−3.2	0.2	0.2
1973–1979	−1.3	−2.1	0.5	0.3	−2.2	−2.9	−0.3	−0.6	−0.7	−1.5	1.2	0.9	−1.2	−2.0	0.6	0.4
1979–1981	−3.9	−5.8	−0.2	0.5	−7.6	−9.4	−4.0	−3.3	−3.4	−5.3	0.3	1.1	−5.0	−6.8	−1.3	−0.6
1981–1985	−3.7	−4.8	−1.4	−1.3	−3.6	−4.7	−1.3	−1.7	−2.0	−3.2	0.2	0.4	−2.6	−3.7	−0.3	−0.1

Table AII.5.3 Energy Intensity in Manufacturing

Panel B: Index: 1979 = 100

	Intensities with respect to output				Intensities with respect to capital				Intensities with respect to labor				Intensities with respect to multifactor input			
Year	Gross energy	Gross nonelec.	Gross elec.	Usable elec.	Gross energy	Gross nonelec.	Gross elec.	Usable elec.	Gross energy	Gross nonelec.	Gross elec.	Usable elec.	Gross energy	Gross nonelec.	Gross elec.	Usable elec.
	(1)	(2)	(3)	(4)	(5)	(6)	(7)	(8)	(9)	(10)	(11)	(12)	(13)	(14)	(15)	(16)
1899	256.9	402.0	21.6	2.0	168.0	262.8	14.1	1.3	41.7	65.3	3.5	0.3	58.7	91.9	4.9	0.5
1903	265.2	406.1	34.9	4.6	166.7	255.2	21.9	2.9	45.0	68.9	5.9	0.8	63.0	96.4	8.3	1.1
1907	302.5	455.7	54.4	7.6	178.4	268.7	32.1	4.5	52.9	79.7	9.5	1.3	73.1	110.1	13.2	1.8
1910	294.1	431.0	69.1	10.7	157.3	230.6	36.9	5.7	53.0	77.7	12.5	1.9	71.4	104.6	16.8	2.6
1913	271.0	388.0	78.0	15.0	144.2	206.4	41.5	8.0	55.5	79.5	16.0	3.1	72.9	104.3	21.0	4.0
1918	280.8	393.9	94.6	23.1	152.7	214.2	51.5	12.6	58.7	82.3	19.8	4.8	77.2	108.3	26.0	6.3
1920	287.0	396.8	106.5	28.5	141.0	194.4	52.2	13.9	60.6	83.6	22.4	6.0	77.8	107.3	28.8	7.7
1923	248.6	343.7	90.8	30.3	138.4	191.4	50.6	16.9	65.3	90.2	23.8	8.0	81.9	113.2	29.9	10.0
1926	225.0	302.5	95.6	38.8	137.9	185.4	58.6	23.8	68.6	92.2	29.1	11.8	85.1	114.4	36.2	14.7
1929	193.6	256.5	88.2	40.9	134.3	178.0	61.2	28.4	66.1	87.5	30.1	13.9	82.2	108.9	37.4	17.4
1937	173.4	221.4	91.7	49.0	144.8	184.8	76.5	40.9	68.4	87.3	36.1	19.3	85.8	109.5	45.4	24.2
1944	115.5	133.8	79.9	48.2	184.8	214.8	128.3	77.3	53.4	62.2	37.1	22.4	74.5	86.7	51.7	31.2
1948	147.1	167.3	109.4	65.2	154.8	176.0	115.1	68.6	67.4	76.7	50.1	29.9	86.4	98.3	64.3	38.3
1953	124.0	141.6	90.9	71.5	143.3	163.7	105.0	82.6	67.5	77.1	49.5	38.9	84.2	96.2	61.7	48.5
1957	126.3	138.0	104.0	95.0	133.3	145.7	108.8	100.2	74.0	80.9	61.0	55.7	87.7	95.9	72.3	66.0
1960	130.5	144.9	103.3	102.9	134.3	149.2	106.4	105.9	80.0	88.8	63.1	63.1	93.0	103.2	73.6	73.3
1969	116.5	124.0	101.8	103.9	126.7	134.9	110.7	113.0	92.8	98.8	81.1	82.7	100.9	107.4	88.2	90.0
1973	112.5	116.3	104.8	105.8	125.7	130.0	117.2	118.3	103.5	107.0	96.5	97.4	108.9	112.6	101.5	102.5
1979	100.0	100.0	100.0	100.0	100.0	100.0	100.0	100.0	100.0	100.0	100.0	100.0	100.0	100.0	100.0	100.0
1981	92.3	89.1	98.9	99.6	82.5	79.6	88.4	89.0	94.2	90.9	101.0	101.7	91.1	88.0	97.7	98.4
1985	72.5	66.0	85.5	87.0	71.0	64.7	83.8	85.3	88.6	80.7	104.5	106.4	83.9	76.4	99.0	100.7

Panel C: Annual percentage growth rates

Period	Intensities with respect to output				Intensities with respect to capital				Intensities with respect to labor				Intensities with respect to multifactor input			
	Gross energy	Gross nonelec.	Gross elec.	Usable elec.	Gross energy	Gross nonelec.	Gross elec.	Usable elec.	Gross energy	Gross nonelec.	Gross elec.	Usable elec.	Gross energy	Gross nonelec.	Gross elec.	Usable elec.
	(1)	(2)	(3)	(4)	(5)	(6)	(7)	(8)	(9)	(10)	(11)	(12)	(13)	(14)	(15)	(16)
1899–1985	-1.5	-2.1	1.6	4.5	-1.0	-1.6	2.1	5.0	0.9	0.2	4.0	7.1	0.4	-0.2	3.6	6.4
1899–1920	0.6	0.0	7.9	13.5	-0.8	-1.4	6.4	12.0	1.8	1.2	9.2	14.9	1.3	0.7	8.8	14.4
1899–1913	0.4	-0.3	9.6	15.5	-1.1	-1.7	8.1	13.8	2.0	1.4	11.4	17.4	1.6	0.9	10.9	16.8
1899–1903	0.9	0.3	13.2	19.4	-0.1	-0.6	12.0	18.2	1.9	1.3	14.3	20.6	1.8	1.3	14.2	20.5
1903–1907	3.6	3.1	11.6	17.5	1.9	1.4	9.8	15.7	4.2	3.7	12.2	18.2	3.8	3.3	11.9	17.8
1907–1910	-1.2	-2.0	8.1	13.7	-4.2	-4.9	4.9	10.4	0.0	-0.8	9.4	15.1	-0.8	-1.6	8.6	14.3
1910–1913	-2.8	-3.6	4.1	9.5	-3.1	-3.8	3.8	9.2	1.5	0.7	8.7	14.4	0.6	-0.2	7.8	13.3
1913–1920	0.9	0.4	4.5	9.8	-0.3	-0.8	3.3	8.5	1.3	0.7	4.9	10.2	0.9	0.4	4.5	9.8
1913–1918	0.8	0.3	3.9	9.2	1.3	0.8	4.4	9.7	1.2	0.7	4.3	9.6	1.2	0.7	4.4	9.6
1918–1920	1.2	0.5	5.9	11.3	-4.0	-4.7	0.5	5.5	1.5	0.8	6.3	11.6	0.3	-0.4	5.0	10.3
1920–1948	-2.4	-3.1	0.1	3.0	0.3	-0.3	2.9	5.9	0.4	-0.3	2.9	5.9	0.4	-0.3	2.9	5.9
1920–1929	-4.3	-4.8	-2.1	4.1	-0.5	-1.0	1.8	8.3	1.0	0.5	3.4	9.9	0.6	0.2	3.0	9.5
1920–1923	-5.0	-5.0	-5.3	2.0	-0.5	-0.5	-0.8	6.8	2.6	2.6	2.3	10.1	1.8	1.9	1.6	9.3
1923–1926	-3.1	-4.0	1.9	9.0	-0.2	-1.2	4.9	12.2	1.7	0.7	6.9	14.4	1.2	0.3	6.4	13.8
1926–1929	-4.8	-5.3	-2.7	1.6	-0.9	-1.3	1.4	5.8	-1.2	-1.7	1.0	5.5	-1.1	-1.6	1.1	5.6
1929–1948	-1.4	-2.2	1.1	2.5	0.8	0.0	3.4	4.8	0.1	-0.7	2.7	4.1	0.3	-0.5	2.9	4.3
1929–1937	-1.4	-1.8	0.5	2.4	1.0	0.5	2.9	4.8	0.4	0.0	2.4	4.2	0.6	0.1	2.5	4.4
1937–1944	-5.7	-7.0	-1.8	-0.3	3.5	2.1	7.7	9.5	-3.5	-4.7	0.5	2.1	-2.0	-3.3	2.0	3.6
1944–1948	6.3	5.7	7.8	7.8	-4.2	-4.8	-2.9	-2.9	6.0	5.4	7.5	7.4	3.8	3.2	5.3	5.2
1948–1985	-1.9	-2.5	-0.7	0.8	-2.1	-2.7	-0.9	0.6	0.7	0.1	2.0	3.5	-0.1	-0.7	1.2	2.6
1948–1960	-1.0	-1.2	-0.5	3.9	-1.2	-1.4	-0.7	3.7	1.4	1.2	2.0	6.4	0.6	0.4	1.1	5.6
1948–1953	-3.4	-3.3	-3.6	1.9	-1.5	-1.4	-1.8	3.8	0.0	0.1	-0.2	5.4	-0.5	-0.4	-0.8	4.8
1953–1957	0.5	-0.6	3.4	7.4	-1.8	-2.9	0.7	4.9	2.3	1.2	5.4	9.4	1.0	-0.1	4.0	8.0
1957–1960	1.1	1.6	-0.2	2.7	0.2	0.8	-0.5	1.9	2.6	3.2	1.2	4.2	2.0	2.5	0.6	3.6

Table AII.5.3 (Continued)

Panel C: Annual percentage growth rates

Period	Intensities with respect to output				Intensities with respect to capital				Intensities with respect to labor				Intensities with respect to multifactor input			
	Gross energy	Gross nonelec.	Gross elec.	Usable elec.	Gross energy	Gross nonelec.	Gross elec.	Usable elec.	Gross energy	Gross nonelec.	Gross elec.	Usable elec.	Gross energy	Gross nonelec.	Gross elec.	Usable elec.
	(1)	(2)	(3)	(4)	(5)	(6)	(7)	(8)	(9)	(10)	(11)	(12)	(13)	(14)	(15)	(16)
1960–1973	-1.1	-1.7	0.1	0.2	-0.5	-1.1	0.7	0.9	2.0	1.4	3.3	3.4	1.2	0.7	2.5	2.6
1960–1969	-1.3	-1.7	-0.2	0.1	-0.6	-1.1	0.4	0.7	1.7	1.2	2.8	3.1	0.9	0.4	2.0	2.3
1969–1973	-0.9	-1.6	0.7	0.5	-0.2	-0.9	1.4	1.2	2.8	2.0	4.4	4.2	1.9	1.2	3.6	3.3
1973–1985	-3.6	-4.6	-1.7	-1.6	-4.6	-5.6	-2.8	-2.7	-1.3	-2.3	0.7	0.7	-2.1	-3.2	-0.2	-0.1
1973–1981	-2.4	-3.3	-0.7	-0.8	-5.1	-5.9	-3.5	-3.5	-1.2	-2.0	0.6	0.5	-2.2	-3.0	-0.5	-0.5
1973–1979	-1.9	-2.5	-0.8	-0.9	-3.7	-4.3	-2.6	-2.8	-0.6	-1.1	0.6	0.4	-1.4	-2.0	-0.2	-0.4
1979–1981	-3.9	-5.6	-0.6	-0.2	-9.2	-10.8	-6.0	-5.7	-2.9	-4.7	0.5	0.8	-4.6	-6.2	-1.2	-0.8
1981–1985	-5.9	-7.2	-3.6	-3.3	-3.7	-5.1	-1.3	-1.1	-1.5	-2.9	0.9	1.1	-2.1	-3.5	0.3	0.6

Notes to Tables AII.5.1, AII.5.2, and AII.5.3

(Following a consistent format for the panels in Tables AII.1 through AII.6, the index numbers are presented in panel B and growth rates in panel C.) cols. 1–16. The *energy intensity* indexes refer to the relation between the input of different forms and measures of energy — gross energy, nonelectric energy, gross electricity, usable electricity — and various measures of output and of the major factor inputs (labor and capital). The energy intensity indexes are calculated by dividing the energy indexes (Tables AII.1 and AII.2) by the output indexes (Tables AII.4) or the labor and capital indexes (Table AII.4).

For the private domestic business economy, energy intensities were calculated by dividing the energy indexes of the nonresidential economy (which includes government energy use) by the output, labor, and capital indexes for the private domestic business economy (which excludes government). Data availability problems led to this approach. However, since government energy use is included but government output and factor inputs are not, energy intensity in the private domestic business economy is overstated.

Table AII.6
Energy and Electricity Price Indexes

Table AII.6   Energy and Electricity Prices: Indexes of Producer Prices

Panel B: Index: 1967 = 100

Year	"Actual" prices		Adjusted "real" prices		Electricity prices relative to energy prices
	Energy	Electricity	Energy	Electricity	
	(1)	(2)	(3)	(4)	(5)
1899	29.5	n.a.	157.8	n.a.	n.a.
1903	43.1	n.a.	212.3	n.a.	n.a.
1907	38.9	n.a.	173.7	n.a.	n.a.
1910	34.0	n.a.	144.1	n.a.	n.a.
1913	43.8	154.3	181.0	637.6	352.3
1918	78.1	139.4	192.8	344.2	178.5
1920	117.0	144.1	222.4	274.0	123.2
1923	69.6	142.1	168.5	344.1	204.2
1926	71.5	137.6	174.0	334.8	192.4
1929	59.4	128.5	145.0	315.7	217.7
1937	55.5	101.2	152.1	277.3	182.3
1944	59.5	95.5	139.7	224.2	160.5
1948	90.5	91.0	137.7	138.5	100.6
1953	92.6	94.9	128.4	131.6	102.5
1957	99.1	97.2	122.2	119.9	98.1
1960	96.1	101.2	111.6	117.5	105.3
1969	100.9	101.8	91.0	91.8	100.9
1973	134.3	129.3	97.4	93.8	96.3
1979	408.1	270.2	186.4	123.4	66.2
1981	694.5	367.2	265.3	140.3	52.9
1985	633.6	453.6	204.0	146.0	71.6

Table AII.6 (Continued)

Panel C: Annual percentage growth rates

Year	"Actual" prices		Adjusted "real" prices		Electricity prices relative to energy prices
	Energy	Electricity	Energy	Electricity	
	(1)	(2)	(3)	(4)	(5)
1899–1985	3.6	n.a.	0.3	n.a.	n.a.
1899–1920	6.8	n.a.	1.6	n.a.	n.a.
1899–1913	2.9	n.a.	1.0	n.a.	n.a.
1899–1903	9.9	n.a.	7.7	n.a.	n.a.
1903–1907	−2.5	n.a.	−4.9	n.a.	n.a.
1907–1910	−4.4	n.a.	−6.0	n.a.	n.a.
1910–1913	8.8	n.a.	7.9	n.a.	n.a.
1913–1920	15.0	−1.0	3.0	−11.3	−13.9
1913–1918	12.3	−2.0	1.3	−11.6	−12.7
1918–1920	22.3	1.7	7.4	−10.8	−16.9
1920–1948	−0.9	−1.6	−1.7	−2.4	−0.7
1920–1929	−7.3	−1.3	−4.6	1.6	6.5
1920–1923	−15.9	−0.5	−8.8	7.9	18.3
1923–1926	0.9	−1.1	1.1	−0.9	−2.0
1926–1929	−6.0	−2.3	−5.9	−1.9	4.2
1929–1948	2.2	−1.8	−0.3	−4.2	−4.0
1929–1937	−0.8	−2.9	0.6	−1.6	−2.2
1937–1944	1.0	−0.8	−1.2	−3.0	−1.8
1944–1948	11.1	−1.2	−0.4	−11.3	−11.0
1948–1985	5.4	4.4	1.1	0.1	−0.9
1948–1960	0.5	0.9	−1.7	−1.4	0.4
1948–1953	0.5	0.8	−1.4	−1.0	0.4
1953–1957	1.7	0.6	−1.2	−2.3	−1.1
1957–1960	−1.0	1.4	−3.0	−0.7	2.4
1960–1973	2.6	1.9	−1.0	−1.7	−0.7
1960–1969	0.5	−0.1	−2.2	−2.7	−0.5
1969–1973	7.4	6.2	1.7	0.5	−1.2
1973–1985	13.8	11.0	6.4	3.8	−2.4
1973–1981	22.8	13.9	13.3	5.2	−7.2
1973–1979	20.4	13.1	11.4	4.7	−6.1
1979–1981	30.5	16.6	19.3	6.6	−10.6
1981–1985	−2.3	5.4	−6.4	1.0	7.9

Note to Table AII.6

(Following a consistent format for the panels in Tables AII.1 through AII.6, the indexes are presented in panel B and growth rates in panel C.)

cols. 1–5. The *energy price* indexes (col. 1) refer to the producers' selling prices of fuels and related products and power; the electricity price indexes (col. 2) refer to the producer selling prices of electric power. The real price indexes (cols. 3, 4) adjust for changes in the general price level by dividing the producer price indexes by the implicit GNP deflator. The relative electricity price refers to the ratio of real electricity prices divided by real energy prices.

ALLOCATION OF ENERGY USE AMONG
VARIOUS COMPONENTS

Tables AII.7

Tables AII.7 draw on earlier tables to show the percentage distribution of gross energy use among its electric and nonelectric components and also to show the nonresidential and manufacturing shares of U.S. energy and electricity consumption.

Table AII.7.1 Electric and Nonelectric Components of Gross Energy

	Total U.S.			Nonresidential U.S.			Manufacturing		
	Nonelectric	Gross electric	Usable electric	Nonelectric	Gross electric	Usable electric	Nonelectric	Gross electric	Usable electric
Year	(1)	(2)	(3)	(4)	(5)	(6)	(7)	(8)	(9)
1899	93.8	6.2	0.2	91.7	8.3	0.2	96.6	3.4	0.1
1903	93.2	6.8	0.2	91.3	8.7	0.3	94.7	5.3	0.2
1907	91.8	8.2	0.3	89.7	10.3	0.4	92.8	7.2	0.3
1910	90.8	9.2	0.4	88.7	11.3	0.5	90.6	9.4	0.4
1913	90.3	9.7	0.5	88.1	11.9	0.7	88.5	11.5	0.6
1918	89.5	10.5	0.7	87.5	12.5	0.9	86.5	13.5	0.9
1920	88.2	11.8	0.9	86.2	13.8	1.1	85.2	14.8	1.1
1923	89.1	10.9	1.0	87.5	12.5	1.2	85.4	14.6	1.4
1926	88.8	11.2	1.3	87.2	12.8	1.5	83.0	17.0	2.0
1929	88.5	11.5	1.5	87.2	12.8	1.7	81.8	18.2	2.5
1937	86.9	13.1	2.0	86.2	13.8	2.1	78.8	21.2	3.3
1944	84.1	15.9	2.7	82.8	17.2	3.0	71.9	28.1	4.9
1948E	82.2	17.8	3.0	81.7	18.3	3.1	70.2	29.8	5.2
1948U	82.2	17.8	3.0	81.7	18.3	3.1	73.5	26.5	4.6
1953	81.0	19.0	4.3	80.5	19.5	4.5	73.9	26.1	6.0
1957	79.1	20.9	5.5	78.4	21.6	5.7	70.7	29.3	7.8
1960	79.2	20.8	6.0	78.5	21.5	6.2	71.8	28.2	8.2
1969	74.6	25.4	7.5	74.5	25.5	7.6	68.9	31.1	9.3
1973U	71.7	28.3	8.3	72.0	28.0	8.2	66.8	33.2	9.8
1973R	71.8	28.2	8.3	72.2	27.8	8.3	69.3	30.7	9.2
1979	68.5	31.5	9.3	68.9	31.1	9.1	67.0	33.0	9.8
1981	65.6	34.4	10.2	66.3	33.6	10.0	64.7	35.3	10.5
1985	62.9	37.1	11.1	63.2	36.8	11.0	61.1	38.9	11.7

Percentage of gross energy

Table AII.7.2 Utility and Nonutility Sources of Electricity

	Percentage of usable electricity consumed					
	Total U.S.		Nonresidential U.S.		Manufacturing	
Year	Utility generated	Industrial generated	Utility generated	Industrial generated	Utility generated	Manufacturing generated
	(1)	(2)	(3)	(4)	(5)	(6)
1899	39.5	60.5	41.7	58.3	49.6	50.4
1903	39.1	60.9	36.4	63.6	41.2	58.8
1907	38.7	61.3	36.4	63.6	39.5	60.5
1910	41.8	58.2	39.3	60.7	40.5	59.5
1913	46.2	53.8	43.5	56.5	42.8	57.2
1918	59.7	40.3	58.0	42.0	52.2	47.8
1920	67.1	32.9	64.7	35.3	55.7	44.3
1923	69.4	30.6	67.0	33.0	54.7	45.3
1926	71.3	28.7	68.8	31.2	59.2	40.8
1929	76.9	23.1	74.3	25.7	63.8	36.2
1937	79.3	20.7	76.0	24.0	63.2	36.8
1944	79.8	20.2	76.4	23.6	68.3	31.7
1948E	82.0	18.0	77.4	22.6	68.9	31.1
1948U	82.0	18.0	77.4	22.6	68.9	31.1
1953	84.7	15.3	80.1	19.9	72.3	27.7
1957	87.2	12.8	82.9	17.1	76.1	23.9
1960	88.6	11.4	84.5	15.5	77.5	22.5
1969	92.2	7.8	88.7	11.3	82.4	17.6
1973U	94.4	5.6	91.5	8.5	86.2	13.8
1973R	94.4	5.6	91.5	8.5	86.2	13.8
1979	96.7	3.3	95.0	5.0	90.9	9.1
1981	97.1	2.9	95.6	4.4	91.5	8.5
1985	96.1	3.9	94.0	6.0	87.8	12.2

Table AII.7.3 Allocation of Energy Services in Manufacturing: Machine
Drive (MD), Materials Conversion (MC), Information Processing (IP), and
Support Services (SS)

Year	*Percentage of manufacturing gross energy used for*		*Percentage of manufacturing nonelectric energy used for*		*Percentage of manufacturing electricity used for*		
	MD	*MC + IP + SS*	*MD*	*MC + IP + SS*	*MD*	*MC*	*IP + SS*
	(1)	*(2)*	*(3)*	*(4)*	*(5)*	*(6)*	*(7)*
1899	28.0	72.0	27.9	72.1	30.0	10.0	60.0
1903	28.2	71.8	27.7	72.3	36.9	11.1	52.0
1907	22.7	77.3	21.2	78.8	42.6	12.3	45.1
1910	22.4	77.6	19.6	80.4	49.2	13.2	37.6
1913	24.6	75.4	20.1	79.9	59.2	13.8	27.1
1918	20.6	79.4	13.0	87.0	69.6	14.8	15.6
1920	19.9	80.1	11.1	88.9	70.9	15.2	13.9
1923	17.3	82.7	8.1	91.9	70.6	15.8	13.6
1926	17.5	82.5	6.6	93.4	70.3	16.4	13.3
1929	16.3	83.7	4.3	95.7	70.0	17.0	13.0
1937	17.1	82.9	3.4	96.6	67.7	20.9	11.4
1944	21.1	78.9	3.7	96.3	65.5	23.8	10.7
1948E	21.5	78.5	3.4	96.6	64.2	25.4	10.4
1948U	19.1	80.9	2.9	97.1	64.2	25.4	10.4
1953	17.9	82.1	2.3	97.7	62.4	27.6	10.1
1957	19.9	80.1	2.4	97.6	62.0	28.0	10.0
1960	19.0	81.0	2.2	97.8	62.0	27.0	11.0
1969	21.1	78.9	2.2	97.8	63.0	26.0	11.0
1973U	22.4	77.6	2.2	97.8	63.0	26.0	11.0
1973R	20.7	79.3	1.9	98.1	63.0	26.0	11.0
1979	22.0	78.0	1.9	98.1	63.0	25.0	12.0
1981	23.5	76.5	2.0	98.0	63.0	24.0	13.0
1985	25.8	74.2	2.0	97.9	63.0	24.0	13.0

Table AII.7.4 Electric and Nonelectric Components of Manufacturing
Energy Services

Year	Percentage of machine drive gross energy		Percentage of materials conversion, information processing, and support services gross energy	
	Nonelectric	*Gross electric*	*Nonelectric*	*Gross electric*
	(1)	*(2)*	*(3)*	*(4)*
1899	96.4	3.6	96.7	3.3
1903	93.0	7.0	95.3	4.7
1907	86.5	13.5	94.7	5.3
1910	79.4	20.6	93.8	6.2
1913	72.2	27.8	93.7	6.3
1918	54.5	45.5	94.8	5.2
1920	47.5	52.5	94.6	5.4
1923	40.0	60.0	94.8	5.2
1926	31.5	68.5	93.9	6.1
1929	21.7	78.3	93.5	6.5
1937	15.9	84.1	91.8	8.2
1944	12.6	87.4	87.7	12.3
1948E	11.1	88.9	86.4	13.6
1948U	11.1	88.9	88.3	11.7
1953	9.3	90.7	88.0	12.0
1957	8.5	91.5	86.1	13.9
1960	8.2	91.8	86.8	13.2
1969	7.0	93.0	85.4	14.6
1973U	6.5	93.5	84.2	15.8
1973R	6.5	93.5	85.7	14.3
1979	5.8	94.2	84.4	15.6
1981	5.5	94.5	82.9	17.1
1985	5.0	95.5	80.6	19.4

Table AII.7.5 Percentage Distribution of Energy and Electricity to Manufacturing and Nonresidential Use

Year	Manufacturing percentage of U.S.			Nonresidential percentage of U.S.			Manufacturing percentage of nonresidential U.S.		
	Gross energy	Gross nonelec.	Gross elec.	Gross energy	Gross nonelec.	Gross elec.	Gross energy	Gross nonelec.	Gross elec.
	(1)	(2)	(3)	(4)	(5)	(6)	(7)	(8)	(9)
1899	41.8	43.1	22.6	72.7	71.1	97.2	57.5	60.6	23.0
1903	37.6	38.2	29.4	75.4	73.8	96.7	49.9	51.7	30.4
1907	38.1	38.5	33.4	76.5	74.7	96.0	49.8	51.6	34.8
1910	36.7	36.7	37.7	77.5	75.7	95.9	47.4	48.4	39.3
1913	35.7	34.9	42.4	78.2	76.3	95.8	45.6	45.8	44.2
1918	39.3	38.0	50.4	79.4	77.6	94.8	49.6	49.0	53.2
1920	39.4	38.1	49.4	79.7	77.9	93.4	49.5	48.9	53.0
1923	35.9	34.4	48.2	80.9	79.5	92.5	44.4	43.3	52.2
1926	35.3	33.0	53.4	80.4	78.9	91.5	43.9	41.8	58.4
1929	33.4	30.9	52.9	80.9	79.8	90.1	41.3	38.8	58.7
1937	32.3	29.3	52.3	81.5	80.9	85.9	39.6	36.2	61.0
1944	34.5	29.5	61.1	79.0	77.8	85.5	43.6	37.9	71.5
1948	32.7	28.0	54.8	77.3	76.9	79.4	42.3	36.4	69.0
1948	36.7	32.8	54.7	77.3	76.8	79.5	47.4	42.7	68.8
1953	38.2	34.8	52.3	74.7	74.3	76.5	51.1	46.9	68.4
1957	36.4	32.5	51.1	72.6	72.0	75.1	50.1	45.1	68.1
1960	35.5	32.2	47.9	70.7	70.1	72.9	50.2	46.0	65.8
1969	34.3	31.7	42.0	68.5	68.4	68.7	50.1	46.3	61.1
1973	33.1	30.9	38.9	67.4	67.6	66.6	49.2	45.7	58.4
1973	35.6	34.3	38.8	67.2	67.6	66.3	53.0	50.8	58.5
1979	33.4	32.7	35.0	67.8	68.3	66.9	49.3	47.9	52.3
1981	31.9	31.4	32.7	67.3	68.0	65.7	47.3	46.2	49.8
1985	29.3	28.4	30.7	66.0	66.4	65.4	44.3	42.8	46.9

Notes to Tables AII.7.1 through AII.7.5

Tables AII.7.1 through AII.7.5 present the percentage distribution of selected components of energy and electricity use, as based on the data shown in Tables AII.1 through AII.3.

End Notes to Tables AII.1 through AII.7

1. Specifically, this was done by multiplying the 1899 to 1948E entries by the ratio of 1948U/ 1948E; multiplying these adjusted 1899 to 1948E entries and the 1948U to 1973U entries by the ratio of 1973R/1973U; and, finally, using these adjusted entries on a 1979 base to estimate continuous indexes for the years 1899 to 1985.

2. There are also some differences between the BOM and EIA definitions of energy use. The energy use estimates that we present are intended to encompass the fuel and nonfuel sources of energy, including mineral fuels (coal and lignite, natural gas, crude oil, and lease condensate), nuclear fuels, geothermal, and hydropower sources, but excluding fuel wood and waste (except as used by electric utilities) and the direct application of wind and water. However, the BOM includes some natural gas transmission losses and adjustments that are not included by EIA; the EIA thermal conversion factors for nuclear power are significantly different from the BOM factors; and, the EIA includes, but the BOM excludes, net coke imports, geothermal power, and electricity produced from wood and waste. Where possible, our methodology makes adjustments for these differences. Another reason for the 1948 discontinuity is that the 1948E thermal efficiency of electricity generation as based on BOM data differed significantly from the EIA-based 1948U thermal efficiency measures.

3. The most important EIA revision that took place resulted in a much larger share of total petroleum products being allocated to the industrial sector in 1973R than in 1973U. While this had little effect on our estimates of total energy consumption, it had a significant effect on manufacturing energy consumption. A second reason for the 1973 discontinuity is that while our 1973R to 1985 estimates of the thermal efficiency of electricity generation were obtained directly from EIA data, the 1948U to 1973U estimates were derived from other sources.

4. These are allocated to the manufacturing sector in proportion to the manufacturing share of total utility sales.

5. Alternative heat rate measures have been derived because the sales-based heat rate is used to calculate gross electricity from utilities, while the production-based heat rate is used to calculate gross electricity from industrial and manufacturing establishments. The heat rates differ not only because one uses electricity sales and the other uses electricity production in the denominator, but also because the sales-based heat rate includes the electricity generated by nuclear and geothermal facilities that have a lower thermal efficiency than fossil fuel steam generating plants.

6. Prime movers installed in manufacturing establishments are rated by their ability to generate the horsepower or kilowatts needed to perform tasks requiring mechanical or electric power. Prime movers are machines that convert the raw energy in falling water, wind, and fuels into a form such as mechanical or electric power, which can do useful work. (Some prime movers — such as water mills, windmills, or hydro turbines — convert raw mechanical energy directly into useful mechanical or electric power; other prime movers — such as steam engines, steam turbines, or internal combustion engines — burn fuels to produce heat or steam, which is then converted into mechanical or electric power.) Included as prime movers are not only those machines that convert nonelectric energy into mechanical drive, but also the electric drive generators that convert mechanical power into the electric power produced within the manufacturing sector itself.

7. Electric drive generators provide the electric power that is used by secondary motors in manufacturing. However, secondary motor capacity (col. 8) differs slightly from the electric drive generator capacity.

8. Since the manufacturing sector not only produces but also purchases power, the measurement of power available to manufacturing (col. 1) includes the prime mover capacity plus the capacity to use purchased power — which is measured by the capacity of primary electric motors. The capacity of secondary motors does not represent an addition to power available in manufacturing provided that the capacity of electric drive generators is included in the prime mover capacity.

9. These discontinuities occur because the definitions and methodology used by the Bureau of Labor Statistics (BLS) to derive their 1948 to 1985 data differed from the 1899 to 1948 estimates, which were based on data from other sources.

10. Although output data are presented for the total U.S. economy, factor input and productivity data are not. This is because for some components of GNP, labor and capital cannot be measured and because for some workers in the total economy productivity gains cannot be measured.

11. Indexes of private domestic business output were derived by the Bureau of Labor Statistics for the years 1909 to 1985 and by Kendrick for the years 1899 to 1909. As measured for the years 1909 to 1985, private domestic business output includes the value of gross domestic product less the imputed value of output from owner occupied housing and the value of activities in general government, government enterprises, nonprofit institutions, and households. As measured for the years prior to 1909, private domestic business output excludes from the GDP only the value of general government activities.

12. The BLS indexes of manufacturing output for years since 1948 are based on the deflated value of manufacturing production less the deflated value of materials, energy, and business services purchases by manufacturing. The Kendrick indexes of manufacturing output for years prior to 1948 were obtained by aggregating weighted indexes of physical output for specific manufacturing industries.

13. For years prior to 1948 we have adopted Kendrick measures of labor input. The Kendrick measures were derived by summing the weighted man-hours worked in the component industries of the business economy where the weights used are average hourly compensation in each industry. This measure differs from the BLS measure for years since 1948 in that it refers to hours worked rather than hours paid for, and it makes a labor quality adjustment by weighting specific industry work hours by the industry's average employee compensation.

14. The BLS defines capital input as the weighted net (i.e., exclusive of depreciation) stock of equipment, nonresidential structures, land, and inventories. The BLS constructs its private domestic business economy measure of capital input by aggregating the weighted net capital stock estimates for the farm, manufacturing, and nonfarm manufacturing producing sectors. The net capital stock for each of these sectors is obtained by using a weighted average of the growth rates for separate categories of equipment, structure, land, and inventories, where the weights are equal to the relative service prices (i.e., user or rental prices) of the different assets. Similarly, the BLS measure of manufacturing capital input is constructed by weighting and then aggregating the growth rates for specific categories of net capital stock held by the separate manufacturing industries.

The Kendrick measures of capital input essentially conform to the BLS measure, although Kendrick uses base period rates of return to capital (rather than service prices) as weights and also uses a different capital stock depreciation formula.

15. As its weights, the BLS uses the average of a given year's and previous year's relative cost share for each of the inputs. The BLS procedure for combining inputs is to weight and then add the growth rates of the individual inputs. Kendrick also uses the input cost shares as weights, but the procedure for combining inputs is to weight and then add indexes of the inputs rather than (as for BLS) the input growth rates.

16. If the production function is characterized by constant returns to scale, Hicks neutral technical change, and perfect competition, then the output growth rate is equal to the sum of the multifactor productivity, labor input, and capital intensity growth rates. Since the sum of the labor input and capital intensity growth rates are equal to the sum of the weighted labor input and weighted capital input growth rates, the output growth rate can be decomposed into three components — the multifactor productivity, weighted labor input, and weighted capital input growth rates.

Works Cited

Jack Alterman, *A Historical Perspective on Changes in U.S. Energy-Output Ratios*, EPRI EA-3997 (Palo Alto, CA: Electric Power Research Institute, June 1985).

Ernst R. Berndt and David O. Wood, "Energy Price Changes and the Induced Revaluation of Durable Capital in U.S. Manufacturing," mimeograph, 1983.

Richard B. DuBoff, "Electric Power in American Manufacturing, 1899–1958," Ph.D. diss., University of Pennsylvania, 1964.

Edison Electric Institute, *Statistical Yearbook, 1975* (Washington, D.C.: EEI, 1975).

Edison Electric Institute, *Statistical Yearbook, 1986* (Washington, D.C.: EEI, 1986).

Edison Electric Institute, *Historical Statistics of the Electric Utility Industry Through 1970*, Pub. no. 73–34 (Washington, D.C.: EEI, 1974).

John W. Kendrick, *Productivity Trends in the United States* (Princeton, NJ: Princeton University Press for the National Bureau of Economic Research, 1961).

John W. Kendrick, *Postwar Productivity Trends in the U.S.* (New York: National Bureau of Economic Research, 1973).

Robert C. Marlay, "Industrial Energy Productivity," Ph.D. diss., Massachusetts Institute of Technology, 1983.

Resource Dynamics Corporation, "Industrial Electrotechnologies and Electrification Program Plan, A Report to the Energy Management and Utilization Division of the Electric Power Research Institute," McLean, VA, August 1984.

Sam H. Schurr, Bruce C. Netschert, Vera Eliasberg, Joseph Lerner, and Hans H. Landsberg, *Energy in the American Economy* (Baltimore, MD: Johns Hopkins Press for Resources for the Future, 1960).

Vivian Eberle Spencer, *Raw Materials in the United States Economy, 1800–1966*, Working Paper no. 30 (Washington, D.C.: U.S. Bureau of the Census, 1963).

U.S. Bureau of the Census, *Historical Statistics of the United States, Colonial Times to 1970* (Washington, D.C.: GPO, 1975).

U.S. Bureau of the Census, *Census of Manufactures, 1963, Fuels and Electric Energy Consumed in Manufacturing Industries: 1962*, MC 63(1)7 (Washington, D.C.: GPO, 1963).

U.S. Bureau of the Census, *Census of Manufactures, 1982* (Washington, D.C.: GPO, 1985).

U.S. Bureau of the Census, *Statistical Abstract of the United States, 1982–1983* (Washington, D.C.: GPO, 1982).

U.S. Energy Information Administration, *Annual Report to Congress, 1980* (Washington, D.C.: GPO, 1981).

U.S. Energy Information Administration, *Monthly Energy Review* (June 1987).

U.S. Bureau of Labor Statistics, *Trends in Multifactor Productivity*, BLS Bulletin 2178 (Washington, D.C.: GPO, 1983).

U.S. Bureau of Labor Statistics, "Multifactor Productivity, a New BLS Measure, *Monthly Labor Review* (December 1983).

U.S. Bureau of Labor Statistics, *Monthly Labor Review* (August 1987).

U.S. Bureau of Labor Statistics, "Productivity and Related Measures, 1948–1986," 87–436 (Washington, D.C.: October 1987).

U.S. Bureau of Labor Statistics, *BLS Handbook* (Washington, D.C.: GPO, 1983 and 1985).

U.S. Bureau of Economic Analysis, *The National Income and Product Account of the United States, 1929–1982, Statistical Tables* (Washington, D.C.: GPO, September 1986).

U.S. Bureau of Economic Analysis, *Survey of Current Business* 66, no. 1 (July 1986).

Author Index

Subject Index

About the Authors and Contributors

SAM H. SCHURR served as Deputy Director, Energy Study Center, Electric Power Research Institute while the research for this book was being performed. He is the author or editor of several books, including *Energy in the American Economy* and *Energy in America's Future,* and was formerly Director of Energy Studies at Resources for the Future in Washington, D.C.

CALVIN C. BURWELL was a staff member of the Institute for Energy Analysis, Oak Ridge Associated Universities while the research for this book was being performed. He is now a private consultant in Clinton, Tennessee, specializing in energy technology and economics.

WARREN D. DEVINE, JR., was a staff member of the Institute for Energy Analysis, Oak Ridge Associated Universities while the research for this book was being performed. He is now a member of a consulting firm in Oak Ridge, Tennessee, specializing in environmental technology and assessment.

SIDNEY SONENBLUM is a private economic consultant in Los Angeles, and is author or editor of several studies dealing with energy, productivity, and economic growth. He was formerly Director of Research at the National Planning Association in Washington, D.C.

MILTON F. SEARL is a consultant to the Electric Power Research Institute. He was a member of the staff of the Institute's Energy Study Center, and was formerly with the Energy Policy Staff of the U.S. Office of Science and Technology.

BLAIR G. SWEZEY is a Senior Economist in the Energy and Environmental Analysis Division of the Solar Energy Research Institute in Golden, Colorado.